CliffsStudySolver™
Biology

By Max Rechtman

WILEY

Wiley Publishing, Inc.

Published by:
Wiley Publishing, Inc.
111 River Street
Hoboken, NJ 07030-5774
www.wiley.com

Copyright © 2004 Wiley, Hoboken, NJ

Published by Wiley, Hoboken, NJ
Published simultaneously in Canada

Library of Congress Cataloging-in-Publication Data
Rechtman, Max, 1947-
  Biology / by Max Rechtman.— 1st ed.
      p. cm. — (CliffsStudySolver)
  ISBN 978-0-7645-5842-9 (pbk.)
  1. Biology—Outlines, syllabi, etc. 2. Biology—Problems, exercises, etc. I. Title. II. Series.
  QH315.5.R43 2004
  570—dc22

                        2004006763

ISBN: 978-0-7645-5842-9

10 9 8 7

1B/QU/QW/QU/IN

For general information on our other products and services or to obtain technical support, please contact our Customer Care Department within the U.S. at 800-762-2974, outside the U.S. at 317-572-3993, or fax 317-572-4002.

Wiley also published its books in a variety of electronic formats. Some content that appears in print may not be available in electronic books.

Note: If you purchased this book without a cover, you should be aware that this book is stolen property. It was reported as "unsold and destroyed" to the publisher, and neither the author nor the publisher has received any payment for this "stripped book."

WILEY

About the Author

Max Rechtman taught high school biology in the New York City public school system for 34 years before retiring in 2003. He was a teacher mentor and holds a New York State certificate in school administration and supervision.

Dedication

Annette, you make my dreams come true.

Stephen, thanks for keeping me on my toes.

Publisher's Acknowledgments

Editorial

Project Editor: Marcia L. Johnson

Senior Acquisitions Editor: Greg Tubach

Technical Editor: Bill McKnight

Editorial Assistant: Amanda Harbin

Composition

Project Coordinator: Kristie Rees

Indexer: Johnna VanHoose

Proofreader: Ann Marie Damian

Wiley Publishing, Inc. Composition Services

Table of Contents

Pretest

Pretest Problems

For problems 1–95, circle the letter corresponding to the correct answer.

1. Which life function provides an organism with energy?

   A. regulation   B. reproduction   C. respiration   D. synthesis

2. To obtain sharp and exact focus when using the low-power lens of a compound microscope, a student must use the:

   A. fine adjustment knobs   B. coarse adjustment knobs
   C. nosepiece   D. diaphragm

3. What is the size of a cell measured under high power (400X magnification) if the size of the cell measured under low power (100X magnification) is 20µm?

   A. 20µm   B. 80µm   C. 200µm   D. 800µm

4. To remove the nucleus from one cell and insert it into another cell, a biologist must use:

   A. a microtome   B. microdissection tools
   C. an electron microscope   D. gene sequencers

5. Which part of an experiment is the comparison group?

   A. independent variable   B. dependent variable   C. hypothesis   D. control

6. The cell is the basic unit of structure and function in all living things. This is a basic concept of:

   A. the heterotroph hypothesis   B. the cell theory
   C. the theory of natural selection   D. cellular respiration

7. A student observes a cell with a microscope. The student can identify the cell as a plant cell and not an animal cell because of the presence of:

   A. cytoplasm   B. vacuoles   C. a cell membrane   D. a cell wall

8. Identify the organelle that is the energy factory of the cell.

   A. Golgi apparatus   B. lysosome   C. mitochondrion   D. ribosome

9. The movement of water into an amoeba through its cell membrane takes place by:

   A. diffusion   B. osmosis   C. endocytosis   D. active transport

10. Organisms are classified on the bases of all the following *except:*

    A. similarity of structure   B. similarity of function
    C. evolutionary development   D. phylogeny

11. *Necator americanus* is the scientific name for a hookworm. What part of the organism's scientific name is *Necator?*

    A. kingdom   B. phylum   C. genus   D. species

12. Which of the following takes control of the host cell's nucleus by destroying its DNA, inserting its own genetic material and taking over the life functions of the host cell?

    A. virus   B. bacterium   C. paramecium   D. slime mold

13. Prokaryotic organisms:

    A. have a nuclear membrane   B. lack a nuclear membrane
    C. lack genetic material   D. are multicellular

14. The animal-like protists contain:

    A. bacteria and blue-green algae   B. euglena and green algae
    C. amoeba and paramecium   D. slime molds and fungi

15. Eukaryotic organisms with cells lacking chloroplasts that are heterotrophs and obtain their food by absorption from decaying vegetation are characteristic of:

    A. algae   B. fungi   C. animal-like protists   D. bryophytes

16. Bryophytes are in the kingdom:

    A. Fungi   B. Protista   C. Anamalia   D. Plantae

17. Exoskeletons composed of chitin are characteristic of:

    A. arthropoda   B. mollusca   C. fishes   D. echinodermata

18. Having lungs, being warm blooded and having a four-chambered heart are characteristics of:

    A. osteichthyes   B. amphibia   C. reptilia   D. aves

19. The formation of a large molecule from two or more smaller molecules by the removal of water is known as:

    A. hydrolysis   B. dehydration synthesis
    C. photosynthesis   D. cellular respiration

20. Lipids are composed of:

    A. glucose and maltose   B. glycerol and fatty acids
    C. amino acids and polypeptides   D. carboxylic acid and glycogen

21. Which of the following elements is found only in proteins?

    A. hydrogen   B. oxygen   C. nitrogen   D. carbon

22. Which of the following pH values indicates the strongest acid?

    A. 3   B. 6   C. 7   D. 8

23. Enzyme action can best be explained by the:

    A. scientific method   B. theory of natural selection
    C. fluid mosaic model   D. lock-and-key model

24. All the following are factors that affect the rate of enzyme action *except:*

    A. oxygen concentration   B. substrate concentration
    C. enzyme concentration   D. pH

25. Identify the structures in the lower epidermis of a leaf that allow carbon dioxide in and water and oxygen out.

    A. guard cells   B. stomates   C. phloem cells   D. palisade cells

26. Which of the following molecules is produced during the dark reactions of photosynthesis?

    A. PGAL   B. ATP   C. DNA   D. CO_2

27. Which cells in a vein are specialized for the upward transport of materials?

    A. phloem   B. xylem   C. cambium   D. meristem

28. Select the cells that have a large surface area and are specialized for the increased absorption of water.

    A. phloem cells   B. root hair cells   C. cortex cells   D. root cap

29. In humans, the end products of digestion enter the:

    A. large intestine   B. small intestine   C. blood   D. liver

30. Which blood vessels have thick, muscular walls that pulse and carry blood away from the heart to the organs?

    A. capillaries   B. arteries   C. veins   D. alveoli

31. Where in the human respiratory system does the exchange of oxygen for carbon dioxide take place?

    A. alveoli   B. bronchioles   C. bronchi   D. stomates

32. Cellular respiration takes place at the:

    A. vacuoles   B. chloroplasts   C. ribosomes   D. mitochondria

33. The structures in the human kidney that filter waste products out of the blood are called:

    A. nephridia   B. nephrons   C. Malpighian tubules   D. tracheal tubes

34. The skin helps maintain body temperature when water is removed from its surface by the process of:

    A. evaporation   B. condensation   C. precipitation   D. transpiration

35. In humans, locomotion is accomplished through the interaction of:

    A. muscles and setae   B. muscles and glands
    C. muscles and chitin   D. muscles and bones

36. Which of the following life functions is most closely associated with the ability of an organism to adapt to changes in its environment?

    A. ingestion   B. excretion   C. circulation   D. regulation

37. The part of a neuron that contains the nucleus, cytoplasm and dendrites is called the:

    A. axon   B. cyton   C. synapse   D. terminal branch

38. The part of the brain used to answer this question is called the:

    A. cerebrum   B. cerebellum   C. medulla   D. hypothalamus

39. Which of the following best describes an endocrine system response?

    A. fast and of short duration   B. slow and widespread, often targeting several organs
    C. fast and of long duration   D. slow and limited, always targeting only one organ

40. Select the endocrine gland known as the *master gland*.

    A. pituitary   B. thyroid   C. hypothalamus   D. adrenal

41. Which of the following chemical substances does *not* have an effect on life functions in animal-like protists?

    A. food   B. oxygen   C. hormones   D. enzymes

42. Regulation in a hydra depends on its:

    A. ganglia   B. spinal cord   C. brain   D. nerve net

43. An earthworm takes in oxygen through its:

    A. skin   B. lungs   C. tracheal tubes   D. mouth

44. If the spiracles of a grasshopper become clogged, the grasshopper has difficulty with the life function of:

    A. circulation   B. excretion   C. digestion   D. respiration

45. During asexual reproduction, daughter cells are formed by:

    A. metamorphosis   B. parthenogenesis   C. mitosis   D. meiosis

46. In mitosis, replication of the chromosomes takes place during:

    A. interphase   B. prophase   C. anaphase   D. telophase

47. Animal-like protists and bacteria usually reproduce by:

    A. budding   B. binary fission   C. regeneration   D. vegetative propagation

48. If a cell has a haploid number of 16, its diploid number of chromosomes is:

    A. 4   B. 8   C. 16   D. 32

49. Two cell divisions is characteristic of the process of:

    A. spore formation   B. cleavage   C. mitosis   D. meiosis

50. In the human male reproductive system, the urethra functions in:

    A. excretion only   B. reproduction only
    C. excretion and reproduction   D. fertilization and development

51. Zygote formation takes place in a part of the female reproductive system known as the:

    A. ovary   B. fallopian tube   C. vagina   D. uterus

52. The menstrual cycle is controlled by the interaction of hormones produced by the:

    A. uterus and ovaries   B. vagina and cervix

    C. fallopian tubes and follicles   D. pituitary gland and ovaries

53. What is the name of the series of cell divisions that results in the formation of an embryo?

    A. cleavage   B. meiosis   C. budding   D. oogenesis

54. Which structure is formed from a combination of tissue composed of the mother's uterine wall and the developing fetus?

    A. umbilical cord   B. placenta   C. amnion   D. fallopian tube

55. Vegetative propagation is:

    A. the formation of a new plant from seeds

    B. a method of producing vegetables

    C. a form of sexual reproduction

    D. the formation of a new plant from part of an already existing plant

56. When a stem is cut off a plant and attached to the stem of a plant that is rooted in the ground, the process is called:

    A. stem cutting   B. grafting   C. layering   D. natural attachment

57. The male gamete produced by a flowering plant is the:

    A. ovule   B. pistil   C. pollen grain   D. seed

58. Which of the following is characteristic of sexual reproduction in flowering plants?

    A. no fertilization   B. single fertilization

    C. double fertilization   D. triple fertilization

59. The epicotyl and hypocotyl are:

    A. parts of a flower   B. food for a developing embryo

    C. the embryo of a seed   D. the male reproductive parts of a plant

60. To reduce competition between a parent plant and germinating plants for scarce resources such as light, soil and water, seeds must:

    A. be dispersed   B. develop   C. be dormant   D. have the same niche

61. The basic hereditary factor that is inherited from one generation to the next is the:

    A. sperm cell   B. egg cell   C. gene   D. zygote

62. The characteristic, or version, of a trait that an organism has is called its:

    A. phenotype   B. genotype   C. gametes   D. DNA

63. In pea plants, the allele for the tall characteristic of height is dominant over the allele for the short characteristic. Tall pea plants are crossed with short pea plants, resulting in 149 tall plants and 153 short plants. Based on these results, the genotype of the tall plant is:

    A. homozygous dominant   B. heterozygous   C. recessive   D. not determinable

64. Which of the following genetic crosses best illustrates Mendel's Law of Segregation?

    A. TT × TT   B. tt × tt   C. Tt × Tt   D. Tt × tt

65. Which of the following genetic crosses can produce a phenotype ratio of 9:3:3:1?

    A. YyRr × YyRr   B. YYRR × YYRR

    C. YYRR × yyrr   D. yyrr × yyrr

66. When two pink Japanese four o'clock plants are crossed, offspring are produced in a ratio of:

    A. 75% white, 25% red   B. 75% red, 25% white

    C. 100% pink   D. 25% red, 50% pink, 25% white

67. In humans, the number of autosomes in a sperm cell is:

    A. 1   B. 22   C. 23   D. 46

68. Which cross can produce a child with blood type A?

    A. $I^B i \times ii$ B. $I^B i \times I^A I^A$ C. $I^A I^A \times I^B I^B$ D. $I^A I^A \times ii$

69. Traits controlled by genes located on the X chromosome are:

    A. sex linked   B. dominant   C. multiple alleles   D. homozygous

70. A normal bean seedling that has the ability to produce chlorophyll doesn't when grown in soil lacking magnesium. This demonstrates:

    A. that a mutation probably occurred, preventing chlorophyll production

    B. natural selection

    C. the influence of the environment on gene expression

    D. the need for light in chlorophyll production

71. Deoxyribose, a phosphate group, and thymine combine to form:

    A. RNA   B. a nucleotide   C. a codon   D. an amino acid

72. The replication of the DNA molecule is known as:

    A. mitosis   B. translation

    C. semiconservative replication   D. asexual reproduction

73. Identify the two steps involved in protein synthesis.

    A. translocation and transpiration   B. evaporation and condensation

    C. DNA replication and ATP production   D. transcription and translation

74. Each set of three nitrogen bases on an mRNA molecule is called a:

    A. peptide   B. nitrogen bond   C. codon   D. nucleotide

75. A tRNA molecule has a structure that allows it to:

    A. carry an amino acid on one end, while having an anticodon on the other end

    B. manufacture mRNA

    C. make DNA molecules

    D. carry ribosomal RNA

76. During protein synthesis, if one nitrogen base is replaced by another, the results is a:

    A. chromosome mutation   B. gene mutation   C. translocation   D. deletion

77. Nondisjunction during meiosis can result in a gamete that is:

    A. $2n$ B. $n + 1$ C. $2n + 1$ D. $2n - 1$

78. Huntington's disease is caused by:

    A. a defective dominant allele   B. a defective recessive allele
    C. nondisjunction of chromosome pair number 21   D. a missing enzyme

79. The process by which genes from the chromosomes of one species are inserted into the chromosomes of another species is called:

    A. addition   B. genomics   C. genetic engineering   D. parthenogenesis

80. Which area of biology explains the change that occurs in a species with the passage of time?

    A. genetics   B. evolution   C. ecology   D. biotechnology

81. The remains of organisms that lived in the distant past are called:

    A. codons   B. fossils   C. alleles   D. Archeae

82. Scientists use the radioactivity of certain elements and their half-life to:

    A. determine the occurrence of genetic mutations   B. create a common ancestor
    C. create variations in a species   D. determine the age of a fossil

83. Structures that are similar in construction and evolutionary development but dissimilar in function are called:

    A. vestigial   B. homologous   C. analogous   D. complementary

84. According to most biologists today, hereditary variations are due to:

    A. the use or lack of use of organs   B. the need to adapt to a changing environment
    C. mutations   D. the inheritance of acquired characteristics

85. According to the Hardy-Weinberg law, for the frequency of an allele to remain constant from generation to generation, all the following conditions must be met *except:*

    A. mutations   B. a large gene pool   C. no migration   D. no natural selection

86. A point on earth north or south of the equator can be measured in degrees of:

    A. attitude   B. longitude   C. altitude   D. latitude

87. An example of a population is all the:

    A. mice living in a field   B. mice and owls living in a field
    C. animals in a field and their surrounding environment   D. animals in a field and their food

88. A relationship between two organisms that live together, where at least one of the organisms benefits from the association is called:

    A. intraspecific competition   B. symbiosis
    C. the predator-prey relationship   D. abiotic interaction

89. The further down we move in a food chain from producers to final consumers, the:

    A. more energy becomes available   B. number of organisms increases
    C. number of plants increases   D. less energy is available

90. Identify the two processes involved in the water cycle.

    A. evaporation and precipitation   B. osmosis and dehydration synthesis
    C. protein synthesis and nitrogen fixation   D. photosynthesis and respiration

91. The process by which one biotic community is replaced by another is known as:

    A. a food chain   B. ecological succession
    C. the flow of energy through the ecosystem   D. population growth

92. A large geographical area characterized by specific abiotic factors and a climax community of plants and animals that are unique to an area is called a(n):

    A. ecosystem   B. habitat   C. biome   D. continent

93. Which of the following factors places the greatest demand on the earth's limited resources?

    A. population growth   B. disease   C. global warming   D. forest fires

94. The tropical rain forests must be protected and conserved because of their:

    A. biodiversity
    B. rich soil that can be used for growing crops
    C. ability to remove carbon dioxide from the atmosphere
    D. ability to produce large amounts of oxygen

95. Which of the following is *not* an alternative source of energy that can replace fossil fuels?

    A. hydrogen fuel   B. wind energy   C. solar energy   D. natural gas

Pretest Answers

1. **C.**

If you missed the preceding problem, study "Life Functions," page 15.

2. **A.**

If you missed the preceding problem, study "Tools of the Biologist," page 18.

3. **A.**

If you missed the preceding problem, study "Microscope Measurement," page 21.

4. **B.**

If you missed the preceding problem, study "Other Tools," page 22.

5. **D.**

If you missed the preceding problem, study "The Scientific Method," page 23.

6. **B.**

If you missed the preceding problem, study "Cell Theory," page 33.

7. **D.**

If you missed the preceding problem, study "Cell Structure and Function," page 35.

8. **C.**

If you missed the preceding problem, study "The Animal Cell As Seen with an Electron Microscope," page 37.

9. **B.**

If you missed the preceding problem, study "Molecular Transport," page 39.

10. **B.**

If you missed the preceding problem, study "Classification Subdivisions," page 47.

11. **C.**

If you missed the preceding problem, study "Binomial Nomenclature," page 48.

12. **A.**

If you missed the preceding problem, study "Viruses," page 50.

13. **B.**

If you missed the preceding problem, study "Domain Bacteria," page 51.

14. **C.**

If you missed the preceding problem, study "Kingdom Protista," page 53.

15. **B.**

If you missed the preceding problem, study "Kingdom Fungi," page 54.

16. **D.**

If you missed the preceding problem, study "Kingdom Plantae," page 56.

17. **A.**

If you missed the preceding problem, study "Kingdom Anamalia," page 58.

18. **D.**

If you missed the preceding problem, study "Vertebrates," page 60.

19. **B.**

If you missed the preceding problem, study "Carbohydrates," page 75.

20. **B.**

If you missed the preceding problem, study "Lipids," page 78.

21. **C.**

If you missed any of the preceding problems, study "Proteins," pages 79–80.

22. **A.**

If you missed the preceding problem, study "Acids and Bases," pages 82–83.

23. **D.**

If you missed the preceding problem, study "Enzyme Action," page 84.

24. **A.**

If you missed the preceding problem, study "Factors That Affect the Rate of Enzyme Action," page 85.

25. **B.**

If you missed the preceding problem, study "Leaf Structure and Function," pages 95–96.

26. **A.**

If you missed the preceding problem, study "The Photosynthetic Process," pages 98–99.

27. **B.**

If you missed the preceding problem, study "Stem Structure and Function," pages 101–102.

28. **B.**

If you missed the preceding problem, study "The Photosynthetic Process," pages 98–99.

29. **C.**

If you missed the preceding problem, study "Nutrition and Digestion," pages 113–114.

30. **B.**

If you missed the preceding problem, study "Circulation," page 116.

31. **A.**

If you missed the preceding problem, study "Organism Respiration," page 119.

32. **D.**

If you missed any of the preceding problems, study "Cellular Respiration," page 121.

33. **B.**

If you missed the preceding problem, study "Structure and Function of the Kidneys," page 125.

34. **A.**

If you missed the preceding problem, study "The Skin As an Organ of Excretion," page 129.

35. **D.**

If you missed the preceding problem, study "Locomotion," page 130.

36. **D.**

If you missed the preceding problem, study "Regulation," page 132.

37. **B.**

If you missed the preceding problem, study "The Structure of a Neuron," page 134.

38. **A.**

If you missed the preceding problem, study "The Brain," page 136.

39. **B.**

If you missed the preceding problem, study "Structure and Function of the Endocrine System," pages 138–139.

40. **A.**

If you missed the preceding problem, study "Endocrine Glands," pages 141–142.

41. **C.**

If you missed the preceding problem, study "Animal-like Protists," pages 153–154.

42. **D.**

If you missed the preceding problem, study "Phylum Cnidaria," pages 155–156.

43. **A.**

If you missed the preceding problem, study "Phylum Annelida," pages 157–158.

44. **D.**

If you missed the preceding problem, study "Phylum Arthropoda," pages 159–160.

45. **C.**

If you missed the preceding problem, study "Asexual Reproduction," page 167.

46. **A.**

If you missed the preceding problem, study "Asexual Reproduction," page 167.

47. **B.**

If you missed the preceding problem, study "Examples of Asexual Reproduction," page 169.

48. **D.**

If you missed the preceding problem, study "Sexual Reproduction," pages 173–174.

49. **D.**

If you missed the preceding problem, study "Sexual Reproduction," pages 173–174.

50. **C.**

If you missed the preceding problem, study "The Male Reproductive System," pages 183–184.

51. **B.**

If you missed the preceding problem, study "The Female Reproductive System," pages 185–186.

52. **D.**

If you missed the preceding problem, study "The Menstrual Cycle," pages 187–188.

53. **A.**

If you missed the preceding problem, study "Embryological Development," pages 189–190.

54. **B.**

If you missed the preceding problem, study "Fetal Development," pages 191–192.

55. **D.**

If you missed the preceding problem, study "Vegetative Propagation," page 199.

56. **B.**

If you missed the preceding problem, study "Artificial Propagation," page 202.

57. **C.**

If you missed the preceding problem, study "Flower Structure and Function," page 205.

58. **C.**

If you missed the preceding problem, study "Pollination and Fertilization," pages 207–208.

59. **C.**

If you missed the preceding problem, study "Seed Structure and Function," pages 209–210.

60. **A.**

If you missed the preceding problem, study "Seed Dispersal and Germination," pages 211-212.

61. **C.**

If you missed the preceding problem, study "Gregor Mendel," page 221.

62. **A.**

If you missed the preceding problem, study "Genetic Vocabulary," page 221–222.

63. **B.**

If you missed the preceding problem, study "Mendel's Law of Dominance," page 223–224.

64. **C.**

If you missed the preceding problem, study "Mendel's Law of Segregation," page 224–225.

65. **A.**

If you missed the preceding problem, study "Mendel's Law of Independent Assortment," page 227.

66. **D.**

If you missed the preceding problem, study "Incomplete Dominance," pages 228–229.

67. **B.**

If you missed the preceding problem, study "Sex Determination," page 237.

68. **D.**

If you missed the preceding problem, study "Blood Types and Multiple Alleles," pages 238–239.

69. **A.**

If you missed the preceding problem, study "Sex Linkage," page 242.

70. **C.**

If you missed the preceding problem, study "The Influence of Environment on Heredity," page 246.

71. **B.**

If you missed the preceding problem, study "DNA Structure," pages 253–254.

72. **C.**

If you missed the preceding problem, study "DNA Replication," page 254.

73. **D.**

If you missed the preceding problem, study "Protein Synthesis," page 256.

74. **C.**

If you missed the preceding problem, study "Transcription," page 256.

75. **A.**

If you missed the preceding problem, study "Translation," pages 257–258.

76. **B.**

If you missed the preceding problem, study "Gene Mutations," page 259.

77. **B.**

If you missed the preceding problem, study "Chromosome Mutations," page 260.

78. **A.**

If you missed the preceding problem, study "Genetic Diseases," pages 262–263.

79. **C.**

If you missed the preceding problem, study "Biotechnology," page 264.

80. **B.**

If you missed the preceding problem, study "Evolution Defined," page 275.

81. B.

If you missed the preceding problem, study "Fossil Evidence for Evolution," page 275.

82. D.

If you missed the preceding problem, study "Dating of Fossils," page 278.

83. B.

If you missed the preceding problem, study "Additional Evidence for Evolution,"
pages 280–281.

84. C.

If you missed the preceding problem, study "Theories of Evolution," pages 282–283.

85. A.

If you missed the preceding problem, study "The Hardy-Weinberg Law," pages 284–285.

86. D.

If you missed the preceding problem, study "Abiotic Factors," pages 297–298.

87. A.

If you missed the preceding problem, study "Biotic Factors," page 299.

88. B.

If you missed the preceding problem, study "Symbiosis," page 301.

89. D.

If you missed the preceding problem, study "The Flow of Energy Through the Ecosystem,"
page 303.

90. A.

If you missed the preceding problem, study "Material Cycles," pages 306–307.

91. B.

If you missed the preceding problem, study "Ecological Succession," page 308.

92. C.

If you missed the preceding problem, study "World Biomes," pages 309–311.

93. A.

If you missed the preceding problem, study "Humans and the Environment," pages 313–314.

94. A.

If you missed the preceding problem, study "Environmental Protection," pages 315–316.

95. D.

If you missed the preceding problem, study "Environmental Protection," pages 315–316.

Chapter 1
Introduction to Biology

Biology is the science that studies living things, their structure and function. All living things (*organisms*) are composed of microscopic (too small to be seen by the human eye) units called cells. The cell is the basic unit of function for all living things. How can we tell if something is alive? Living things perform various *life functions* or *processes*.

Life Functions

The following sections introduce the major life functions.

Nutrition

Nutrition is the process by which an organism takes in food and uses this food for energy, growth, and repair. First the organism takes in food from its environment. This process is called *ingestion*. These large food molecules are *digested*, broken down into many small molecules, so that they can enter the cells through a process called absorption. When digested, large food molecule AB is broken down to produce small food molecules A and B (AB → A + B). When in the cell, some end products of digestion can be used for energy production.

Is a green plant alive? We all know that plants don't eat food. Plants make their own food by photosynthesis. Plants take in carbon dioxide and water from their environment, producing a sugar called glucose. This glucose serves as food for the plant.

Transport

Transport is the life function by which materials such as food, water, and oxygen from the environment are distributed to all cells of the organism. The transport system also carries waste products away from the cells. In humans and other animals, the blood circulatory system is responsible for carrying out this life function. In many plants, specialized structures called vascular bundles carry materials throughout the organism.

Respiration

Respiration provides the organism with the energy needed to carry out all the other life processes. During the process of respiration, oxygen is brought into the organism and is used to chemically release energy that is stored in food. Respiration in animals takes place on two different levels. On the organism level, oxygen is taken in and carbon dioxide is released. We usually refer to this process as breathing. On the cellular level, food and oxygen interact chemically to produce energy and the waste products carbon dioxide and water. This process is generally referred to as cellular respiration.

Excretion

As organisms perform their various life functions, waste products are produced. These waste products are often harmful or poisonous to the organism. Excretion is the removal of cellular waste products such as water, carbon dioxide, and nitrogen waste from an organism.

Elimination

Not all the food that is eaten by an organism is capable of being digested and can be classified as waste. The removal of undigested food as a semisolid waste material is called elimination.

Reproduction

Reproduction is the life function by which organisms produce new individuals of the same kind (species). Reproduction also produces new cells in an organism that are necessary for growth and repair. Reproduction is necessary for the survival of the species. If no individual members of a species reproduce, the species becomes extinct. However, reproduction is not necessary for the survival of an individual member of a species.

Growth

Growth is the increase in the size of the organism. Growth results from the reproduction of new cells and from the increase in cell size.

Repair

Repair is the ability of an organism to fix or mend a damaged part. Some organisms have the ability to replace a lost or damaged part. For example, a starfish that loses an arm can grow back the missing arm.

Synthesis

Synthesis is the process by which two or more smaller molecules are combined to form a larger molecule. For example, $A + B \rightarrow AB$. Small molecule A combines with small molecule B to form large molecule, AB. Synthesis is a building process (photosynthesis is an example of a synthesis reaction). The new, larger molecules produced by synthesis can be used to form cell parts, or they can be incorporated into the body of the organism (*assimilation*).

Locomotion

Locomotion is the ability of an organism to move from one place to another. Animals depend on locomotion for finding food, locating a mate, and avoiding predators. Most plants have very limited locomotion. Their leaves can grow toward light and their roots toward water.

Regulation

Regulation is the ability of an organism to respond to a *stimulus* (a change in the environment); the reaction of the organism is the *response*. A stimulus generally upsets the *homeostasis* (the stable internal environment) of an organism; the response returns the organism to homeostasis. Homeostasis occurs when all systems within the organism are in balance and working properly.

For example, human body temperature is approximately 37°C. If the temperature of the room in which the person is in rises from 22°C to 38°C (this is a stimulus), homeostasis is upset. The person begins to sweat (this is the response); sweating cools the body allowing the person to maintain a constant internal body temperature of 37°C (homeostasis). A person can also return to homeostasis by turning on an air conditioner or by going into a cool room. Regulation is the life function that allows an organism to adapt to a changing environment.

Two systems in humans help regulate the body and maintain homeostasis; they are the nervous system and the endocrine system. The nervous system is composed of the brain, spinal cord, and nerve cells. The endocrine system is composed of ductless glands (glands that do not have tubes). Examples of some endocrine glands are the pituitary gland, thyroid gland, and adrenal glands.

Metabolism is a term that is used to refer to all the chemical reactions that take place within an organism. Metabolism includes all the life functions that are performed by an organism.

Example Problems

The following problems are based on the life functions.

1. How does a biologist determine whether something is living or nonliving?

    Answer: A biologist determines whether something is living or nonliving by observing whether or not most of the life functions are performed.

2. How is nutrition in plants different from nutrition in animals?

    Answer: Plants make their own food, but animals must eat other organisms for food. Plants make food by the process of photosynthesis. Animals must hunt, capture, ingest, and digest food.

3. Place the following life functions into a logical sequence: digestion, transport, ingestion, absorption, excretion, and respiration. *Explain* why you selected this sequence.

    Answer: ingestion → digestion → transport → absorption → respiration → excretion

    ingestion—food is taken into the organism.

    digestion—food is broken down into small particles that can enter a cell.

    transport—food is distributed to all cells of the organism.

    absorption—food molecules enter the cell.

    respiration—the digested food is used for the production of energy.

    excretion—the waste products of digestion (carbon dioxide and water) are removed from the organism.

Work Problems

Use these problems on life functions for additional practice.

1. Why must an organism possess a transport system?

2. Explain why homeostasis is necessary for the survival of the organism.

3. Explain the following statement: "Reproduction is necessary for the survival of the species, but not for the survival of the individual."

Worked Solutions

1. In large organisms, most cells are not in direct contact with the environment. Such organisms require a transport system so that materials such as food, water, and oxygen from the environment can be distributed to all cells.

2. The life function of regulation allows the organism to adapt to changes in its environment. An organism that cannot respond to changes in its environment might become damaged or die.

    Stimulus → Homeostasis is upset. → Response → Homeostasis is restored.

    A stimulus (change in the environment of the organism) occurs.

    Homeostasis is upset (out of balance).

    The organism responds to the stimulus.

    Homeostasis is restored (balance).

3. Reproduction is necessary for the survival of the species. If no members of a species reproduce, the species becomes extinct. However, reproduction is not necessary for the survival of an individual member of a species. For example, if no humans reproduced, the human species would become extinct. But, an individual person does not have to reproduce to survive.

Tools of the Biologist

To study living things and carry out experiments, biologists use a variety of tools. One of the most important tools of the biologist is the microscope. Several different kinds of microscopes exist, and each one is specialized for a specific purpose. Two other tools used by the biologist are the microtome and the centrifuge.

Microscopes

Microscopes are fundamental to the study of biology. The following sections introduce these basic tools.

Simple Microscope

The simple microscope consists of a single convex lens that is capable of magnifying an object. When magnified, an object looks bigger, but its true size remains the same. A magnifying glass and a hand lens are examples of simple microscopes. A good simple microscope can magnify up to 300X.

Compound Microscope

The compound microscope uses two convex lenses at the same time. A good compound microscope can have a magnification up to 2,000X. The following diagram shows a typical compound microscope.

The Compound Microscope

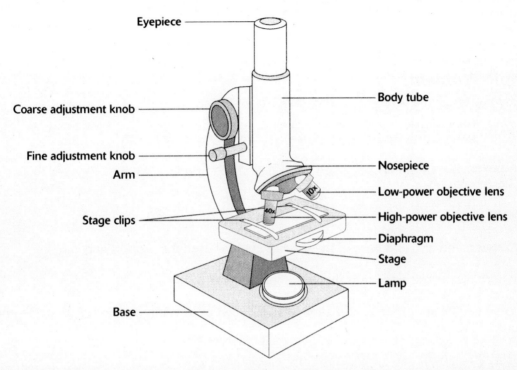

The following table is based on the diagram of the compound microscope. The column on the left lists the structure, and the column on the right gives its function.

The Compound Microscope	
Structure (Part)	**Function (Job)**
Ocular (eyepiece)	A convex lens used to magnify the image (usually 10X).
Tube (barrel)	Supports the eyepiece and the nosepiece.
Coarse adjustment knob	Moves the tube up and down and produces a rough focus of the image.
Fine adjustment knob	Moves the tube up and down by very small amounts and produces a sharp focus of the image.
Arm	Used to carry the microscope.
Nosepiece	Contains the high- and low-power objective lenses.
Low-power objective lens (LP)	A convex lens used to magnify the image (usually 10X).
High-power objective lens (HP)	A convex lens used to magnify the image (usually 40X).
Stage	Supports the glass slide and contains the specimen being observed.

(continued)

The Compound Microscope (continued)	
Structure (Part)	Function (Job)
Stage clips	Hold the slide in place.
Diaphragm	Adjusts the amount of light passing through the stage.
Light source (lamp or mirror)	Illuminates the specimen.

Example Problems

The following problems review the compound microscope.

1. Which parts of the compound microscope magnify the specimen?

    Answer: The eyepiece, low-power and high-power objective lenses magnify the specimen.

2. Identify the parts of the microscope that are used for support and carrying.

    Answer: The arm and base of the microscope are used for support and carrying.

3. Where on the microscope is the specimen placed?

    Answer: The specimen is on a slide that is placed on the stage of the microscope under one of the objective lenses.

Work Problems

Use these problems on the compound microscope for additional practice.

1. Which part of the microscope produces sharp focus?

2. How many lenses does a compound microscope use at one time?

3. Why must the specimen that is placed on a slide be thin?

Worked Solutions

1. The fine adjustment knob produces sharp focus by moving the tube up and down by very small amounts.

2. The compound microscope uses two lenses at a time. One is the ocular, and the other is either the low-power or high-power objective lens.

3. The compound microscope is a light microscope, and to see a specimen, light must be able to pass through it. Therefore, the specimen must be thin.

Microscope Measurement

If the eyepiece of a microscope magnifies 10X and the low-power objective magnifies 10X, the total magnification using these two lenses together is 100X (10×10). If the eyepiece magnifies 10X and the high-power objective magnifies 40X, the total magnification of the microscope is 400X (10×40). When using the high-power lens, your field of view is smaller, but your specimen looks larger. As magnification increases, the field of view of a microscope decreases proportionately. For example, if the magnification of a microscope is doubled, the field of view is cut in half. Also, the specimen appears to be upside down when viewed through a microscope, and the slide might need to be moved around to locate the specimen.

A microscope can also be used to measure the size of a specimen such as a cell. The micrometer (μm) is the unit used for cellular measurement. One millimeter (mm) is the equivalent of 1,000 μm. If the field of view of a microscope measures 1.5mm, it is equal to 1,500 μm ($1.5 \times 1,000$). A cell that takes up half of this field of view measures 750 μm ($1,500 \div 2$).

Stereomicroscope or Dissecting Microscope

This microscope uses dual eyepieces and an objective lens producing a three-dimensional image of a specimen. This microscope is used to study small organisms, either plant (moss, duckweed) or animal (insects, earthworms). Typically, these microscopes magnify from 10X to 40X. Also, as the name suggests, they can be used to dissect small organisms.

Electron Microscope

The electron microscope differs from the other microscopes in that it does not use light to illuminate the specimen. Instead, a beam of electrons is used to produce an image resulting in upward of 300,000X magnification. Electron microscopes can be used to study viruses and bacteria. The major advantage of the electron microscope is its high magnification. Its main disadvantage is that the specimen cannot be alive because it is placed in a vacuum inside the microscope.

Example Problems

The following problems are based on microscope measurement.

1. What is the size of a cell when the diameter of the field of view of a microscope is 1,000 μm and 10 equal-sized cells fit across the field of view?

    Answer: 100 μm

    Field of view size ÷ number of cells = cell size.

    1,000 μm ÷ 10 = 100 μm

2. The size of a cell when measured under low power (100X) is 240 μm. Find the size of the same cell when measured under high power (400X).

    Answer: 240 μm

    The size of a cell does not change as the magnification of the microscope increases. The cell only looks bigger.

3. How many micrometers long is a cell that measures 1.25 mm?

   Answer: 1,250 μm

   1 mm = 1,000 μm

   1.25 × 1,000 = 1,250 μm

   An alternate solution is to move the decimal three places to the right. This is the same as multiplying by 1,000.

Work Problems

Use these problems on microscope measurement for additional practice.

1. What is the *maximum* magnification of a microscope having a 10X eyepiece, a 43X objective and a 90X objective?

2. How many millimeters long is a cell that measures 375 μm in length?

3. How does the letter *P* look when viewed through a microscope?

Worked Solutions

1. 900

   Some compound microscopes have three lenses on the nosepiece. To find the *maximum* magnification of the microscope, multiply the magnifying power of the eyepiece (10X) with that of the most powerful objective lens (90X).

   10 × 90 = 900

2. 0.375 μm

   1mm = 1,000 μm

   375 mm ÷ 1,000 = 0.375 μm

   An alternate solution is to move the decimal three places to the left. This is the same as dividing by 1,000.

3. The letter *d* appears to be upside down when viewed through a microscope because the microscope inverts the image. If you turn your book upside down, you can see how the image should look.

Other Tools

In addition to microscopes, biologists use the following tools:

❏ **Microtome:** This instrument is used to slice a specimen very thin so that it can be observed with a light or electron microscope.

❏ **Centrifuge:** The centrifuge is used to spin specimens at very high speeds. The heaviest and densest parts settle to the bottom, and lighter parts remain at the top. If human blood is subjected to centrifugation, plasma (the liquid part of the blood) stays at the top, and the blood cells settle to the bottom.

❏ **Microdissection Tools:** These tools are extremely small and are used with the aid of a microscope to remove or add parts to cells. Cloning, which will be discussed in a later chapter, is one use for these tools.

The Scientific Method

The scientific method is an organized series of steps used by biologists to solve a problem. Although different biologists use different approaches to solve a problem, they all use certain steps in common.

1. **Problem Statement:** identify a question to be answered.

2. **Hypothesis:** Form an educated guess that provides a possible answer to the problem statement.

3. **Experiment:** Perform the actual procedures used to support or reject a hypothesis. The item being tested or changed is called the *independent variable*. The *dependent variable* is the change that occurs due to the procedures performed on the independent variable. The independent variable controls or determines the dependent variable. Generally, for an experiment to be considered valid, a *control* must be present. The control serves as a comparison point or group.

4. **Results:** Collect and record the outcomes and information (*data*) obtained as a result of the experiment. A biologist determines the results by observing, measuring and weighing. The data that a biologist gathers are often presented in a table or a graph. This makes it easier to interpret the results of the experiment.

5. **Conclusions:** Using the results of the experiment, a biologist either supports or rejects the hypothesis, thus answering the original question.

An experiment must be capable of being repeated and producing substantially similar results. Otherwise, the original experiment must have been flawed in some way and is therefore not valid. The greater the number of experimental test subjects used, the more statistically valid the results. Only one independent variable should be used in an experiment. Everything else in the experiment must be kept constant.

Example Problems

The following paragraph illustrates the use of the scientific method. Read the paragraph and answer the example problems that follow.

A student at the beach noticed that as she walked further and further into the water, the temperature of the water seemed to get colder. She wondered if any correlation existed between water temperature and water depth. The student performed the following experiment: She measured the temperature of the water at five different depths and compared her results to the surface temperature of the water. The following table indicates her findings:

Water Temperatures at Various Depths	
Water Depth (Meters)	Temperature (°C)
0	20
10	18
20	15
30	12
40	10
50	7

1. What is the problem statement for this investigation?

    Answer: How is water temperature affected by water depth?

    A problem statement should always be in the form of a question. In this investigation the student was measuring the temperature of the water at different depths.

2. State an appropriate hypothesis for this investigation.

    A. As the depth of the water increases, temperature increases.
    B. As the depth of the water increases, temperature decreases.
    C. As the depth of the water increases, temperature remains the same.

    A hypothesis is an educated guess that might provide an answer to the question. At this point, the student does not know the answer to the question. However, she noticed that as she went further into the water, the temperature of the water seemed to get colder. This is why hypothesis **B** is the best choice. It is possible that as depth increases, water temperature increases or remains the same. Hypotheses **A** and **C** are also acceptable because the hypothesis can be accepted or rejected based on the results of the experiment.

3. Identify the independent variable.

    Answer: The independent variable is water depth.

4. What is the control in this investigation?

    Answer: The temperature of the water at the surface (0 meters) is the control for this experiment.

    A control can serve as a comparison point in an experiment. We are measuring the temperature of the water at different depths and comparing the results to the temperature of the water at the surface. Without knowing the surface temperature of the water, we would not be able to determine whether water temperature increases, decreases, remains the same, or varies with water depth.

5. Write a conclusion for this experiment based on the results given in the preceding table.

    A. As water depth increases, water temperature decreases.

    B. As water temperature decreases, water depth increases.

    C. As water temperature increases, water depth decreases.

    An examination of the table shows an inverse (opposite) relationship existing between the two variables in this experiment; as one increases, the other decreases. Although all three answers are correct, answer **A** is preferred because it clearly demonstrates that the independent variable (water depth) is responsible for determining or controlling the dependent variable (water temperature).

Work Problems

Work problems 1–5 are based on the following passage and the accompanying table.

Some people believe that taking high doses of vitamin C reduces the number of days that a person suffers from the common cold. To test this belief, two groups of 100 people per group were selected. Group A (cold sufferers) was given 1,000 mg of vitamin C each day for one week beginning with the first day of cold symptoms. Group B (cold sufferers) was given a placebo (a substance of no remedial value, such as sugar) each day for one week beginning with the first day of cold symptoms. The following table indicates the results of the experiment:

	Effects of Vitamin C on Cold Sufferers	
Number of Days with a Cold	*Group A* *Number of Cold Sufferers Receiving 1,000 mg Vitamin C per Day*	*Group B* *Number of Cold Sufferers Receiving Placebo*
1	100	100
2	98	97
3	53	56
4	41	38
5	10	8
6	2	3
7	0	0

1. What is the problem statement for this experiment?

2. State the hypothesis.

3. Identify the dependent variable.

4. Which is the control group?

5. State the conclusion for this experiment.

Worked Solutions

1. Does 1,000 mg per day of vitamin C reduce the number of days a person suffers with the common cold?

   We are looking to see the effect of vitamin C on the common cold. We want to know whether the vitamin reduces the number of days that a person has a cold.

2. Taking 1,000 mg per day of vitamin C reduces the number of days a person suffers with the common cold.

   A hypothesis is an educated guess. We are wording the hypothesis in this manner because we are hopeful that vitamin C works. However, if we do not believe that vitamin C works, we can word the hypothesis in the negative: Taking 1,000 mg per day of vitamin C does not reduce the number of days a person suffers with the common cold.

3. The dependent variable is the change (number of individuals) that occurs due to the procedure performed. The use of 1,000 mg of vitamin C per day is the independent variable.

   The dependent variable in this experiment is the number of individuals with the common cold.

4. Group B is the control group.

   A control group is a comparison group. In this experiment we are comparing the results of giving vitamin C to group A to the results of giving a placebo to group B (the control group). If we did not have a control group, it would appear that vitamin C is effective because with each passing day fewer people had a cold. However, when we look at the control group, we see approximately the same result.

5. Vitamin C has no effect on the duration of the common cold.

   We can come to this conclusion because the result of taking vitamin C is the same as not taking vitamin C. Note that our conclusion rejects the hypothesis.

Chapter Problems and Answers

Problems

The following is a description of the life functions of the African antelope. For problems 1–10, fill in the name of the life function that is suggested.

A herd of antelope running across the African plains. _____ is an exciting sight to see.

1

The grass the antelope consume is broken down chemically in the stomach and the small

intestine by the action of enzymes _____. The broken-down food substances eventually

2

reach all the cells of the antelope by way of the blood stream _____. In the cells, the
3
nutrient molecules undergo chemical reactions being built up _____, after which they
4
become living parts of the cell. Powerful lungs supply oxygen, which when combined

with food results in the production of energy _____. Waste products from the body's
5
activities pass to the skin and kidneys to be removed from the body _____. Meanwhile,
6
undigested food solids are _____. A sharp sense of smell warns the animal of
7
approaching danger _____. Young antelope are usually born singly in the springtime
8
_____. The increase in the size of the young antelope over a year is surprising _____.
9   10

For problems 11–18, answer the question.

11. What is the size of a cell being observed through a microscope if the low-power field of view size is 1,200 µm and the number of cells counted across the diameter of the field of view is 10? _____

12. What is the size of a cell if the low-power field of view measues 1,300 µm and the number of cells counted across the diameter of the field of view is 10? _____

13. How do we find the size of a cell, if we know the size of the field of view and the number of cells that fit across the diameter of the field of view? _____

14. What is the size of a cell measured under high power (400X magnification) if the size of the cell measured under low power (100X magnification) is 80 µm? _____

15. When switching from low to high power, the real size of the cell _____.

16. Under high power a cell only looks _____.

17. How many amoeba (small one-celled animals) can be seen under high power if the high-power field of view of a microscope measures 400 µm and the amoeba measures 200 µm? _____

18. How many micrometers long is a cell that measures 0.75 mm? _____

Problems 19–21 are based on the following diagram of a microscope.

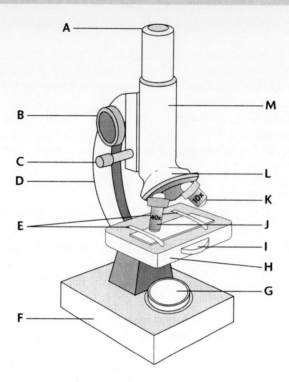

19. Which part of the microscope is the high-power objective lens? _____

20. Which part of the microscope is used to select the desired objective lens?

      _____

21. Which *two* parts of the microscope are used for focusing? _____

For problems 22–25, give the name of the biological tool that is associated with each description.

22. This tool produces a three-dimensional image. _____

23. This tool separates materials based on their density. _____

24. This tool is used to remove parts of cells. _____

25. This tool does not use light for magnification. _____

Answers

1. **Locomotion.** The antelope are running and moving from one place to another.

2. **Digestion.** Digestion is a process by which large food molecules are broken down.

3. **Transport.** This is the life function by which materials such as food, water, and oxygen from the environment are distributed to all cells of the organism. In animals, the blood circulatory system is responsible for carrying food substances to all parts of the organism.

4. **Synthesis.** This is a building process by which smaller molecules are combined to form larger molecules.

5.   **Respiration.** This life function provides the antelope with energy. During the process of respiration, oxygen is brought into the antelope and is used to chemically release energy that is stored in food.

6.   **Excretion.** This is the removal of waste products (water, carbon dioxide, nitrogen waste) from an organism. The skin and kidneys are organs of excretion.

7.   **Eliminated.** Not all the food that is eaten by the antelope is capable of being digested. The removal of undigested food as a semisolid waste material is called elimination.

8.   **Regulation.** Regulation is the ability of an organism to respond to a stimulus; the reaction of the organism is the response. The antelope's keen sense of smell can detect a stimulus such as the scent of a lion and warns the animal of the approaching danger. The antelope can respond to this situation by running away. Regulation is the life function that allows the organism to survive by enabling it to adapt to changes in its environment.

9.   **Reproduction.** This is the life function by which organisms produce new individuals of the same kind (species).

10. **Growth.** This is the increase in the size of the organism.

11. **120 μm.** 1,200 μm ÷ 10 = 120 μm. You must indicate the correct unit of measurement, micrometers (μm), in your answer.

12. **130 μm.** 1,300 μm ÷ 10 = 130 μm. You must indicate the correct unit of measurement, micrometers (μm), in your answer.

13. **The field of view size divided by the number of cells equals the cell size.** Keep in mind that this works only if the cells are approximately the same size.

14. **80 μm.** The size of a cell does not change when magnified. Magnification just makes the cell look bigger. You must indicate the correct unit of measurement, micrometers (μm), in your answer.

15. **Remains the same.** Increasing the magnification does not change the actual size of a cell.

16. **Bigger.** Increased magnification results in a cell that looks bigger.

17. **Two.** Divide the field of view size by the size of the amoeba: 400 μm ÷ 200 μm = 2 amoeba.

18. **750 mm.** 1,000 μm = 1 mm. Multiply 0.75 by 1,000, or move the decimal three places to the right. You must indicate the correct unit of measurement, millimeters (mm), in your answer.

19. **J.** Structure **J** is the high-power objective lens. This lens is longer; it has greater magnification (40X) than the low-power objective lens (10X).

20. **L.** This is the nosepiece, which holds the objective lenses. The nosepiece is capable of being turned to select either the low-power or high-power objective lens.

21. **B and C.** Letter **B** indicates the coarse adjustment knob, which moves the tube up and down to produce a rough focus of the image. Letter **C** indicates the fine adjustment knob, which moves the tube up and down by very small amounts to produce a sharp focus of the image.

22. **Stereomicroscope.** This tool uses dual eyepieces and an objective lens, producing a three-dimensional image of the specimen. The effect obtained is similar to using binoculars.

23. **Centrifuge.** This tool is used to spin specimens at very high speeds. The heaviest and densest parts settle to the bottom, and lighter parts stay at the top.

24. **Microdissection tools.** These tools are extremely small and are used with the aid of a microscope to remove or add parts to cells.

25. **Electron microscope.** This tool uses a beam of electrons to produce an image, resulting in upward of 300,000X magnification.

Supplemental Chapter Problems

Problems

For problems 1–20, select the *best* answer.

1. The taking in of food by an organism is called:

   A. digestion   B. ingestion   C. assimilation   D. reproduction

2. The sum total of all the life functions is known as:

   A. homeostasis   B. regulation   C. synthesis   D. metabolism

3. All the following are necessary for the survival of the organism *except:*

   A. homeostasis   B. metabolism   C. reproduction   D. regulation

4. The life functions of an organism work together to maintain stability and constant conditions within an organism. This concept is referred to as:

   A. assimilation   B. homeostasis   C. transport   D. respiration

5. A general function of a transport system is to:

   A. receive stimuli from the external environment

   B. bring materials from the external environment into contact with all the cells of the organism

   C. break down nutrients so that the cells can use them

   D. remove solid waste materials from the digestive system

6. The change of digested food into the living matter of the cell is called:

   A. ingestion   B. digestion   C. absorption   D. assimilation

7. The process by which an organism obtains the energy it needs by releasing the chemical energy in nutrients is:

   A. ingestion   B. synthesis   C. respiration   D. digestion

8. The cells of an organism are capable of chemically combining simple substances to form more complex substances. This process is called:

   A. synthesis   B. digestion   C. assimilation   D. respiration

9. Which of the following life functions allows an organism to respond to changes in its environment?

    A. transport   B. regulation   C. respiration   D. synthesis

10. The removal of metabolic wastes from an organism is known as:

    A. excretion   B. repair   C. transport   D. synthesis

11. Which instrument is used to observe the structure of a virus?

    A. simple microscope   B. compound microscope
    C. ultracentrifuge   D. electron microscope

12. Which tool is used to move a cell part from one cell to another?

    A. dissecting microscope   B. ultracentrifuge
    C. microdissection tools   D. electron microscope

13. Insects, earthworms, and other small organisms can best be studied by using:

    A. a dissecting microscope   B. an ultracentrifuge
    C. microdissection tools   D. an electron microscope

14. This microscope uses one convex lens.

    A. simple microscope   B. compound microscope
    C. ultracentrifuge   D. electron microscope

15. A student looking at his finger through a microscope sees a totally black field of view. The most probable explanation for this is that:

    A. the microscope is not properly focused   B. the microscope is too powerful
    C. the microscope is not powerful enough   D. the student's finger is too thick

16. The total magnification of a microscope using a 10X ocular and a 60X objective lens is:

    A. 50X   B. 70X   C. 600X   D. 6000X

17. Which part of the microscope regulates light?

    A. diaphragm   B. ocular   C. fine adjustment knob   D. coarse adjustment knob

18. A student measures the length of a paramecium to be 100 µm under low power (100X magnification). He switches to high power (400X magnification) and measures the paramecium. What is the size of the paramecium under high power?

    A. 25µm   B. 100µm   C. 400µm   D. 4,000µm

19. How many millimeters is a cell that measures 675 micrometers?

    A. 0.675mm   B. 6.75mm   C. .000675mm   D. 675mm

20. The field of view of a microscope is 1,000 micrometers. A student counts five cells of equal size going across the field of view. How long is each cell?

    A. 20µm   B. 200µm   C. 500µm   D. 5,000µm

Answers

1. **B.** "Life Functions," p. 15

2. **D.** "Life Functions," p. 15

3. **C.** "Life Functions," p. 15

4. **B.** "Life Functions," p. 15

5. **B.** "Life Functions," p. 15

6. **D.** "Life Functions," p. 15

7. **C.** "Life Functions," p. 15

8. **A.** "Life Functions," p. 15

9. **B.** "Life Functions," p. 15

10. **A.** "Life Functions," p. 15

11. **D.** "Tools of the Biologist," p. 18

12. **C.** "Tools of the Biologist," p. 18

13. **A.** "Tools of the Biologist," p. 18

14. **A.** "Tools of the Biologist," p. 18

15. **D.** "Tools of the Biologist," p. 18

16. **C.** "Tools of the Biologist," p. 18

17. **A.** "Tools of the Biologist," p. 18

18. **B.** "Tools of the Biologist," p. 18

19. **A.** "Tools of the Biologist," p. 18

20. **B.** "Tools of the Biologist," p. 18

Chapter 2
The Unity of Life

The millions of different kinds of living things that exist on earth today have the following characteristics in common: They are composed of cells, function as a result of their cells and produce new offspring by cell reproduction. The concepts of *the cell theory* serve to *unite* all living things.

The Cell Theory

The development of the cell theory was closely tied to improvements in the microscope. In the 1590s two Dutch lens makers, Zacharias Janssen and his father, made a compound microscope by placing two convex lenses at each end of a tube. Robert Hooke (1665) used a compound microscope to observe cork from the bark of the cork oak tree and saw small, empty box-like structures that he named cells. The cells that Hooke studied were nonliving. However, Anton van Leeuwenhoek (working during the 1670s and 1680s) was the first to observe and describe magnified living things with simple microscopes. Leeuwenhoek was the first to see bacteria from teeth scrapings and animal-like protists (small one-celled animals) from pond water. Robert Brown (1831), working with orchid and other plant cells, saw a small structure inside the cells that he named the nucleus. Matthias Schleiden (1838) worked with plants and noticed that all plants are made of cells. The following year Theodor Schwann (1839), who worked with animals, stated that all animals are made of cells. Rudolf Virchow (1858) proposed that new cells arise only from previously existing cells, thus disputing the idea of spontaneous generation. The combined contributions of these scientists is known as *the cell theory:*

- ❏ The cell is the basic unit of structure in all living things.
- ❏ The cell is the basic unit of function in all living things.
- ❏ New cells arise only from previously existing cells.

The following is a list of some exceptions to the cell theory:

- ❏ The *first cell* is an exception to the cell theory because it could not have come from a previously existing cell.
- ❏ *Chloroplasts* and *mitochondria* are parts of cells that have their own DNA and reproduce independently of the cell; this makes them more basic than a cell in structure and function.
- ❏ *Viruses* are exceptions because they do not have the cell structures common to most cells. Most biologists do not consider them living things.

The following flowchart shows the *structural hierarchy* that is found in all organisms. Note that the cell is the basic unit of structure in all living things. Similar cells combine to form tissues, similar tissues combine to form organs, organs work together to form organ systems, and all the different organ systems combine to make up the organism.

cells → tissues → organs → organ systems → organism

Example Problems

These problems review cell theory and the unity of life.

1. Who was the first person to study living cells?

   Answer: Anton van Leeuwenhoek was the first to see living things such as bacteria and animal-like protists. Although Zacharias Janssen preceded Leeuwenhoek and made a compound microscope, he did not use it to study living things.

2. Who named the cell?

   Answer: Robert Hooke used a compound microscope to observe cork from the bark of the cork oak tree and saw small box-like structures that he named cells.

3. Who discovered the nucleus?

   Answer: Robert Brown discovered the nucleus while working with plant cells.

Work Problems

Use these problems on the cell theory and the unity of life for additional practice.

1. Why is a microscope necessary to study cells?

2. State the main concepts of the cell theory.

3. Explain the concept of *unity*.

4. Outline the structural hierarchy that is found in all organisms.

5. Why are viruses considered exceptions to the cell theory?

Worked Solutions

1. A microscope is needed to see cells because they are too small to be seen by the naked eye. The tools most often used to magnify cells are the compound and electron microscopes.

2. The cell theory states that the cell is the basic unit of structure and function in all living things. Also, new cells arise only from previously existing cells.

3. All organisms are composed of cells, function as a result of their cells, and produce new offspring by cell reproduction. These concepts of the cell theory serve to *unite* all living things.

4. The structural hierarchy that is found in all organisms builds on the concept that the cell is the basic unit of structure in all living things. Similar cells combine to form tissues, similar tissues combine to form organs, organs work together to form organ systems, and all the different organ systems combine to form the organism.

5. **Viruses** do not have the cell structures common to most cells. Most biologists do not consider them living things.

Cell Structure and Function

Animal and plant cells have many parts in common. The parts of a cell are called *organelles* (little organs). The following diagrams show typical animal and plant cells as they look when viewed with a compound microscope.

Typical Animal and Plant Cells

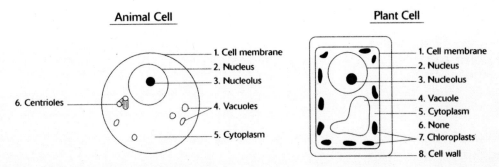

The following table is based on the diagrams of typical animal and plant cells. The numbers next to each organelle correspond to the numbers in the diagrams. The function of each organelle is given in the column on the right.

Organelles and Their Function	
Organelle	**Function (Job)**
1. Cell membrane	The outer boundary of the cell protects the cell, controls what goes in and out of the cell (*selective permeability*), and holds the cell together.
2. Nucleus	Contains *chromosomes,* which have DNA and control the heredity of the cell. Controls the reproduction of the cell. A nuclear membrane surrounds the chromosomes.
3. Nucleolus	Found inside the nucleus; associated with the production of ribosomes.
4. Vacuoles	Storage sites for food or water.
5. Cytoplasm	The liquid part of the cell found within the cell membrane. The organelles float inside the cytoplasm.
6. Centrioles	Centrioles are cylindrical in shape and function in the reproduction of the cell.
7. Chloroplasts	Small oval-shaped structures containing the green pigment, chlorophyll. They produce glucose during photosynthesis.
8. Cell wall	Found in plant cells outside the cell membrane. The cell wall is made of cellulose, which is rigid and nonliving. This organelle gives the plant cell support and shape.

All cells have the following parts: cell membrane, nucleus, nucleolus, vacuoles and cytoplasm.

Several key differences distinguish animal cells from plant cells: Animal cells have two centrioles, and plant cells do not. Plant cells have chloroplasts and cell walls; animal cells do not. Animal cells have many small vacuoles, but plant cells tend to have one or two large vacuoles. Both animal and plant cells have different shapes and sizes. Thus, shape and size are not a difference between these cells.

Example Problems

These problems review cell structure and function.

1. Define the term organelle.

    Answer: Organelles are the parts that make up a cell.

2. Name the organelles that are found in most cells?

    Answer: The organelles that are found in most cells are cell membrane, nucleus, nucleolus, vacuoles, and cytoplasm. Note: Mature red blood cells do not have a nucleus and are an exception.

3. What is the function of the nucleus?

    Answer: The nucleus contains chromosomes, which have DNA and control the heredity and reproduction of the cell.

4. What is the function of the cytoplasm?

    Answer: The cytoplasm is the liquid within the cell membrane in which all parts of the cell float.

Work Problems

Use these problems on cell structure and function for additional practice.

1. How are animal cells different from plant cells?

2. Why must the cell membrane be selectively permeable?

3. How is the cell membrane different from the cell wall?

4. Why must a plant cell have chloroplasts?

Worked Solutions

1. Animal cells have two centrioles and many small vacuoles. Plant cells do not have centrioles, but they have chloroplasts and one or two large vacuoles.

2. Selective (or semi) permeability is the ability of a cell membrane to control the movement of materials into and out of the cell. Materials that enter the cell are water, food, and oxygen, which are needed by the cell to perform its life functions. Materials that leave through the cell membrane are wastes such as carbon dioxide. Also, the cell membrane prevents harmful organisms and substances such as poisons, viruses, and bacteria from entering the cell.

3. The cell membrane is a living part of the cell and is found in all cells. The cell wall is nonliving and made of cellulose. It is found only in plant cells.

4. Chloroplasts contain the green pigment, chlorophyll, and function in photosynthesis for the production of food.

The Animal Cell As Seen with an Electron Microscope

If an electron microscope is used to magnify a cell, organelles can be seen that are not visible with a compound microscope.

Typical Animal Cell As Seen with an Electron Microscope

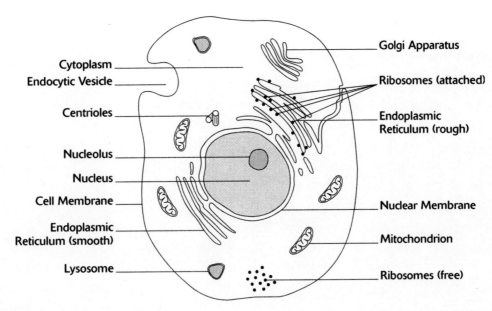

The following table lists the structures of the cell shown in the preceding figure and their function.

Typical Animal Cell As Seen with an Electron Microscope: Organelles and Their Function	
Organelle	*Function (Job)*
Cell membrane	Increased magnification shows that the cell membrane is composed of two layers of lipid with protein molecules suspended inside. This model of the cell membrane is called the fluid mosaic model.
Endoplasmic reticulum (ER)	The ER forms a canal-like network within the cell that functions in the transport of materials. ER with ribosomes attached is called rough ER.
Ribosome	This organelle makes proteins for the cell. Proteins are used for growth, repair, and reproduction of new cells.
Mitochondrion	This organelle is the energy factory of the cell and is associated with cellular respiration. It has an elliptical shape with outer and inner membranes. Mitochondria contain their own DNA and reproduce independently of the cell.
Golgi apparatus (body)	Flattened membranes that look like plates stacked on top of each other. The Golgi apparatus packages proteins.

(continued)

Typical Animal Cell As Seen with an Electron Microscope *(continued)*	
Organelle	*Function (Job)*
Lysosomes	Small, irregularly shaped structures that have their own membrane, containing digestive enzymes for the breakdown of food.
Endocytotic vesicle	Sometimes the cell membrane can form a small pocket or pouch that can engulf particles of food (endocytosis) that are too large to pass through the membrane. This pocket is called the endocytotic vesicle.

The following diagram shows the fluid mosaic model of the cell membrane.

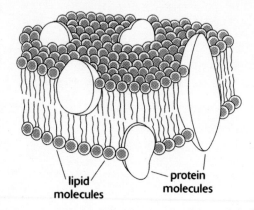

lipid molecules

protein molecules

Example Problems

These problems review cell structure and function.

1. How does the cell membrane differ from the nuclear membrane?

   Answer: The cell membrane is the outer boundary of the cell; the nuclear membrane surrounds and protects the chromosomes of the cell.

2. What is the function of ribosomes?

   Answer: Ribosomes are sites for protein synthesis. Proteins are needed for growth, repair, and reproduction of new cells.

3. Why are mitochondria important for a cell?

   Answer: Mitochondria are the energy factories of a cell and are associated with cellular respiration. In the mitochondria, oxygen is used to chemically release energy that is stored in food. During this process, the waste products carbon dioxide and water are produced.

Work Problems

Use these problems on cell structure and function for additional practice.

1. Explain the fluid mosaic model of the cell membrane.

2. How is the Golgi body different from the endoplasmic reticulum?

3. Why does the breakdown of the lysosome membrane result in the death of a cell?

4. Describe the process of endocytosis.

Worked Solutions

1. According to the fluid mosaic model, the cell membrane consists of two layers of lipid with proteins spread throughout the layers and in between the layers.

2. The Golgi body has flattened membranes that look like plates stacked on top of each other, and their function is to package proteins. The membranes of the endoplasmic reticulum form a canal-like network within the cell that functions in the transport of materials. Some endoplasmic reticulum membranes have ribosomes attached to them and are referred to as rough endoplasmic reticulum. Ribosomes are never attached to the Golgi body.

3. Lysosomes contain digestive enzymes for the breakdown of food. If the lysosome membrane breaks, the enzymes it contains leak into the cytoplasm and digest the organelles of the cell. This results in the death of the cell. Thus, lysosomes are sometimes called the "suicide sacs" of the cell.

4. Endocytosis is a process by which the cell membrane forms a small pocket or pouch that surrounds and engulfs food particles that are too large to pass through the membrane; this pocket is called the endocytotic or pinocytotic vesicle.

Molecular Transport

Molecules of food, oxygen, and water enter the cell through the cell membrane. Waste products such as carbon dioxide and urea must pass out of the cell through the cell membrane, or the cell dies. Molecular transport is the movement of molecules across a membrane. Two kinds of molecular transport are used in an organism: *passive transport* and *active transport*.

Passive transport is the movement of molecules from areas of high concentration to areas of low concentration (from where the molecules are many to where they are few). No energy is required for passive transport. Two kinds of passive transport exist:

❏ **Diffusion:** The movement of *any* molecule (*except water*) from an area of high concentration to an area of low concentration.

❏ **Osmosis:** The movement of *the water molecule only* from an area of high concentration to an area of low concentration.

Active transport is the movement of molecules from areas of low concentration to areas of high concentration (from where the molecules are few to where they are many). Energy is required for active transport.

Example Problems

These problems review molecular transport.

1. What is molecular transport?

   Answer: Molecular transport is the movement of molecules across a membrane.

2. Define passive transport.

   Answer: Passive transport is the movement of molecules from areas of high concentration to areas of low concentration. No energy is required.

3. Define active transport.

   Answer: Active transport is the movement of molecules from areas of low concentration to areas of high concentration. Energy is required.

Work Problems

Use these problems on molecular transport for additional practice.

1. How is molecular transport different from circulation?

2. Why can some molecules pass through a cell membrane while others cannot?

3. How is diffusion different from osmosis?

Worked Solutions

1. Circulation is concerned with the movement of materials such as food, oxygen, and water from the environment to all cells of the organism. Molecular transport explains how these materials get into the cell.

2. Small molecules can pass through a cell membrane; large molecules cannot.

3. Diffusion is the movement of any molecule (except water) from areas of high concentration to areas of low concentration. Osmosis is the movement of water molecules from areas of high concentration to areas of low concentration.

Chapter Problems and Answers

Problems

The following is a brief paragraph based on cells, their structure and function. For problems 1–10, provide the missing terms.

A student was looking at a cell with a compound microscope. The student observed the

presence of small green structures called _____. These structures are important because
                                            1

they provide the cell with _____ through the process of _____. A large, dense circular
                           2   3

structure called the _____ was also observed. A major part of the cell was occupied by
4

two large, clear storage structures called _____. All the cell parts were floating within a
5

clear liquid substance known as the _____. After careful observation, the student was
6

not able to find cylindrical structures called _____. The outer boundary of the cell is
7

the _____, which holds the cell together. This structure is important because it controls
8

the _____ of materials in and out of the cell. The cell the student was looking at was a
9

_____ cell.
10

For problems 11–15, provide the name of the scientist associated with the discovery mentioned.

11.   New cells arise from previously existing cells. _____

12.   All plants are composed of cells. _____

13.   All animals are composed of cells. _____

14.   First to see bacteria and animal-like protists. _____

15.   Made the first compound microscope. _____

For problems 16–20, use the following answer choices to select the organelle that is associated
with the description given. A choice can be used more than once or not at all.

   A. Centriole   B. Chloroplast   C. Endoplasmic reticulum   D. Golgi body   E. Lysosome

16.   This canal-like network of the cell functions in intracellular transport. _____

17.   These flattened membranes look like plates stacked on top of each other and package
      proteins. _____

18.   This structure, when combined with a food vacuole, helps with the process of digestion.
      _____

19.   This structure functions in the reproduction of animal cells. _____

20.   This organelle uses carbon dioxide and water to produce glucose during photosynthesis.
      _____

For problems 21–25, use the following answer choices to select the life function that is associated
with the given organelle and description. A choice can be used more than once or not at all.

   A. Respiration   B. Transport   C. Ingestion   D. Digestion   E. Reproduction

21.   *Chromosomes* have DNA and control the heredity of the cell. _____

22.   The *lysosome* contains enzymes capable of breaking large food molecules into many
      smaller molecules. _____

23. In *mitochondria*, food and oxygen interact chemically to produce energy. _____

24. The movement of the *cytoplasm* within the cell helps in the distribution of materials within a cell. _____

25. The *cell membrane* controls what goes in and out of the cell. _____

Answers

1. **Chloroplasts.** Chloroplasts are green because they contain the green pigment chlorophyll. Chloroplasts are found only in plant cells.

2. **Food.** The food produced by chloroplasts is glucose.

3. **Photosynthesis.** Photosynthesis is the process by which plant cells use water and carbon dioxide to produce glucose. This process takes place in the chloroplasts.

4. **Nucleus.** The nucleus of a cell usually appears as a dark circular structure in the cell. Often, the cell must be stained for its nucleus to be seen.

5. **Vacuoles.** Plants cells have one or two large vacuoles; animal cells have many small vacuoles.

6. **Cytoplasm.** Cytoplasm is the liquid part of the cell. All the cell's organelles move within the cytoplasm.

7. **Centrioles.** These structures function in the reproduction of animal cells.

8. **Cell membrane** and/or **cell wall.** The cell membrane and cell wall are so close to each other that they appear as one structure when observed with a compound microscope. Cell membrane is the best answer because the next sentence states that this structure controls the movement of materials in and out of the cell.

9. **Movement.** The cell membrane is semipermeable, allowing for the movement of materials in and out of the cell.

10. **Plant cell.** Several clues in the paragraph lead to the conclusion that the student is looking at a plant cell. The cell has small, green structures (chloroplasts) and two large vacuoles. Chloroplasts and large vacuoles are characteristic of plant cells.

11. **Virchow.**

12. **Schleiden.**

13. **Schwann.**

14. **Leeuwenhoek.**

15. **Janssen.**

16. **C. Endoplasmic reticulum.** Look carefully again at the diagram of the cell. The membranes of endoplasmic reticulum interconnect; the membranes of the Golgi body do not. Also, some endoplasmic reticulum membranes have ribosomes attached to them. Ribosomes are never attached to the Golgi body.

17. **D. Golgi body.** The membranes of the Golgi body never interconnect, and ribosomes are never attached to the Golgi body.

18. **E. Lysosome.** Lysosomes are structures that contain digestive enzymes.

19. **A. Centriole.** Plant cells do not have centrioles and can reproduce without them.

20. **B. Chloroplast.** Only plant cells have chloroplasts for the production of glucose.

21. **E. Reproduction.** Chromosomes also take part in cellular reproduction.

22. **D. Digestion.** When a lysosome combines with a food vacuole, its enzymes digest the food.

23. **A. Respiration.** Energy production is always associated with the life function of respiration.

24. **B. Transport.** Within a cell, the movement of the cytoplasm is the transport system.

25. **C. Ingestion.** The cell membrane allows food, oxygen and water to enter the cell.

Supplemental Chapter Problems

Problems

For problems 1–13, select the *best* answer.

1. All the following are concepts of cell theory *except:*

   A. The cell is the basic unit of structure in all living things.

   B. The cell is the basic unit of function in all living things.

   C. New cells arise only from previously existing cells.

   D. All living things have chloroplasts.

2. Which of the following *best* explains the concept of *unity of life?*

   A. metabolism   B. the cell theory   C. organ systems   D. fluid mosaic model

3. Which of the following represents the correct order from simplest to most complex?

   A. organism → organ systems → organs → tissues → cells

   B. cells → tissues → organs → organ systems → organism

   C. cells → organs → tissues → organ systems → organism

   D. organ systems → organism → organs → tissues → cells

4. A student observed a cell under the microscope. The student identified the cell as an animal cell and not a plant cell because of the presence of:

   A. chloroplasts   B. centrioles   C. a cell membrane   D. vacuoles

5. Which of the following organelles can be seen only with the aid of an electron microscope?

   A. ribosomes   B. cell membrane   C. nucleus   D. vacuoles

6. Which organelles are responsible for respiration and energy production?

 A. mitochondria   B. ribosomes   C. Golgi body   D. lysosomes

7. Which structure provides support for a plant cell?

 A. cell wall   B. cell membrane   C. endoplasmic reticulum   D. nucleus

8. The structure in the cell that stores food is the:

 A. Golgi body   B. mitochondrion   C. vacuole   D. centriole

9. Which structure of the cell is capable of engulfing a large food particle?

 A. Golgi body   B. endoplasmic reticulum

 C. lysosome   D. endocytotic vesicle or pinocytotic vesicle

10. Select the description that best illustrates the structure of the cell membrane.

 A. three layers: protein/lipid/protein

 B. three layers: lipid/protein/lipid

 C. two layers of protein

 D. two layers of lipid with proteins spread throughout the layers and in between the layers

11. Proteins are needed by a cell for:

 A. growth and repair   B. energy   C. transport   D. respiration

12. Semipermeability describes the ability of the cell membrane to:

 A. transport materials from one location to another within the cell

 B. control the movement of materials into and out of the cell

 C. actively manufacture vacuoles and lysosomes

 D. hold the cell together

13. A cell with many small vacuoles is *most* probably a(n):

 A. virus   B. bacteria   C. plant cell   D. animal cell

Problems 14–22 are based on the following diagram. Select the *best* answer.

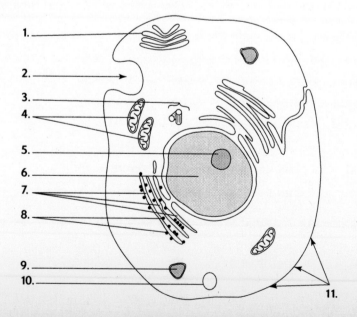

14. Which structure serves as the transportation network of the cell?

    A. 1   B. 3   C. 6   D. 7

15. This structure is a storage site.

    A. 2   B. 4   C. 9   D. 10

16. Protein synthesis takes place here.

    A. 1   B. 4   C. 8   D. 9

17. This structure is the site for cellular respiration.

    A. 1   B. 4   C. 8   D. 10

18. Which structure contains DNA and genes?

    A. 6   B. 7   C. 8   D. 9

19. Which structure functions in the reproduction of a cell?

    A. 2   B. 4   C. 6   D. 9

20. Digestive enzymes are always found here.

    A. 1   B. 4   C. 5   D. 9

21. This structure is an endocytotic or pinocytotic vesicle.

    A. 1   B. 2   C. 3   D. 4

22. The cell shown is a(n):

    A. plant cell   B. animal cell   C. paramecium   D. amoeba

For problems 23–25, select the *best* answer.

23. Which type of molecular transport requires energy?

    A. passive transport   B. active transport   C. diffusion   D. osmosis

24. The movement of molecules from areas of low concentration to areas of high concentration is characteristic of:

    A. passive transport   B. active transport   C. diffusion   D. osmosis

25. Water can enter a cell by the process of:

    A. diffusion   B. active transport   C. endocytosis   D. osmosis

Answers

1. **D.** "The Cell Theory," p. 33

2. **B.** "The Cell Theory," p. 33

3. **B.** "The Cell Theory," p. 33

4. **B.** "Cell Structure and Function," p. 35

5. **A.** "The Animal Cell As Viewed with an Electron Microscope," p. 37

6. **A.** "The Animal Cell As Viewed with an Electron Microscope," p. 37

7. **A.** "Cell Structure and Function," p. 35

8. **C.** "Cell Structure and Function," p. 35

9. **D.** "The Animal Cell As Viewed with an Electron Microscope," p. 37

10. **D.** "The Animal Cell As Viewed with an Electron Microscope," p. 37

11. **A.** "The Animal Cell As Viewed with an Electron Microscope," p. 37

12. **B.** "Molecular Transport," p. 39

13. **D.** "Cell Structure and Function," p. 35

14. **D.** "The Animal Cell As Viewed with an Electron Microscope," p. 37

15. **D.** "The Animal Cell As Viewed with an Electron Microscope," p. 37

16. **C.** "The Animal Cell As Viewed with an Electron Microscope," p. 37

17. **B.** "The Animal Cell As Viewed with an Electron Microscope," p. 37

18. **A.** "The Animal Cell As Viewed with an Electron Microscope," p. 37

19. **C.** "The Animal Cell As Viewed with an Electron Microscope," p. 37

20. **D.** "The Animal Cell As Viewed with an Electron Microscope," p. 37

21. **B.** "The Animal Cell As Viewed with an Electron Microscope," p. 37

22. **B.** "The Animal Cell As Viewed with an Electron Microscope," p. 37

23. **B.** "Molecular Transport," p. 39

24. **B.** "Molecular Transport," p. 39

25. **D.** "Molecular Transport," p. 39

Chapter 3
The Diversity of Life

D*iversity* refers to the millions of different kinds of living things that exist on earth. *Classification* is a method of organizing these diverse living things into a system of logical groupings that makes it easier for biologists to study them. *Taxonomy* is the branch of biology that is concerned with the classification and naming of organisms. Carolus Linnaeus, an 18th-century Swedish botanist, is credited with having developed our modern system of classification. Linnaeus is often referred to as the father of classification.

Classification

Today, one modern classification system organizes living things into three domains: Bacteria, Archaea and Eukarya, which is subdivided into four kingdoms: Protista, Fungi, Plantae, and Animalia. Another classification system is based on six kingdoms: Eubacteria, Archaeabacteria, Protista, Fungi, Plantae, and Animalia. The following table shows the relationship between a three-domain system of classification and a six-kingdom system of classification, arranged from simplest (on the top) to most complex (on the bottom).

Modern Systems of Classification	
Three-Domain System	*Six-Kingdom System*
Bacteria	Eubacteria
Archaea	Archaebacteria
Eukarya (includes the kingdoms:)	Protista
	Fungi
	Plantae
	Animalia

The domain system of classification replaces the older five-kingdom system (Monera, Protista, Fungi, Plantae, and Animalia). Domains Bacteria and Archaea replace the kingdom Monera, and domain Eukarya now includes the kingdoms Protista, Fungi, Plantae, and Animalia.

Classification Subdivisions

Each domain can be subdivided into the classification *taxa* (groups) shown in the following list. A domain is a diverse group of organisms whose characteristics have the *least* in common and is listed on the top. A species is a group containing organisms whose characteristics have the *most* in common and is listed on the bottom. If the definition of a species is known, the definitions for all other groups in the classification system follow.

Domain: a group of similar kingdoms

Kingdom: a group of similar phyla

Phylum: a group of similar classes

Class: a group of similar orders

Order: a group of similar families

Family: a group of similar genera

Genus: a group of similar species

Species: organisms that can mate with each other and produce fertile offspring

Organisms are classified on the basis of similarity of structure, evolutionary development and history (*phylogeny*). *Evolution* explains the origin and changes in a species from the distant past to the present.

Following is the classification of humans:

Domain: Eukarya

Kingdom: Animalia

Phylum: Chordata

Class: Mammalia

Order: Primates

Family: Hominidae

Genus: *Homo*

Species: *Homo sapiens*

Binomial Nomenclature

Linnaeus also developed *binomial nomenclature,* a two-name system for organisms. All organisms have a scientific name composed of their genus and specific epithet names. The first name of an organism is its genus name and is always capitalized. The second is its specific epithet name, which is written in lowercase. Both of these are either italicized or underlined to indicate that they are in a foreign language, in this case Latin. Linnaeus selected Latin because most scientists at the time knew the language, and it was not the language of any nation. Linnaeus understood the importance of being politically correct. The scientific name for humans is *Homo sapiens.*

The following table demonstrates how to determine the scientific name or species name of an organism. Notice that the scientific name of an organism is composed of its genus (first name) and a specific epithet (second name). The specific epithet often describes a distinguishing characteristic of the organism.

Scientific Naming			
Common Name	*Genus*	*Specific Epithet*	*Scientific Name*
Tiger	*Felis*	*tigris*	*Felis tigris*
Lion	*Felis*	*leo*	*Felis leo*
House cat	*Felis*	*domestica*	*Felis domestica*
Dog	*Canis*	*familiaris*	*Canis familiaris*

The tiger, lion, and house cat are all in the same *genus* (*Felis*). This tells us that these organisms have many characteristics in common. However, they are members of different species. Organisms that are in the same species have the greatest number of characteristics in common. Notice that *Canis familiaris*, the dog, is in a different genus (*Canis*) from cats. This immediately indicates that the dog shares fewer characteristics with cats.

Example Problems

These problems review classification.

1. Why do biologists classify living things?

    Answer: Living things are classified into logical groupings to make it easier for biologists to study them.

2. What is meant by diversity?

    Answer: Diversity refers to the millions of different kinds of living things that exist on earth.

3. What is phylogeny?

    Answer: Phylogeny refers to the evolutionary development and history of a species of organism.

4. How is the three-domain system of classification different from the six-kingdom system of classification?

    Answer: The three-domain system of classification includes Bacteria, Archaea, and Eukarya. Domain Eukarya is subdivided into the following four kingdoms: Protista, Fungi, Plantae, and Animalia. The six-kingdom system of classification includes Eubacteria, Archaeabacteria, Protista, Fungi, Plantae, and Animalia.

5. List the taxa of classification from the largest grouping to the smallest.

    Answer: domain → kingdom → phylum → class → order → family → genus → species

Work Problems

Use these problems on classification for additional practice.

1. How do biologists classify organisms?

2. Explain how organisms are named using binomial nomenclature.

3. How can we define a species?

4. *Sequoia sempervirens* is the scientific name for a redwood tree. Identify the genus and the species epithet for this organism.

5. *Ursus arctos* is a grizzly bear; *Ursus americanus* is a black bear. How is their classification similar and different?

Worked Solutions

1. Organisms are classified into groups on the basis of similarity of structure and evolutionary development. For example, an oak tree with structures such as roots, stems and leaves is classified into the plant kingdom and not the animal kingdom. Animal structures are totally different from those found in plants.

2. Binomial nomenclature is a two-name system for all organisms that was developed by Linnaeus. The first name of an organism is its genus name, and the second is its specific epithet. The genus name of the organism is always capitalized, and the specific epithet is written in lowercase. The genus name combined with the specific epithet forms the scientific or species name of the organism.

3. A species is defined as a group of organisms that can mate with each other and produce fertile offspring. Two conditions must be met for organisms to be considered members of the same species. First, the organisms must be able to mate with each other. (An oak tree and a bear are members of different species because they cannot mate with each other.) Second, the organisms must be able to produce fertile offspring (offspring capable of reproduction). A horse and a donkey are able to mate with each other and produce a mule. However, the horse and donkey are in different species because the mule is sterile (not capable of reproduction).

4. The genus name for the redwood tree is *Sequoia,* and the specific epithet is *sempervirens.* In binomial nomenclature, the genus name is written first, and the specific epithet is written second.

5. The grizzly bear and the black bear are in the same genus, *Ursus,* but in different species. The grizzly bear is in the species *Ursus arctos;* the black bear is in the species *Ursus americanus.*

Viruses

Biologists do not classify viruses into any domain. Viruses can be classified into an order, family, genus, and species. Most biologists do not consider viruses living things. At best, they seem to occupy a position between living and nonliving. When outside a cell, the virus does not perform any life functions. However, when inside a cell, the virus takes control of the host cell's nucleus by destroying its DNA, inserting its own genetic material (DNA or RNA) and taking over the life functions of the cell. The virus begins to reproduce or replicate (make copies of itself) within the host cell. Eventually, the cell bursts and dies, releasing the newly formed viruses so that the process can begin again. A virus consists mostly of genetic material surrounded by a protein coat called a *capsid.* Some viruses have a second protein coat. Viruses can infect the cells of bacteria, plants or animals. Viruses that infect one species or cell type are usually unable to infect other species or cell types. Many viruses are harmful to humans such as the cold virus, influenza, poliovirus, and HIV.

Example Problems

These problems review viruses.

1. Why are viruses not classified into any domain?

    Answer: Viruses are not classified into a domain because many biologists do not consider them living things. Outside a cell, viruses do not perform life functions such as respiration, ingestion, transport, excretion, and locomotion.

2.   Which life function is a virus capable of performing?

Answer: A virus is capable of reproduction only when inside a host cell.

3.   What is the structure of a virus?

Answer: A virus consists mostly of genetic material (DNA or RNA) surrounded by one or two protein coats.

Work Problems

Use these problems on viruses for additional practice.

1.   Describe the life cycle of a virus.

2.   Why can't a human catch a cold from a dog?

Worked Solutions

1.   A virus invades a cell's nucleus destroying the cell's DNA. The virus then inserts its own genetic material and takes over the life functions of the cell. The virus begins to replicate within the cell. Eventually, the cell bursts and dies—releasing the newly formed viruses, and the process can begin again.

2.   A human cannot catch a cold from a dog because the virus that causes a cold in dogs is unable to insert itself into human cells.

Domain Bacteria

Bacteria are unicellular (one-celled) organisms and are considered to be primitive living things. Bacteria first appeared 3.5 billion years ago. Also, they are the most numerous and widely spread organisms on earth. Bacteria have a ring of genetic material (one circular chromosome), a cell membrane, and a cell wall containing peptidoglycan (a sugar and polypeptide compound). However, they are *prokaryotic* cells (cells that lack a nuclear membrane).

Most bacteria are *heterotrophs;* they get their energy by consuming organic molecules found in their environment. Most bacteria are harmless to humans, but some are harmful such as anthrax, tuberculosis, and *Streptococcus,* which cause strep throat. Antibiotics can kill many harmful bacteria. Some bacteria are beneficial, such as nitrogen-fixing bacteria that make nitrogen available to plants, and the bacteria of decay, which help recycle materials in the environment.

In a six-kingdom system of classification, organisms in the domain Bacteria are classified in the kingdom Eubacteria.

Domain Archaea

Archaea are unicellular, prokaryotic organisms that are often found in the harshest environments on earth: vents in the sea, salt lakes, and hot springs. Archaea have cell walls that lack peptidoglycan, and they do not respond to antibiotics. Many biologists believe that the Archaea evolved into the early *eukaryotes* (cells that have a nuclear membrane).

In a six-kingdom system of classification, organisms in the domain Archaea are classified in the kingdom Archaeabacteria.

Example Problems

These problems review Bacteria and Archaea.

1. List the major characteristics of the domain Bacteria.

   Answer: Single-celled organisms (unicellular), primitive cells dating back at least 350 million years, with prokaryotic cells.

2. Define prokaryotic.

   Answer: Prokaryotic cells lack a nuclear membrane surrounding their genetic material.

3. How are heterotrophs different from autotrophs?

   Answer: Heterotrophs are organisms that obtain food for energy, growth, repair, and reproduction of new cells. Autotrophs are usually green plants that make their own food by the process of photosynthesis.

Work Problems

Use these problems on Bacteria and Archaea for additional practice.

1. How can we distinguish between members of the domain Bacteria and the domain Archaea?

2. Why are Bacteria considered primitive organisms?

3. A student found a colony of organisms in a salt lake and noticed that they had the following characteristics: no nuclear membrane, no peptidoglycan in their cell walls and they lacked the ability to manufacture their own food. Where should the student classify these organisms?

4. Which evolved first, viruses or Bacteria? (Review the section on viruses before answering this question.)

Worked Solutions

1. Bacteria have cell walls that contain peptidoglycan, and they respond to antibiotics. Archaea have cell walls that lack peptidoglycan, and they do not respond to antibiotics.

2. Bacteria are considered primitive organisms because the current-day Bacteria are similar to those that lived 350 million years ago. These organisms are prokaryotic, which is a primitive cellular characteristic.

3. These organisms should be classified in the domain Archaea. The lack of a nuclear membrane indicates that the organisms are prokaryotic. Archaea is the only domain with organisms that are prokaryotic, lack peptidoglycan, and are found in salt lakes.

4. Bacteria evolved before viruses. At first it might seem that viruses evolved before Bacteria because they are not cells and do not perform life functions except for reproduction. However, this is not the case. To reproduce, a virus must have a host cell to invade. The host cells must be there *first* for viruses to reproduce.

Domain Eukarya

The domain Eukarya is subdivided into the following four kingdoms: Protista, Fungi, Plantae, and Animalia. All organisms in these kingdoms have eukaryotic cells.

Kingdom Protista

In a three-domain system of classification, organisms in the kingdom Protista are in the domain Eukarya. All protists have *eukaryotic* cells because their chromosomes are surrounded by a nuclear membrane. The cell walls of some protists contain cellulose (a polysaccharide), and some protists have chloroplasts. The protists are a very diverse group of organisms. Most are unicellular with animal, plant-like, or fungus-like characteristics. Some form colonies, and others are multicellular. Protists can be heterotrophic or autotrophic. The three major categories of Protista and their most important phyla are described in the following sections.

Animal-Like Protists

Phylum Rhizopoda consists of organisms that have *pseudopods* (false feet—cellular extensions that constantly change shape), which are used for locomotion and to capture food. The Rhizopoda are found in fresh and salt water. In animals they cause disease, such as amebic dysentery. The amoeba is a member of the Rhizopoda. We will study the life functions of the amoeba in Chapter 7.

Phylum Ciliophora consists of organisms that have *cilia* (hair-like projections that beat back and forth), which are used for locomotion. These organisms are found in fresh and salt water. The paramecium is a member of the Ciliophora. We will study the life functions of the paramecium in Chapter 7.

Phylum Zoomastigina contains organisms that have a *flagellum* (tail), which is used for locomotion. These organisms are found in fresh water or as parasites in animals. *Trypansoma* is a member of the Zoomastigina that causes the disease called African sleeping sickness.

Older systems of classification use the term protozoa when discussing animal-like protists.

Plant-Like Protists

Phylum Euglenophyta consists of organisms that are unicellular and have both animal and plant characteristics. Euglena use flagella for locomotion and are found in fresh water. They are capable of photosynthesis when exposed to light and are heterotrophic in the dark.

Phylum Chlorophyta are green algae. They are unicellular, colonial or multicellular and are capable of photosynthesis. *Spirogyra* are members of Chlorophyta.

Fungus-Like Protists

Phylum Myxomycota consists of unicellular organisms that have many nuclei. Their means of locomotion is *amoeboid movement* (streaming of the cytoplasm). They are heterotrophic—eating bacteria, yeast, and decaying vegetation. Examples of Myxomycota are acellular slime molds.

Phylum Acrasiomycota consists of organisms that have one nucleus. Under starvation conditions, the single cells combine to form a group of cells called a pseudoplasmodium. Their means of locomotion is *amoeboid movement*. They are heterotrophic—eating bacteria, yeast, and decaying vegetation. Examples of Acrasiomycota are the cellular slime molds.

Example Problems

These problems review the kingdom Protista.

1. List the major characteristics of the kingdom Protista.

    Answer: Protista are mostly unicellular organisms with animal-like, plant-like, or fungus-like characteristics. Protists can be heterotrophic or autotrophic. All protists are eukaryotic.

2. How do amoeba, paramecia, and trypanosoma differ in their mode of locomotion?

    Answer: Amoeba move with the help of pseudopods, false feet that are cellular extensions of the organism and constantly change shape. As a result, the amoeba is an organism that does not have a definite shape. The paramecium is covered with tiny hair-like structures called cilia that are used to propel the organism back and forth in the water. Trypanosoma has a tail called a flagellum that it whips back and forth to propel itself.

3. What are the characteristics of slime molds?

    Answer: Slime molds are unicellular organisms. Some have only one nucleus (cellular slime molds); others have many nuclei (acellular slime molds). In many ways slime molds resemble a huge amoeba. All slime molds are nongreen and heterotrophic.

Work Problems

Use these problems on the kingdom Protista for additional practice.

1. How are prokaryotic cells different from eukaryotic cells?

2. How are the Euglenophyta and Chlorophyta different from animal-like protists?

3. How can we distinguish between green algae and Archaea?

Worked Solutions

1. Prokaryotic cells do not have a nuclear membrane and are characteristic of organisms in the domains Bacteria and Archaea. Eukaryotic cells have a nuclear membrane and are characteristic of organisms in the domain Eukarya.

2. Euglenophyta and Chlorophyta are phyla, containing organisms that have plant-like characteristics. They are green, autotrophic, and are capable of photosynthesis. The animal-like protists are nongreen and heterotrophic.

3. Green algae are in the kingdom Protista, and they are eukaryotic with cell walls that have cellulose. Archaea are prokaryotic and have cell walls lacking cellulose.

Kingdom Fungi

All fungi are eukaryotic: They have cell walls made of chitin (a complex carbohydrate), but their cells do not have chloroplasts. They are heterotrophic, mostly terrestrial and obtain their food by absorption from decaying vegetation or as parasites of animals and plants. As they decompose dead plant materials, fungi return important nutrients to the soil. Most are multicellular; however, yeast cells are unicellular. Many fungi are composed of thread-like filaments called hyphae that are attached to rhizoids on the bottom end. Rhizoids anchor the organism to its food source. The top of the organism often forms a reproductive structure called a sporangium. Several phyla of fungi exist, and they are classified according to methods of reproduction. The suffix -*mycota* indicates a phylum in the kingdom Fungi. Examples are: Zygomycota, Ascomycota, Basidiomycota, and Deuteromycota.

Beneficial fungi include yeasts, which are used in the baking (bread) and brewing (alcohol) industries and molds, some of which produce antibiotics. For example, *penicillium* is a mold that produces the antibiotic penicillin. Mushrooms are edible fungi, but many are poisonous. Some harmful fungus diseases are ringworm and athlete's foot in humans, and rust and striped smut in plants. These fungi are *parasitic* (organisms that live on or in other organisms for food). We will study yeasts and molds in Chapter 8.

Lichen is a combination of two organisms that live together, a fungus and a green alga or blue-green bacteria. The alga or bacteria live inside the fungus. The alga or bacteria provide the fungus with food produced by photosynthesis. The fungus provides the alga or bacteria with a place to live and the moisture necessary for photosynthesis. *Symbiosis* is a relationship between two organisms that live together where at least one of the organisms benefits from the association. This example of symbiosis is known as mutualism. Both organisms depend on each other for their survival. We will study symbiosis in more detail in Chapter 15.

Example Problems

These problems review the kingdom Fungi.

1. List the major characteristics of the kingdom Fungi.

    Answer: Fungi are eukaryotic, have cell walls made of chitin and their cells lack chloroplasts. They are heterotrophic, decomposers, and unicellular and multicellular organisms. Fungi are broken down into several phyla, with organisms classified according to methods of reproduction.

2. How do Fungi obtain food?

    Answer: Many fungi are heterotrophic and obtain their food by absorption from decaying vegetation.

3. Describe the structure of a fungus.

    Answer: Fungi are composed of thread-like filaments called hyphae that are attached to rhizoids on the bottom end. Rhizoids anchor the organism to its food source. The top of the organism often forms a reproductive structure called a sporangium.

4. What is a parasite? Give an example of a parasite that is a fungus.

    Answer: A parasite is an organism that lives off another organism called the host. The parasite often harms or kills its host. Examples of fungi that are parasitic in humans are those that cause ringworm and athlete's foot.

5. What is symbiosis? Give an example of symbiosis.

    Answer: Symbiosis is a relationship between two organisms that live together where at least one of the organisms benefits from the association. An example of symbiosis can be seen in lichen.

Work Problems

Use these problems on the kingdom Fungi for additional practice.

1. How are Fungi different from plant-like protists?

2. How are Fungi important to soil health and maintenance?

3. How do the two organisms that make up lichen depend on one another?

4. Why can't Fungi make their own food?

Worked Solutions

1. Fungi cells lack chloroplasts; as a result, fungi are heterotrophs. The plant-like protists are green, have chloroplasts and are autotrophic (making their own food by photosynthesis).

2. Fungi are important to soil health and maintenance because they decompose dead plant material, returning important nutrients to the soil.

3. Lichen is a combination of a green alga (or blue-green bacteria) living inside a fungus. The alga or bacteria provide the fungus with food produced by photosynthesis. The fungus provides the alga or bacteria with a place to live and the moisture necessary for photosynthesis.

4. Fungi can't make their own food because they are nongreen, lacking the photosynthetic pigment chlorophyll.

Kingdom Plantae

The members of the Plantae kingdom are green plants that probably evolved from algae. They are autotrophic, making their own food by photosynthesis. Their green color derives from the green pigment chlorophyll, which is found in their cells. Plants are multicellular and eukaryotic. The classification of green plants is complex, and many different approaches exist. For the sake of simplicity, we divide this kingdom into phyla. Many plant taxonomists use the term *division* instead of phylum. The suffix *-phyta* means plant.

Some phyla or divisions in the plant kingdom (*Bryophyta*) are nonvascular plants because they lack vascular tissue, which are specialized structures containing veins that transport materials throughout the plant. Other phyla or divisions (*Pterophyta Coniferophyta Anthophyta*) have vascular tissue, true roots, stems, and leaves and are called vascular plants. Vascular plants can grow very tall, with some trees exceeding heights of 100 meters. Some classification systems refer to vascular plants as Tracheophytes. We will study plant structure and function in Chapter 5.

The following section lists four of the key phyla or divisions found in the plant kingdom.

Phylum Bryophyta can best be described by what they *do not* have. Bryophytes do not have vascular tissue, true roots, stems, or leaves. As a result, Bryophytes are low-growing plants, reaching a height of only a few centimeters. Bryophytes reproduce by spores. Mosses are Bryophytes.

Phylum Pterophyta contains the ferns and is the second largest division in the plant kingdom comprised of some 20,000 species. Ferns reproduce by spores and are seedless plants.

Phylum Coniferophyta is made up of nonflowering plants that usually reproduce by seeds found in cones. Examples of Coniferophyta are evergreen trees such as pine, spruce, cedar, hemlock, and fir trees. The leaves of these trees are needle-like and stay on the tree all year long. If a tree looks like a Christmas tree, it is a conifer.

Phylum Anthophyta contains the angiosperms (hidden seeds from the Greek words *angios* [hidden] and *sperma* [seed]) and is the largest division in the plant kingdom with over 250,000 species. Anthophyta are flowering plants that usually reproduce by seeds hidden inside fruit. The phylum Anthophyta is divided into two classes: Monocotyledons and Dicotyledons.

- ❏ **Monocotyledons** (monocots) have one seed part (cotyledon) that cannot be split and one seed leaf with parallel veins. Examples are corn, grass, and wheat.

- ❏ **Dicotyledons** (dicots) have two seed parts (cotyledons) that can be split in two and two seed leaves with netted veins. Examples are peas, beans, and maple trees.

Example Problems

These problems review kingdom Plantae.

1. List the major characteristics of the kingdom Plantae.

   Answer: Plantae are autotrophic and make their own food by photosynthesis. They have the green pigment chlorophyll. Green plants are multicellular and eukaryotic and can be classified into several phyla or divisions.

2. How do green plants obtain food?

   Answer: Green plants have chlorophyll and can make their own food by photosynthesis. During photosynthesis, plants take in carbon dioxide and water from their environment, producing a sugar called glucose. Glucose serves as food for the plant.

3. What are the characteristics of Bryophytes?

   Answer: Bryophytes are low-growing green plants that do not have vascular tissue for the transport of materials. Also, they do not have true roots, stems or leaves. Examples of Bryophytes are liverworts and mosses.

4. What are the characteristics of vascular plants?

   Answer: Vascular plants have vascular tissue (for the transport of materials), true roots, stems, and leaves. As a result, they can grow tall. Examples of vascular plants are trees and shrubs.

5. How can we tell if a plant is a conifer?

   Answer: Most conifers have needle-like leaves, are evergreen, and reproduce from seeds found in cones.

Work Problems

Use these problems on the kingdom Plantae for additional practice.

1. How are green plants different from fungi?

2. Why can't Bryophytes grow as tall as vascular plants?

3. How is reproduction in Coniferophyta different from Anthophyta?

4. Compare the seeds of a Monocot with the seeds of a Dicot.

Worked Solutions

1. Green plants have chlorophyll, are autotrophic, and make their own food by photosynthesis. Fungi cells lack chloroplasts, are heterotrophic, and obtain their food from decaying vegetation.

2. Bryophytes do not have the vascular tissue needed to transport materials upward in a plant. As a result, they can't grow as tall as vascular plants.

3. Coniferophyta reproduce from seeds found in cones. Anthophyta reproduce from seeds found inside fruit.

4. Monocots have one seed part (cotyledon) that cannot be split. Dicots have two seed parts (cotyledons) that can be split into two halves.

Kingdom Animalia

Animals are multicellular, heterotrophic organisms with eukaryotic cells. The organisms in the animal kingdom have the greatest diversity. We can divide the animal kingdom into two groups: invertebrates and vertebrates.

Invertebrates

Invertebrates are animals without backbones. The invertebrate phyla discussed in this section are in order of increasing complexity.

Phylum Porifera contains sponges. Sponges have pores that draw in water for their cells to filter and absorb food. Sponges do not have a mouth. These organisms have two cell layers and live in water (mostly marine). All sponges have *radial symmetry*. (A cut through the axis of the organism produces halves that are mirror images of each other.) Sponges are sessile (not able to move).

Phylum Cnidaria's members are known also as the coelenterates. These organisms have two cell layers, an opening (mouth) for ingestion, a hollow body cavity for digestion, stinging cells (nematocysts), and tentacles. Many cnidaria are *motile* (able to move). Some examples of cnidaria are: hydra, jellyfish, coral, and sea anemone. We will study the hydra in Chapter 7.

Phylum Platyhelminthes contains flatworms and have *bilateral symmetry*. (The right and left sides of the organism are similar.) Some examples of flatworms are planaria and tapeworms.

Phylum Nematoda consists of roundworms. These organisms have bilateral symmetry and are parasitic. Trichinella (pork worm), hookworm, and ascaris are all roundworms.

Phylum Annelida are segmented worms, have a tube-within-a-tube body organization, and bilateral symmetry. The sandworm, leech, and earthworm are examples of annelids. We will study the earthworm in Chapter 7.

Phylum Mollusca are also known as mollusks and have soft bodies and a hard shell. Examples are: snails, clams, oysters, and octopi. The shell of an octopus is inside its soft body.

Phylum Arthropoda contain organisms that have jointed appendages and an *exoskeleton* (an outside skeleton made of chitin, a hard shell-like substance). Three of the classes that make up the phylum Arthropoda are Crustacea, Arachnida, and Insecta.

❏ **Crustaceans** live in marine or fresh water. They have two body parts: a *cephalothorax* (head and thorax combined) and an abdomen. They have gills for breathing and antennae. Their appendages are used for locomotion, food gathering, and eating. (Shrimp, lobsters, and crabs are examples of this class.)

❏ **Arachnids** have two body parts (cephalothorax and abdomen), eight legs, and no wings. (Examples are spiders, scorpions, and ticks.)

❏ **Insects** have three body parts: a head, thorax, and abdomen. Insects have three pairs of legs. Most have two pairs of wings, and many have the ability to perform *metamorphosis*. Metamorphosis is a change in body form from young to adult. For example, a caterpillar becomes a butterfly. Insects have a system of tracheal tubes for respiration. This class contains the largest number of species in the animal kingdom. (Some examples of insects are flies, mosquitoes, moths, ants, and grasshoppers.) We will study the grasshopper in Chapter 7.

Phylum Echinodermata contains spiny-skinned organisms. All echinoderms live in marine water and have radial symmetry. Starfish and sea urchins are examples of echinoderms.

Example Problems

These problems review invertebrate classification.

1. List the major characteristics of kingdom Animalia.

   Answer: Animals are multicellular, heterotrophic organisms. Animals use locomotion to find food. Animal cells are eukaryotic. Some animals are classified as invertebrates; others are vertebrates.

2. How can we distinguish between the Porifera and the Cnidaria?

   Answer: Porifera are sponges; they do not have a mouth but have pores that draw in water for cells to filter and absorb food. Sponges have two cell layers, and most are marine and sessile. Cnidaria are the coelenterates and include hydra, jellyfish, coral, and sea anemones. Cnidaria have two cell layers, a mouth for ingestion, and a hollow body cavity for digestion. They also have stinging cells, tentacles, and many are motile.

3. Define bilateral symmetry.

   Answer: In bilateral symmetry, the right and left sides of the organism are similar.

4. Distinguish between Platyhelminthes, Nematoda, and Annelida. Give examples of each.

   Answer: Platyhelminthes are flatworms. Examples are planaria and tapeworms. Nematodes are roundworms. Examples are trichinella and hookworm. Annelida are segmented worms. Examples are the earthworm and sandworm.

5. What are the major characteristics of the Annelida?

   Answer: Annelida are segmented worms and have a tube-within-a-tube body construction.

Work Problems

Use these problems on invertebrate classification for additional practice.

1. How are animals different from green plants?

2. Why is an octopus classified as a mollusk?

3. How are insects different from spiders?

4. What is metamorphosis?

5. How is respiration in a crustacean different from respiration in an insect?

Worked Solutions

1. Animals are heterotrophic organisms that depend on locomotion to find food. Green plants have chlorophyll and can make their own food by photosynthesis. During photosynthesis, plants take in carbon dioxide and water from their environment—producing a sugar called glucose. Glucose serves as food for the plant.

2. Mollusks have soft bodies and a hard shell. In the case of the octopus, the shell is inside a soft body.

3. Insects are members of the class Insecta. Insects have three body parts: head, thorax, and abdomen. Insects also have three pairs of legs. Most have two pairs of wings, and many are capable of metamorphosis. Spiders are members of the class Arachnida. Spiders have two body parts (cephalothorax and abdomen), eight legs, and no wings.

4. Metamorphosis is a complete change in the body structure of an organism from young to adult. The transformation of a caterpillar into a butterfly is an example of metamorphosis.

5. Crustaceans use gills for breathing, and insects use tracheal tubes.

Vertebrates

Vertebrates are animals with backbones. All *chordates* have a dorsal (back side) *notochord,* which is a flexible rod-like structure found at some time in their embryonic development. In the more complex chordates, the notochord forms a vertebrate. Five classes under phylum Chordata are shown here in order of increasing complexity.

❑ **Osteichthyes** are called *bony fishes.* Bony fishes live in marine or fresh water. They have gills, scales, fins, a two-chambered heart, and are *cold blooded.* (Their body temperature is the same as that of the environment.) Some examples of bony fishes are tuna, flounder, trout, bass, salmon, and goldfish. The shark is a *boneless* fish with a notochord and hard fibrous tissue called *cartilage* instead of bones. As a result, sharks are not in the class Osteichthyes.

❑ **Amphibia** can live in water or on land. Most have the ability of metamorphosis. (For example, a tadpole becomes a frog.) Young amphibians have gills, but adults have lungs, a three-chambered heart, and are cold blooded. Examples of amphibians are frogs, salamanders, and newts.

❑ **Reptilia** is made up of mostly land animals that have lungs for breathing. Reptiles have scales, a four-chambered heart, and are cold blooded. Examples are snakes, lizards, turtles, alligators, and dinosaurs.

❏ **Aves** are birds. They have feathers, wings, beaks, hollow bones (for flying), scales on their legs, a four-chambered heart, and are *warm blooded*. (They maintain a constant body temperature.) Sparrows, owls, pigeons, and penguins are examples of birds.

❏ **Mammalia** have mammary glands to feed milk to their young, body hair, a four-chambered heart, and are warm blooded. Humans, dolphins, pigs, whales, bats, and polar bears are mammals.

Example Problems

These problems review vertebrate classification.

1. What is meant by cold blooded? Which classes of vertebrates are cold blooded?

    Answer: In a cold-blooded animal, the body temperature of the organism is the same as that of the environment. Fishes are cold blooded. As the temperature of the water that a fish is swimming in changes, the body temperature of the fish changes accordingly. Other vertebrate classes that are cold blooded are amphibians and reptiles.

2. How are invertebrates different from vertebrates?

    Answer: Invertebrates are animals without backbones; vertebrates are animals with backbones.

3. Why are turtles classified as reptiles and not fishes?

    Answer: Although many turtles spend much of their lives in water, they have lungs for breathing and they return to land to reproduce. Fishes breathe through gills and are basically marine or fresh water organisms.

Work Problems

Use these problems on chordate classification for additional practice.

1. Why is a bat a mammal and not a bird?

2. How are fishes different from amphibians?

3. Why are sharks not in the class Osteichthyes?

4. In which class of chordates are humans classified?

5. Which two classes in the animal kingdom have the ability to perform metamorphosis?

Worked Solutions

1. A bat is classified as a mammal because it has hair, and the females have mammary glands that can feed milk to their young. Although a bat can fly, it does not have feathers, which is a characteristic that is exclusive to birds.

2. Fishes have scales, fins, and gills for breathing. Amphibians have lungs as adults and lack fins and scales. Amphibians have the ability to perform metamorphosis, but fishes do not.

3. Osteichthyes are bony fishes; sharks are boneless fishes containing cartilage.

4. Humans are mammals (in the class Mammalia) because we have mammary glands and body hair.

5. Insects are invertebrates that have the ability to perform metamorphosis. Amphibians are vertebrates that can perform metamorphosis.

Chapter Problems and Answers

Problems

The following is a brief paragraph based on classification that describes an organism that lived in the past. For problems 1–10, fill in the missing terms.

Scientists have found the remains of an animal that lived about 150 million years ago.

They named this organism *Archaeopteryx lithographica*, following the naming system

known as _____. The genus name of the organism is _____. The organism had bones,
　　　　　　　　1　　　　　　　　　　　　　　　　　　2

claws, teeth, legs, scales, and a tail. The most likely phylum for *Archaeopteryx*

lithographica is _____. However, *Archaeopteryx lithographica* also had two wings,
　　　　　　　　　3

feathers, and the ability to fly—suggesting that it should be a member of the class _____.
　　4

Other biologists feel that *Archaeopteryx lithographica* should be placed into the

class _____. If *Archaeopteryx lithographica* were able to maintain a constant body
　　　　　5

temperature, it would be considered _____. Biologists are certain that *Archaeopteryx*
　　　　　　　　　　　　　　　　　　　　6

lithographica is *not* in the class _____ because it was exclusively a land animal and could
　　　　　　　　　　　　　　　　　7

not perform metamorphosis. The lack of hair and mammary glands indicate that

Archaeopteryx lithographica was not in the class _____. In attempting to classify
　　　　　　　　　　　　　　　　　　　　　　　　　　8

Archaeopteryx lithographica, scientists must consider its _____ and _____.
　　　　　　　　　　　　　　　　　　　　　　　　　9　　　　　　10

For problems 11–14, select the classification group (from the following list) that is associated with the description given in each question. A choice can be used more than once or not at all.

　　A. Anthophytes　　B. Bryophytes　　C. Coniferophytes　　D. Fungi　　E. Dicotyledons

11. These organisms produce cones for reproduction. _____

12. Which group contains the flowering plants? _____

13. Members of this group of plants do not have true roots, stems, or leaves. _____

14. These plants produce seeds that can be split in two. _____

For problems 15–20, select the classification group (from the following list) that is associated with the organisms given in each question. A choice can be used more than once or not at all.

> A. Chordata   B. Arthropoda   C. Annelida   D. Cnidaria   E. Anthophyta

15. Fish, amphibians, reptiles, birds, and mammals _____

16. Insects, spiders, and scorpions _____

17. Coral and jellyfish _____

18. Vertebrates _____

19. Earthworms, sandworms, and leeches _____

20. Trees and shrubs _____

For problems 21–25, fill in the term that *best* completes each statement.

21. Ringworm is a disease caused by a _____.

22. The change of a tadpole into a frog is known as _____.

23. An organism that is sessile cannot perform the life function of _____.

24. Starfish and sea urchins are classified as _____.

25. Plants that lack true roots, stems, and leaves are classified in the phylum _____.

Answers

1. **Binomial nomenclature.** All organisms have two names that make up their scientific name. The first name of an organism is its genus, and the second is its specific epithet.

2. **Archaeopteryx.** The first name of the organism is its genus name. The genus name of the organism is always capitalized and italicized; the specific epithet name is written in lowercase.

3. **Chordata.** The paragraph states that *Archaeopteryx lithographica* had bones. The only phylum in the animal kingdom that has organisms with bones is Chordata.

4. **Aves or birds.** Feathers are a characteristic that is exclusive to birds. Wings and the ability to fly were also found in reptiles known as dinosaurs that are now extinct.

5. **Reptiles.** The presence of teeth, scales, and a tail suggest that *Archaeopteryx lithographica* could be considered a reptile.

6. **Aves or birds.** Warm-blooded animals maintain a constant body temperature. Birds are warm blooded, but reptiles are not.

7. **Amphibian.** In the phylum Chordata, the only class of animals with the ability to perform metamorphosis is Amphibia.

8. **Mammalia.** In the phylum Chordata, the only class of animals that have hair and mammary glands is Mammalia.

9. **Structure or evolutionary development.** When considering the classification of *Archaeopteryx lithographica,* scientists looked at the following structures: bones, claws, teeth, legs, scales, a tail, two wings, and feathers. Many scientists consider *Archaeopteryx lithographica* a transitional organism that bridges the gap between the evolution of reptiles to birds. Scientists do not consider function when classifying organisms. Butterflies have wings that function in flight, as do birds. However, the *structure* of the wings serves as the basis for classifying these two organisms. A butterfly's wing is a thin membrane, while a bird's wing has feathers and contains bones. We cannot classify these two organisms in the same group!

10. **Evolutionary development or structure.** *See the explanation to problem 9.*

11. **C. Coniferophyta.** Coniferophyta are nonflowering plants that reproduce by seeds found in cones. Examples of Coniferophyta are coniferous (evergreen) trees such as pine, spruce, cedar, hemlock, and fir.

12. **A. Anthophyta.** Anthophyta are flowering plants that reproduce by seeds hidden inside fruit.

13. **B. Bryophytes.** Bryophytes do not have vascular tissue; as a result they do not have true roots, stems, or leaves and are low-growing plants.

14. **E. Dicotyledons.** Dicots are plants that produce seeds that have two cotyledons or seed parts that can be split in half.

15. **A. Chordata.** These organisms represent different classes in the phylum Chordata.

16. **B. Arthropoda.** These organisms represent different classes in the phylum Arthropoda.

17. **D. Cnidaria.** These organisms are in the phylum Cnidaria, which is also known as the colenterates.

18. **A. Chordata.** Vertebrates are animals with backbones and are classified as chordates.

19. **C. Annelida.** The earthworm, sandworm, and leech are segmented worms in the phylum Annelida.

20. **E. Anthophyta.** Trees and shrubs are green plants that have vascular tissue and are classified as Anthophytes.

21. **Fungus.** Don't let the name fool you; ringworm is a fungal disease.

22. **Metamorphosis.** Metamorphosis is the ability of an organism to undergo a complete change in body structure from young to adult. This is exactly what happens when a tadpole becomes a frog.

23. **Locomotion.** Organisms that are sessile do not have the ability to move around.

24. **Echinoderms.** Starfish and sea urchins have spiny skins and are in the phylum Echinodermata.

25. **Bryophyta.** Bryophytes lack true roots, stems, leaves, and vascular tissue. As a result, bryophytes are low-growing plants.

Supplemental Chapter Problems

Problems

For problems 1–30, select the *best* answer.

1. Scientific classification is primarily based on:

   A. development   B. size
   C. structure and evolutionary development   D. function

2. The binomial system of classification was developed by:

   A. Leeuwenhoek   B. Linnaeus   C. Hooke   D. Brown

3. The scientific name of an organism includes its:

   A. genus and specific epithet   B. class and genus
   C. genus and phylum   D. phylum and specific epithet

4. The greatest similarity in structure occurs between members belonging to the same:

   A. kingdom   B. species   C. phylum   D. class

5. Organisms capable of mating and producing fertile offspring belong to the same:

   A. phylum   B. genus   C. class   D. species

6. The gray wolf is classified as *Canis lupus* and the red wolf as *Canis rufus*. These two animals belong to:

   A. the same genus but different species   B. the same species but different genus
   C. the same species but different phyla   D. the same genus but different kingdoms

7. *Acer rubrum* is the scientific name of the red maple. This organism belongs to the species:

   A. Acer   B. rubrum   C. Acer rubrum   D. plants

8. Which of the following represents the correct order of classification from largest grouping to smallest grouping?

   A. species → genus → phylum → class
   B. phylum → class → genus → species
   C. genus → species → phylum → class
   D. phylum → genus → class → species

9. Which group includes the other four?

   A. genus   B. species   C. kingdom   D. phylum

10. According to the binomial naming system of classification, the organism named *Acetabularia mediterrania* is most closely related to:

    A. *Crenulta acetabularia*   B. *Mediterrania acetabularia*
    C. *Acetabularia crenulta*   D. *Mediterrania crassa*

11. All of the following are domains *except:*

    A. Archaea   B. Bacteria   C. Eukarya   D. Protista

12. Which domain contains primitive cells that are lacking a nuclear membrane?

    A. Bacteria B. Protista C. slime molds D. Fungi

13. Eukaryotic organisms:

    A. lack a nuclear membrane B. have a nuclear membrane

    C. lack chromosomes D. lack DNA

14. Paramecia, amoeba and trypanosoma are classified as:

    A. animal-like protists B. plants C. fungi D. Archaea

15. Although euglena has a tail, it is classified as a plant-like protist because it has the ability to perform:

    A. photosynthesis B. reproduction C. digestion D. excretion

16. The paramecium is a one-celled, eukaryotic, heterotrophic organism with cilia. This organism is best classified in the kingdom:

    A. Protista B. Fungi C. Plantae D. Animalia

17. All the following are characteristics of fungi *except:*

    A. green plants B. eukaryotic C. heterotrophic D. decomposers of dead plants

18. Which two phyla can be found in the plant kingdom?

    A. Eubacteria and Fungi B. algae and blue-green algae

    C. Coniferophyta and Anthophyta D. protozoa and slime molds

19. Organisms that are capable of manufacturing their own food are found in the group:

    A. Cnidaria B. Fungi C. Anthophyta D. Porifera

20. Which of the following are cnidaria?

    A. slime molds and green algae B. amoeba and paramecia

    C. hydra and jellyfish D. ferns and grasses

21. All the following are annelids *except:*

    A. leech B. earthworm C. planarian D. sandworm

22. Select the pair of organisms that have an exoskeleton.

    A. planarian and leech B. hydra and coral

    C. starfish and sea urchin D. insect and spider

23. The octopus is classified as a mollusk because it has:

    A. tentacles B. a soft body and a hard interior shell

    C. an exoskeleton D. jointed appendages

24. A starfish is classified in the phylum:

    A. Cnidaria B. Annelida C. Arthropoda D. Echinodermata

25. Of the following, the closest relative to the whale is the:

    A. shark   B. tuna   C. eel   D. cat

26. Animals that have backbones can be found in the phylum:

    A. Invertebrates   B. Chordata   C. Cnidaria   D. Animalia

27. Chordates all have:

    A. an exoskeleton   B. a dorsal nerve cord
    C. chloroplasts   D. a two-cell layered body wall

28. Organisms capable of metamorphosis are found in the class:

    A. Osteichthyes   B. Aves   C. Amphibia   D. Reptilia

29. An organism that has hair, mammary glands and feeds milk to its young can be classified as a:

    A. bird   B. mammal   C. reptile   D. amphibian

30. Why is a virus *not* classified into a kingdom?

    A. Most biologists do not consider them living things.   B. They are primitive cells.
    C. They can reproduce.   D. Many cause disease in plants and animals.

Answers

1. **C.** "Classification," p. 47

2. **B.** "Classification," p. 47

3. **A.** "Classification," p. 47

4. **B.** "Classification," p. 47

5. **D.** "Classification," p. 47

6. **A.** "Classification," p. 47

7. **C.** "Classification," p. 47

8. **B.** "Classification," p. 47

9. **C.** "Classification," p. 47

10. **C.** "Classification," p. 47

11. **D.** "Classification," p. 47

12. **A.** "Domain Bacteria," p. 51

13. **B.** "Kingdom Protista," p. 53

14. **A.** "Kingdom Protista," p. 53

15. **A.** "Kingdom Protista," p. 53

16. **A.** "Kingdom Protista," p. 53

17. **A.** "Kingdom Fungi," p. 53

18. **C.** "Kingdom Plantae," p. 53

19. **C.** "Kingdom Plantae," p. 53

20. **C.** "Invertebrates," p. 58

21. **C.** "Invertebrates," p. 58

22. **D.** "Invertebrates," p. 58

23. **B.** "Invertebrates," p. 58

24. **D.** "Invertebrates," p. 58

25. **D.** "Vertebrates," p. 60

26. **B.** "Vertebrates," p. 60

27. **B.** "Vertebrates," p. 60

28. **C.** "Vertebrates," p. 60

29. **B.** "Vertebrates," p. 60

30. **A.** "Viruses," p. 50

Chapter 4
The Chemical Basis of Life

One of the main concepts of the cell theory is that all living things are made of cells and that cells contain organelles. Organelles are composed of certain key chemical compounds. These compounds are in turn made up of atoms of elements.

organism → organ systems → organs → tissues → cells → organelles → compounds → atoms

Chemistry Review

Chemistry is the science that studies matter and the changes in matter. Matter is anything that occupies space and has mass. Books, rocks, water, oil, air, and hydrogen are some examples of matter. An organism is also considered to be matter. Anything that is not matter is energy. Examples of energy that are important in the study of biology are: heat, light, electrical, and chemical energy.

All matter is made up of elements. An *element* is a substance that contains only one kind of atom. A symbol is an "abbreviation" for the name of an element and consists of one or two letters.

The following table lists elements that are found in all living things and their corresponding symbols.

Examples of Elements and Their Symbols	
Element	*Symbol*
Hydrogen	H
Oxygen	O
Nitrogen	N
Carbon	C

An *atom* is the smallest part of an element (that cannot be broken down further by ordinary chemical means). The center of an atom is called the *nucleus*, which contains protons and neutrons. *Electrons* surround the nucleus. *Protons* are positively charged, *neutrons* do not have a charge (neutral), and *electrons* are negatively charged.

A *compound* is a substance composed of two or more kinds of atoms chemically combined. A chemical bond is the electron arrangement (or force) that holds the atoms in a compound together. Ionic or covalent bonding can form compounds. In *ionic bonding*, atoms gain or lose electrons to form a compound. In *covalent bonding*, atoms share electrons to form a compound. In a compound, atoms of elements lose their original properties and take on the properties of the compound. For example, sodium is a soft gray metal and chlorine is a green poisonous gas. When these two elements combine, sodium chloride (table salt) is formed. Sodium chloride is a white crystal and can be used to flavor food.

The following table lists compounds that are found in living things and their corresponding formulas.

Examples of Compounds and Their Formulas	
Compound	**Formula**
Water	H_2O
Salt	NaCl
Carbon dioxide	CO_2
Glucose	$C_6H_{12}O_6$

A *molecule* is two or more atoms of the same element combined together (for example, O_2, Cl_2, N_2) *or* two or more atoms of different elements combined together (for example, H_2O, CO_2, $C_6H_{12}O_6$). Molecules are the smallest part of a compound.

Interpreting a Chemical Formula

A chemical formula is an abbreviation for a compound. A chemical formula gives the elements and the number of atoms for each element in a compound. Look at the following formula for water:

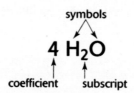

A *coefficient* is a number written in front of a formula. This is the number of molecules for the compound. In this example, water has four molecules. This is the same as saying H_2O + H_2O + H_2O + H_2O. It is more efficient to simply write $4H_2O$. If a coefficient is not present, the number of molecules for that compound is one.

Symbols indicate the elements in a compound. Water is made up of the elements hydrogen and oxygen.

A *subscript* (written after and below a symbol) indicates the number of atoms for each element in a compound. If no subscript is present, the number of atoms for that element is one. Water has two atoms of hydrogen and one atom of oxygen for each molecule.

Organic and Inorganic Compounds

Two major branches of chemistry exist: *inorganic chemistry* and *organic chemistry.* Inorganic chemistry deals with inorganic compounds—those that usually *do not* contain carbon and that are formed by ionic or covalent bonding. Organic chemistry deals with organic compounds—those that contain carbon. Almost all organic compounds contain carbon and hydrogen together and are usually formed by covalent bonding.

The following table shows examples of typical inorganic and organic compounds. Notice that all these organic compounds have carbon and hydrogen together. Many organic compounds contain carbon, hydrogen, *and* other elements.

Examples of Inorganic and Organic Compounds			
Inorganic Compounds		**Organic Compounds**	
Water	H_2O	Methane	CH_4
Salt	NaCl	Glucose	$C_6H_{12}O_6$
Nitric acid	HNO_3	Glycerol	$C_3H_5(OH)_3$
Sodium hydroxide	NaOH	Acetic acid	CH_3COOH

Example Problems

These problems review basic chemistry.

1. How is an atom different from an element?

    Answer: An element is a substance that contains only one kind of atom. The element sulfur contains only sulfur atoms, iron contains only iron atoms and so on. The atom is the smallest part of an element that still retains all the properties of the element.

2. Define *compound* and give some examples of compounds.

    Answer: A compound is a chemical combination of two or more kinds of atoms. In a compound, atoms of elements lose their original properties and take on the properties of the compound. Examples of compounds are water, salt, and methane.

3. When do we use symbols and formulas?

    Answer: A symbol is an abbreviation for an element, and a formula is an abbreviation for a compound.

4. What is a chemical bond?

    Answer: A chemical bond is the electron arrangement (or force) that holds atoms together. The two kinds of chemical bonds are the ionic bond and the covalent bond.

5. Look at the chemical formula for stearic acid: $C_{17}H_{35}COOH$. What kind of compound is stearic acid?

    Answer: Stearic acid is an organic compound. Stearic acid contains carbon and hydrogen together; compounds of carbon and hydrogen are usually organic.

Work Problems

Use these problems on basic chemistry for additional practice.

1. How is a compound different from the elements that make up the compound?

2. How are inorganic compounds different from organic compounds?

3. Explain the difference between ionic and covalent bonding.

4. How is the oxygen molecule (O_2) different from the carbon dioxide molecule (CO_2)?

5. Look at the following chemical formula: $C_6H_{12}O_6$.

    A. Name the compound.

    B. How many molecules of this compound are there?

    C. How many atoms of carbon are in the compound?

    D. How many atoms of hydrogen are in the compound?

    E. How many elements are in the compound?

Worked Solutions

1. In a compound, atoms of elements lose their original properties and take on the properties of the compound. For example, water, which is a liquid, is composed of two elements that are gases—hydrogen and oxygen.

2. Inorganic compounds are compounds that do not contain carbon and are usually formed by ionic bonding. Organic compounds generally contain carbon and hydrogen together and are usually formed by covalent bonding.

3. In ionic bonding, atoms gain or lose electrons to form inorganic compounds. In covalent bonding, atoms share electrons to form the compound, which is usually organic.

4. Oxygen is a molecule that is composed of two atoms of the same element. Carbon dioxide is a molecule composed of two different elements: carbon and oxygen.

5. **A. Glucose** is the name of the compound with the chemical formula $C_6H_{12}O_6$.

    B. One. There is no coefficient in front of the chemical formula for glucose. This means that only one molecule is present, the one we are looking at.

    C. Six. The subscript after carbon is six. A subscript indicates the number of atoms for each element in the compound.

    D. Twelve. The subscript after hydrogen is twelve. A subscript indicates the number of atoms for each element in the compound.

    E. Three. This compound has the elements: carbon, hydrogen, and oxygen.

Drawing Organic Compounds

To correctly draw organic compounds, some simple bonding rules must be followed. Each element in the following table cannot have less than or more than the number of bonds indicated. The table also shows the four key elements found in all living things. Remember *HONC*, and you will know the correct number of bonds for each of these elements. Drawing the structure of a compound allows us to see the relationships between the atoms and is a useful way to distinguish between different compounds that have the same chemical formula.

Bonding Rules		
Element	**Symbol**	**Bonds**
Hydrogen	H	1
Oxygen	O	2
Nitrogen	N	3
Carbon	C	4

Consider the *chemical formula* for methane (CH_4). This formula indicates that one atom of carbon and four atoms of hydrogen are in the compound. Methane can be drawn three different ways, as shown in the following figure. Figure A shows an atomic drawing of methane. In Figure B, a dot represents a hydrogen electron, an *X* represents a carbon electron, and each pair of shared electrons represents a covalent bond. Figure C is the fastest and easiest method of drawing methane. Each line between the atoms represents a covalent bond. Notice that the bonding rules have been followed; carbon has four bonds and each hydrogen has one bond.

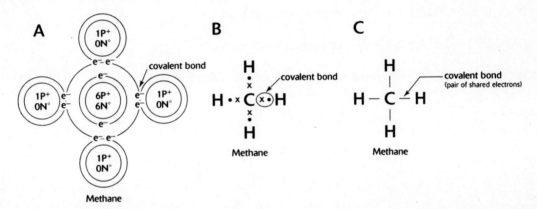

Figure C shows the structural formula for methane. A *structural formula* indicates the number of molecules, elements, atoms, and their arrangement for each element in a compound.

Notice that in each structural formula in the following figure, every hydrogen atom has one bond, and every carbon atom has four bonds.

Ethane, C_2H_6

Propane, C_3H_8

Example Problems

These problems review drawing organic compounds.

1. Name the elements found in all living things.

   Answer: The elements that are found in all living things are hydrogen, oxygen, nitrogen, and carbon. Remember the pneumonic device *HONC*.

2. How many bonds can a carbon atom form?

    Answer: Carbon forms four bonds when creating a compound.

3. Look at the structural formulas for ethane and propane in the preceding figures. What kind of compound are these?

    Answer: Ethane and propane are organic compounds. Their structural formulas show that they contain carbon and hydrogen.

Work Problems
Use these problems on drawing organic compounds for additional practice.

1. How is a chemical formula different from a structural formula?

2. What does the line between C and H in a structural formula represent?

3. Draw butane, which has the chemical formula C_4H_{10}.

4. Draw octane, which has the chemical formula C_8H_{18}.

5. What is the chemical formula for the compound in the following figure?

$$
\begin{array}{ccccccc}
  & H & H & H & H & H & H \\
  & | & | & | & | & | & | \\
H- & C- & C- & C- & C- & C- & C & -H \\
  & | & | & | & | & | & | \\
  & H & H & H & H & H & H \\
\end{array}
$$

Worked Solutions

1. A chemical formula shows the elements and the number of atoms for each element in a compound. A structural formula adds to this the arrangement of the atoms in the compound.

2. In a structural formula, the line between the carbon and hydrogen atoms represents a covalent bond, which is a pair of shared electrons.

3. Begin by drawing a chain of 4 carbon atoms connected with a bond between each one. Next, distribute the 10 hydrogen atoms. *Note:* Each carbon atom must have four bonds, and each hydrogen atom must have one bond.

$$
\begin{array}{ccccc}
  & H & H & H & H \\
  & | & | & | & | \\
H- & C- & C- & C- & C & -H \\
  & | & | & | & | \\
  & H & H & H & H \\
\end{array}
$$

Butane, C_4H_{10}

4. Begin by drawing a chain of 8 carbon atoms connected with a bond between each one. Next, distribute the 18 hydrogen atoms. *Note:* Each carbon atom must have four bonds, and each hydrogen atom must have one bond.

Octane, C_8H_{18}

5. C_6H_{14} (This is the formula for hexane.)

    First, count the number of carbon atoms in the chain. Then select a hydrogen atom as a starting point and begin counting the number of hydrogen atoms around the molecule. When writing a chemical formula, carbon is written first, and hydrogen is written second.

The Molecules of Life

Water, carbohydrates, lipids, and proteins are molecules that are found in all living things and are called the molecules of life.

Water

Water is an inorganic molecule and is the most common molecule found in all living things. Water is stable, can dissolve many different substances, and allows chemical reactions to take place within a cell. The cytoplasm of a cell is mostly water, and most organisms are 70%–95% water. Life on earth cannot exist without water.

Carbohydrates

Carbohydrates are organic molecules that include sugars and starches. Carbohydrates are a major source of energy for an organism. Examples of sugars are glucose, maltose, and sucrose. Notice that all sugars end in the suffix *-ose*. All carbohydrates contain the elements carbon, hydrogen, and oxygen. In a carbohydrate, the ratio of hydrogen atoms to oxygen atoms is 2:1.

Monosaccharides are simple sugars. Glucose, fructose, and galactose are examples of monosaccharides. All three have the same chemical formula ($C_6H_{12}O_6$), but their structural formulas are different.

Glucose   Fructose   Galactose

Glucose ($C_6H_{12}O_6$) is a good example of a carbohydrate. If we divide the number of atoms for each element by the highest common denominator (6), we get the following:

$$\frac{C_6}{6} \quad \frac{H_{12}}{6} \quad \frac{O_6}{6} = CH_2O$$

This demonstrates that for carbohydrates, the ratio of hydrogen atoms to oxygen atoms is 2:1. Notice that sugar is carbon with water attached to it.

Disaccharides are double sugars formed from the combination of two simple sugars. If we combine two glucose molecules, we get the disaccharide maltose.

$$\text{Glucose} + \text{Glucose} \rightleftharpoons \text{Maltose} + \text{Water}$$
$$C_6H_{12}O_6 + C_6H_{12}O_6 \rightleftharpoons C_{12}H_{22}O_{11} + H_2O$$

Notice that a molecule of water had to be removed to make room for the carbon-oxygen-carbon bonds needed to hold the two molecules of glucose together. *Dehydration synthesis* is the formation of a large molecule from two smaller molecules by the removal of water. Maltose can be broken into two molecules of glucose by adding water. *Hydrolysis* is the breakdown of a large molecule into two or more smaller molecules by adding water. The double arrows in the following figure show that these chemical reactions are reversible.

The Formation of Maltose by Dehydration Synthesis

A *polysaccharide* is formed when many glucose molecules are combined. Starch is a large polysaccharide and is the storage form of glucose in plants. Glycogen, another large polysaccharide, is the storage form of glucose in animals.

Example Problems

These problems review carbohydrates.

1. What is the most abundant compound found in all living things? Why is this compound important?

    Answer: Water, which is inorganic, is the most abundant compound found in all living things. Without water, life on earth could not exist because the cytoplasm of a cell contains mostly water. Most organisms are composed of 70%–95% water.

2. What elements are found in carbohydrates?

   Answer: The elements found in carbohydrates are carbon, hydrogen, and oxygen. The suffix *-ate* indicates the presence of oxygen.

3. What is the ratio of hydrogen atoms to oxygen atoms in a carbohydrate?

   Answer: The ratio of hydrogen atoms to oxygen atoms in a carbohydrate is 2:1.

4. Why are carbohydrates biologically important?

   Answer: Carbohydrates are an important source of energy for organisms.

5. How are monosaccharides different from disaccharides?

   Answer: Monosaccharides are simple (or single) sugars. When two monosaccharides are combined, we get a double sugar called a disaccharide.

Work Problems

Use these problems on carbohydrates for additional practice.

1. What is dehydration synthesis?

2. Explain how the maltose molecule is formed.

3. Explain how maltose can be broken down.

4. How are polysaccharides formed?

5. Why are starch and glycogen important molecules in living things?

6. Glucose has the chemical formula $C_6H_{12}O_6$. When two glucose molecules are combined, we get maltose, which has a formula of $C_{12}H_{22}O_{11}$. Why isn't the formula for maltose $C_{12}H_{24}O_{11}$?

Worked Solutions

1. Dehydration synthesis is the formation of a large molecule from two smaller molecules by the removal of water.

2. Maltose is a disaccharide that is formed by combining two glucose molecules via dehydration synthesis.

3. The maltose molecule can be broken down into two glucose molecules by adding water. This type of chemical reaction is called hydrolysis. Hydrolysis is the opposite of dehydration synthesis.

4. A polysaccharide consists of many glucose molecules combined. Polysaccharides can be formed by dehydration synthesis.

5. Starch is a storage form of sugar in plants, and glycogen is a storage form of sugar in animals. It is much more efficient to store one molecule of starch than hundreds of glucose molecules. When a plant needs sugar for energy production, starch molecules can be broken down by hydrolysis.

6. Maltose is formed from two glucose molecules by dehydration synthesis. When H_2O is removed from $C_{12}H_{24}O_{11}$, the formula for maltose becomes $C_{12}H_{22}O_{11}$.

Lipids

Lipids are organic molecules that include fats, oils, and waxes. Lipids are used for energy and, along with proteins, help form the membranes of a cell. A lipid is composed of one molecule of glycerol combined with three molecules of fatty acid. The molecule glycerol $C_3H_5(OH)_3$ is an alcohol; all alcohols end in the suffix -ol. The presence of a hydroxyl group (OH^-) on a carbon chain makes a compound an alcohol. Note that glycerol has three OH^- groups. A fatty acid molecule has a long carbon chain that ends with a carboxyl group (COOH). Note the double bond between carbon and oxygen in the carboxyl group. (Carbon must have four bonds; oxygen must have two bonds.) The letter R is often substituted for the long carbon chain. (R = the rest of the molecule.)

glycerol

Fatty acid (stearic acid)

or

Notice that three molecules of water have to be removed to make room for the carbon-oxygen-carbon bonds needed to hold the fatty acid molecules and the glycerol together. A lipid is formed by dehydration synthesis and can be broken down by hydrolysis.

The Formation of a Lipid by Dehydration Sysnthesis

Glycerol   3 Fatty Acids   1 Lipid   3 Water

Example Problems

These problems review lipids.

1. Describe the structure of glycerol.

    Answer: Glycerol is an alcohol that has a three-carbon chain. Each carbon has a hydroxyl group (OH^-) attached to it.

2. Describe the structure of a fatty acid.

    Answer: A fatty acid is an organic molecule with a long carbon chain that ends with a carboxyl group (COOH).

3. Give several examples of lipids.

    Answer: Lipids are organic molecules that include fats, oils, and waxes.

4. Why are lipids biologically important?

    Answer: Lipids are used for energy and, along with proteins, help form the membranes of cells.

Work Problems

Use these problems on lipids for additional practice.

1. Describe the formation of a lipid.

2. How are lipids broken down?

3. Both lipids and sugars provide energy. From your examination of their structural formulas, can you discover which molecule is capable of providing the most energy?

Worked Solutions

1. Lipids are formed by dehydration synthesis. One glycerol molecule is combined with three fatty acid molecules.

2. Lipids are broken down by hydrolysis. One lipid molecule + three water molecules *yield* one glycerol molecule + three fatty acid molecules.

3. A molecule of lipid can release more energy than a molecule of sugar. A careful examination of the structural formulas of the two molecules shows that the lipid molecule is larger than the sugar molecule. When broken down, the lipid releases the most energy.

Proteins

Proteins are organic molecules that are used for growth, repair, and reproduction of new cells. All proteins contain the elements carbon, hydrogen, oxygen, and nitrogen. Proteins are generally not used for energy. Proteins are made of 40 or more amino acids combined. In humans thousands of different proteins can be formed by the combination of 20 kinds of amino acids. Some examples of amino acids are glycine, alanine, and valine. Proteins are the most common organic molecules found in living things. An amino acid molecule has an amino group (NH_2) on one end and a carboxyl group (COOH) on the other. The letter *R* represents the rest of the molecule and varies from amino acid to amino acid.

An Amino Acid

amino   carboxyl
group   group

A molecule of water has to be removed to make room for the C-N bond needed to hold the two amino acid molecules together. A carbon-nitrogen bond between two amino acids is known as a *peptide* bond. The molecule that forms is called a *dipeptide*. When several amino acid molecules are combined, a *polypeptide* is formed, and when 40 or more amino acid molecules are combined, a protein is formed. A protein is formed by dehydration synthesis and can be broken down by hydrolysis. Protein synthesis takes place at cellular organelles called ribosomes.

The Formation of a Dipeptide by Dehydration Synthesis

The following is a list of some of the most important proteins. Humans depend on all of these proteins for survival.

❑ **Hemoglobin**—An iron containing pigment that gives red blood cells their color and aids in the transport of oxygen to the cells of the organism.

❑ **Antibodies**—Proteins in the blood that can help fight disease by destroying harmful substances such as viruses.

❑ **Nucleoproteins**—The nucleus of the cell contains deoxyribonucleic acid (DNA), which is important for heredity and reproduction, and ribonucleic acid (RNA), which is important for protein synthesis. The nucleoproteins will be discussed further in a later chapter.

❑ **Hormones**—Chemical messengers that are produced by endocrine glands and are important in the life function of regulation.

❑ **Enzymes**—Organic catalysts that speed up the rate of chemical reactions.

Example Problems
These problems review proteins.

1.   What elements are found in proteins?

     Answer: Carbon, hydrogen, oxygen, and nitrogen are the elements found in all proteins. Only proteins have the element nitrogen.

2. What are the building blocks of proteins?

    Answer: Amino acids are the building blocks of proteins.

3. Give several examples of proteins and state their importance.

    Answer:

    Hemoglobin—Aids in the transport of oxygen to the cells of an organism.

    Antibodies—Proteins in the blood that can help fight disease.

    Nucleoproteins—The DNA molecule is important for heredity and reproduction. RNA aids in protein synthesis.

    Hormones—Chemical messengers that aid in the life function of regulation.

    Enzymes—Organic catalysts that speed up the rate of chemical reactions.

4. What is a peptide bond?

    Answer: A peptide bond is a carbon-nitrogen bond that holds two amino acid molecules together. The peptide bond is formed by dehydration synthesis.

Work Problems

Use these problems on proteins for additional practice.

1. Describe the formation of a protein.

2. Where are proteins formed?

3. State the difference between a dipeptide and a protein.

4. How are proteins broken down?

5. How many kinds of proteins can be formed from 20 different amino acids?

Worked Solutions

1. Proteins are formed by the dehydration synthesis of 40 or more amino acids. Many proteins are complex molecules containing millions of amino acids.

2. Protein synthesis takes place at the ribosome.

3. A dipeptide is a compound of two amino acids combined by one peptide bond. A protein is the combination of 40 or more amino acids.

4. Proteins are broken down into amino acids by hydrolysis.

5. The number of different proteins that can be made is infinite. Think of each of the 20 different amino acids in the human body as a letter of the alphabet. The smallest word these letters form can have any combination of 40 letters, and many words can be millions of letters long. An infinite number of words can be produced. Now just imagine that each word is a protein! Of the infinite number of proteins possible, the human body has approximately 50,000 different proteins.

Acids and Bases

Many substances form acids or bases when dissolved in water. Acids and bases affect chemical reactions in organisms. Some chemical reactions require acidic conditions, such as those involved with the digestion of food in the stomach. Other chemical reactions require basic conditions. Substances that are neither acid nor base are called neutral.

Acids

Acids are compounds that form the hydrogen ion (H^+) when dissolved in water.

Examples of Acids

❏ Hydrochloric acid (HCl) is an acid because of the presence of the hydrogen ion ($HCl \rightarrow H^+ + Cl^-$).

❏ Hydrofluoric acid (HF) is an acid because of the presence of the hydrogen ion ($HF \rightarrow H^+ + F^-$).

❏ Sulfuric acid (H_2SO_4) dissolves in water to form two hydrogen ions ($H_2SO_4 \rightarrow 2H^+ + SO_4^=$).

Bases

Bases are compounds that form the hydroxyl ion (OH^-) when dissolved in water.

Examples of Bases

❏ Potassium hydroxide (KOH) is a base because of the presence of the hydroxyl ion ($KOH \rightarrow K^+ + OH^-$).

❏ Sodium hydroxide (NaOH) is a base because of the presence of the hydroxyl ion ($NaOH \rightarrow Na^+ + OH^-$).

❏ Calcium hydroxide ($Ca(OH)_2$) dissolves in water to form two hydroxyl ions ($Ca(OH)_2 \rightarrow Ca^{++} + 2(OH)^-$).

Water

Water (H_2O) can also be written as HOH. If water is separated ($HOH \rightarrow H^+ + OH^-$), one positive hydrogen ion and one negative hydroxyl ion are produced. As a result, water is *neutral* (not acidic or basic).

The pH Scale

The pH scale is a logarithmic scale (from 0 to 14) that is used to indicate the relative strengths of acids and bases.

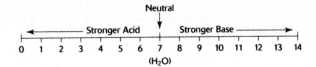

The further away a substance is from neutral (pH = 7) on the acid side, the stronger the acid. The further away a substance is from neutral on the base side, the stronger the base.

Example Problems

These problems review acids and bases.

1. What does the pH scale measure?

    Answer: The pH scale is used to indicate the strengths of acids and bases.

2. How are acids different from bases?

    Answer: Acids are compounds that form the hydrogen ion (H^+) when dissolved in water. Bases are compounds that form the hydroxyl ion (OH^-) when dissolved in water.

3. Give several examples of acids and bases.

    Answer: Some examples of acids are hydrochloric acid (HCl), hydrofluoric acid (HF), and sulfuric acid (H_2SO_4). Some examples of bases are potassium hydroxide (KOH), sodium hydroxide (NaOH), and calcium hydroxide ($Ca(OH)_2$).

Work Problems

Use these problems on acids and bases for additional practice.

1. Why is water neutral?

2. What kind of compound is H_2SO_4? How do we know?

3. What kind of compound is $Al(OH)_3$? How do we know?

4. Which is a stronger acid, pH 3 or pH 5?

5. Which is a stronger base, pH 8 or pH 10?

6. Why is NaCl not an acid?

Worked Solutions

1. Water (H_2O or HOH) contains one positive hydrogen ion (H^+) and one negative hydroxyl ion (OH^-). These ions cancel each other out, making water neutral.

2. H_2SO_4 is sulfuric acid. This compound is an acid because it can produce the hydrogen ion (H^+).

3. $Al(OH)_3$ is aluminum hydroxide. This compound is an acid because it can produce the hydroxyl ion (OH^-).

4. pH 3 is the stronger acid because it is further away from neutral on the acid side of the pH scale. The lower the acid pH number, the stronger the acid.

5. pH 10 is a stronger base because it is further away from neutral on the base side of the pH scale. The higher the base pH number, the stronger the base.

6. NaCl is not an acid because it does not produce the hydrogen ion (H^+).

Enzyme Action

Enzymes are large protein molecules that speed up the rate of chemical reactions but are not changed during the reaction. Without enzymes, the chemical reactions that take place in an organism would not work, and the organism would die. Maltase, lipase, and protease are examples of enzymes. Notice that the name of an enzyme ends in the suffix *-ase*. The substance that an enzyme works on is called the *substrate*. Dehydration synthesis and hydrolysis are chemical reactions that require the presence of an enzyme to work. Look at the chemical reaction that shows the dehydration synthesis of two glucose molecules to form one maltose molecule. The two glucose molecules are the substrate, and maltase is the enzyme.

$$C_6H_{12}O_6 + C_6H_{12}O_6 \xrightarrow{\text{maltase}} C_{12}H_{22}O_{11} + H_2O$$

If we reverse the reaction and break maltose down by hydrolysis, maltose is the substrate and maltase is still the enzyme. Maltase can be used to make or break down maltose. An enzyme can be used again and again because it is not changed by the reaction that it is helping to speed up.

The Lock-and-Key Model of Enzyme Action

The lock-and-key model is used to explain how enzymes work. If we examine a lock and a key, we notice that the shape of the key is designed to fit into the lock. The key can be used to open or close the lock, and the key can be used again and again. Only one key can fit into the lock. In organisms, the enzyme is the key, and the substrate is the lock. Only one kind of enzyme can be used for each type of chemical reaction. In other words, *enzyme action is specific*. Look at the following two diagrams. Enzyme 1 can only work to speed up reaction 1, and enzyme 2 can only work to speed up reaction 2. The shape (or active site) of the enzyme determines which chemical reaction it can speed up.

Enzyme Action: Lock-and-Key Theory

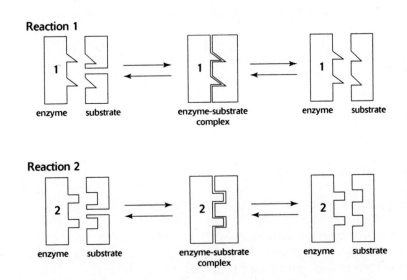

Factors That Affect the Rate of Enzyme Action

Four factors affect the rate of enzyme action: temperature, pH, enzyme concentration, and substrate concentration.

❑ **Temperature**—At low or high temperatures, the active site of most enzymes becomes distorted, and the enzyme no longer works. In humans enzymes work best at a normal body temperature of 37°C.

❑ **pH**—At low or high pH levels, the shape of most enzymes becomes distorted, and the enzyme no longer works. In humans most enzymes work best at a pH near neutral (7).

❑ **Enzyme concentration**—As more enzymes are added, enzyme action increases and eventually levels off when all the substrate is being worked on.

❑ **Substrate concentration**—As more substrate is added, enzyme action increases and eventually levels off when all the enzymes are being used.

Example Problems

These problems review enzyme action.

1. Why are enzymes important?

   Answer: Enzymes speed up the rate of chemical reactions. Without enzymes, the chemical reactions that take place in an organism do not work, and the organism dies.

2. Define the term *substrate*.

   Answer: The substrate is the substance that an enzyme works on. For example, in the formation of a lipid, the substrate is glycerol and three fatty acids.

3. What is the *enzyme-substrate complex?*

    Answer: An enzyme-substrate complex forms when an enzyme combines with a substrate.

4. Explain what is meant by saying "enzyme action is specific."

    Answer: Only one kind of enzyme can be used for each type of chemical reaction. For example, maltase can be used to make or break down maltose; lipase can be used to make or break down a lipid. However, maltase cannot be used to make or break down a lipid, and lipase cannot be used to make or break down maltose.

5. What kind of molecule is kinase?

    Answer: Kinase is an enzyme. All enzymes end in the suffix *-ase*. The ending on the name of a compound can be used to identify the type of compound. Compounds with names ending in *-ose* are sugars, compounds with names ending in *-ol* are alcohols, and most compounds with names ending in *-ine* are amino acids. Note that some amino acids have names that do not end in *-ine*, such as aspartic acid.

Work Problems

Use these problems on enzyme action for additional practice.

1. Explain the lock-and-key model of enzyme action.

2. Why can't the enzyme protease speed up the formation of maltose from two glucose molecules?

3. How do temperature and pH affect the rate of enzyme action?

4. How do enzyme and substrate concentration affect the rate of enzyme action?

5. Why does a high fever make a person feel sick?

Worked Solutions

1. The lock-and-key model of enzyme action is used to show that only one kind of enzyme can be used to speed up each type of chemical reaction, just as only one key can fit into a lock. An enzyme can be used to help make or break down a compound, and an enzyme is never used up. A key can be used to open or close a lock again and again. In organisms, the enzyme is the key, and the substrate is the lock.

2. Protease cannot speed up the formation of maltose from glucose because the active site of protease is not compatible with the shape of the glucose molecules, and therefore an enzyme-substrate complex cannot form. However, maltase can be used to make maltose, and protease can be used to make a protein. Enzyme action is specific: One enzyme is used for each type of chemical reaction, and it must be the correct enzyme!

3. Very high and low temperatures distort the active site of an enzyme. As a result, the rate of enzyme action decreases. Very low or very high pH has a similar effect on the rate of enzyme action.

4. As more enzymes are added, enzyme action increases and eventually levels off when all the substrate is being worked on. As more substrate is added, enzyme action increases and eventually levels off when all the enzymes are being used.

5. When a person has a fever, the higher temperature distorts the active sites of enzymes in the body. As a result, the rate of many chemical reactions decreases, making a person feel sick.

Chapter Problems and Answers

Problems

Observe the following chemical reaction. For problems 1–10, analyze the steps shown and answer the questions.

1. Name the process shown in the preceding steps. _____

2. Identify molecule A. _____

3. Identify molecule B. _____

4. Name the group shown inside circle C. _____

5. What is the name of molecule E? _____

6. How many water molecules are formed in this chemical reaction? _____

7. What is the name of the chemical bond indicated by letter D? _____

8. Where in the cell does this chemical reaction take place? _____

9. Name the enzyme that is needed to speed up this chemical reaction. _____

10. What is the opposite of this chemical process called? _____

For problems 11–16, select the chemical structure from the following list that is associated with each description. A choice can be used more than once or not at all.

   A. Amino group   B. Carboxyl group   C. Hydrochloric acid (HCl)   D. Lipase   E. Lipid

11. Contains one glycerol and three fatty acids. _____

12. An inorganic acid. _____

13. Which structure is found only in proteins? _____

14. Speeds up the rate of a chemical reaction. _____

15. This group is found in amino and fatty acid molecules. _____

16. An inorganic compound formed by ionic bonding. _____

For problems 17–20, select the compound from the following diagram that is associated with each description. A choice can be used more than once or not at all.

(A)   (B)   (C)

(D)   (E)   (F)

17. Which of these molecules is an amino acid? _____

18. Identify the molecule that can be broken down by hydrolysis. _____

19. Which molecule is glycerol? _____

20. Which molecule is a building block found in enzymes? _____

Problems 21–25 are based on the following graphs.

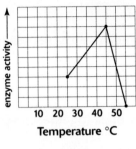

Temperature °C

(A)

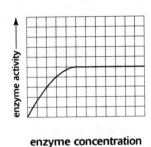

enzyme concentration
increases

(B)

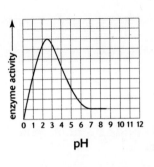

pH

(C)

21. At what temperature does enzyme activity begin to decrease in graph A?
    _____

22. Why does enzyme activity decrease in graph A? _____

23. At what temperature does all enzyme activity stop? _____

24. Why does enzyme activity increase and then level off in graph B? _____

25. At what pH level is enzyme activity best in graph C? _____

Problems 26–28 are based on the following diagram of the pH scale.

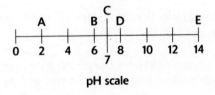

pH scale

26. Which molecule is a very strong base? _____

27. Which molecule is a weak acid? _____

28. Which molecule is probably water? _____

Answers

1. **Dehydration synthesis.** Dehydration synthesis is the formation of a large molecule from two smaller molecules by the removal of water. In step one, small molecules (A and B) are being combined. In step two, water is being removed. In step three, the large molecule (E) is being formed.

2. **Molecule A is an amino acid.** Amino acid molecules have an amino group (NH_2) on one end and a carboxyl group (COOH) on the other.

3. **Molecule B is an amino acid.** Amino acid molecules have an amino group on one end and a carboxyl group on the other. Note that molecule B is different from A because it has a methyl group (CH_3) instead of hydrogen.

4. **Carboxyl group.** A carboxyl group has the chemical formula COOH.

5. **Molecule E is a dipeptide.** Dipeptides are composed of two amino acids combined.

6. **One.** Step two of the chemical reaction shows the removal of one molecule of water (HOH).

7. **Peptide bond.** A carbon-nitrogen bond that holds two amino acids together is a peptide bond.

8. **Ribosome.** The ribosome is the site of protein synthesis in a cell.

9. **Protease.** Protease is an enzyme that can be used to make or break down a protein.

10. **Hydrolysis.** Hydrolysis is the breakdown of a large molecule into two or more smaller molecules by the addition of water. The addition of water to a dipeptide results in the formation of two amino acid molecules.

11. **E. Lipid.** A lipid is composed of one molecule of glycerol combined with three molecules of fatty acid.

12. **C. Hydrochloric acid (HCl).** All acids contain the hydrogen ion (H^+).

13. **A. Amino group.** The amino group contains the element nitrogen. Nitrogen is found only in proteins.

14. **D. Lipase.** Enzymes speed up the rate of chemical reactions. We know that lipase is an enzyme because its name ends in the suffix *-ase*.

15. **B. Carboxyl group.** Both amino acid and fatty acid molecules contain a carboxyl group on one end. However, only an amino acid has an amino group on the other end.

16. **C. Hydrochloric acid (HCl).** This compound is inorganic because it does not have carbon and hydrogen together.

17. **C.** This compound is an amino acid because it has an amino group on one end and a carboxyl group on the other.

18. **F.** Molecule F is the disaccharide maltose, which can be broken down into two glucose molecules by hydrolysis.

19. **B.** Glycerol is a three-carbon compound with three OH^- groups attached.

20. **C.** Enzymes are proteins. The building blocks of proteins are amino acids. Molecule C is an amino acid.

21. **45°C.** Graph A indicates that the highest rate of enzyme activity takes place at 45°C, after which enzyme activity begins to decrease.

22. Enzyme activity decreases because the high temperature changes the active sites of the enzyme, so it is no longer able to act upon the substrate.

23. All enzyme activity stops at 55°C because at this point we are at zero on the graph.

24. As more enzymes are added, enzyme activity increases and eventually levels off when all the substrate is being worked on.

25. For this enzyme, pH 2.5 is best because at this pH level, the rate of enzyme activity is at its highest.

26. **E.** The further away a molecule is from neutral on the *base* side of the pH scale, the stronger the base.

27. **B.** The closer a molecule is to neutral on the *acid* side of the pH scale, the weaker the acid.

28. **C.** Water, which has a pH of 7, is neutral.

Supplemental Chapter Problems

Problems

For problems 1–30, select the *best* answer.

1. A great number of natural organic molecules can exist because carbon atoms can:

   A. form four ionic bonds   B. form four covalent bonds

   C. lose four electrons to form compounds   D. react with enzymes

2. A structural formula indicates the:

   A. number of atoms in a molecule

   B. kind of atoms in a molecule

   C. number *and* kind of atoms in a molecule

   D. number, kind *and* arrangement of atoms in a molecule

3. Which inorganic substance is present in the greatest quantity inside cells?

   A. salt   B. sugar   C. protein   D. water

4. Fructose and pentose are examples of a class of organic compounds known as:

   A. proteins   B. lipids   C. carbohydrates   D. enzymes

5. Which of the following correctly represents the formula for a monosaccharide?

   A. H_2CO_3 B. $C_3H_5(OH)_3$ C. $C_5H_{10}O_5$ D. $C_{17}H_{35}COOH$

6. A very large polysaccharide can also be called a:

   A. sugar   B. starch   C. fatty acid   D. protein

7. All the following are examples of polysaccharides *except:*

   A. glycogen   B. cellulose   C. starch   D. maltase

8. The breakdown of a large molecule into two or more smaller molecules by adding water is known as:

    A. photosynthesis   B. regulation   C. hydrolysis   D. dehydration synthesis

9. Which element *is* found in a protein, but *is not* found in a carbohydrate or lipid?

    A. hydrogen   B. oxygen   C. nitrogen   D. carbon

10. Amino acids are required in a person's diet for the synthesis of:

    A. lipids   B. proteins   C. carbohydrates   D. water

11. Proteins are needed for all the following *except*:

    A. energy   B. growth   C. repair   D. reproduction

12. Which of the following can release the greatest amount of energy?

    A. carbohydrates   B. lipids   C. proteins   D. sugars

13. The cellular site for protein synthesis is the:

    A. ribosome   B. lysosome   C. mitochondrion   D. chloroplast

14. The element nitrogen can be found in:

    A. amino acids   B. proteins
    C. amino acids and proteins   D. neither amino acids nor proteins

15. The release of energy from glucose takes place at the:

    A. mitochondria   B. ribosomes   C. chloroplasts   D. lysosomes

16. Which organelle can store enzymes for the hydrolysis of starch?

    A. mitochondria   B. ribosomes   C. chloroplasts   D. lysosomes

17. The hydrolysis of one lipid molecule results in the formation of:

    A. three water molecules   B. two glucose molecules
    C. three fat molecules   D. one glycerol molecule and three fatty acid molecules

18. The enzyme protease acts on:

    A. maltose and glucose   B. polysaccharides
    C. glycerol and fatty acids   D. amino acids

19. Which of the following substances are proteins?

    A. amino acids   B. fatty acids   C. enzymes   D. nitric acid

20. Identify the two groups that are found in all amino acids.

    A. an amino group and a carboxyl group
    B. an amino group and a hydroxyl group
    C. a carboxyl group and a hydroxyl group
    D. a hydroxyl group and a methyl group

21. The substance that an enzyme acts on is known as the:

    A. specificity   B. subscript   C. substrate   D. concentrate

22. Which of the following substances can be used to speed up a chemical reaction?

    A. hemoglobin   B. glycerol   C. amylase   D. sodium hydroxide

23. Which of the following pH values indicates the strongest base?

    A. 1   B. 6   C. 7   D. 14

24. All the following are acids *except:*

    A. HCl   B. HF   C. H_2SO_4 D. HOH

25. A solution having a large number of hydrogen ions is:

    A. acidic   B. basic   C. neutral   D. specific

Problems 26–30 are based on the following diagram.

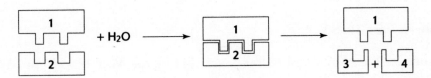

26. What is the name of the process shown?

    A. dehydration synthesis   B. hydrolysis   C. photosynthesis   D. protein synthesis

27. Which molecule is the enzyme?

    A. 1   B. 2   C. 3   D. 4

28. Which combination of molecules represents an enzyme-substrate complex?

    A. 1 and 2   B. 2 and 3   C. 3 and 4   D. 1, 3 and 4

29. If molecule 2 is maltose, then molecule 3 is most likely:

    A. maltase   B. glucose   C. an amino acid   D. a fatty acid

30. What happens to the rate of this reaction if it is subjected to high temperature (above 50°C)?

    A. The reaction speeds up.

    B. The reaction stops.

    C. The rate of reaction remains the same.

    D. The rate of reaction increases and then levels off.

Answers

1. **B.** "Chemistry Review," p. 69

2. **D.** "Chemistry Review," p. 69

3. **D.** "Water," p. 75.

4. **C.** "Carbohydrates," pp. 75

5. **C.** "Carbohydrates," pp. 75–76

6. **B.** "Carbohydrates," pp. 75–76

7. **D.** "Proteins," pp. 79–80

8. **C.** "Carbohydrates," pp. 75–76

9. **C.** "Proteins," pp. 79–80

10. **B.** "Proteins," pp. 79–80

11. **A.** "Proteins," pp. 79–80

12. **B.** "Lipids," p. 78

13. **A.** "Proteins," pp. 79–80

14. **C.** "Proteins," pp. 79–80

15. **A.** Chapter 2, "The Animal Cell As Seen with an Electron Microscope," pp. 37–38

16. **D.** Chapter 2, "The Animal Cell As Seen with an Electron Microscope," pp. 37–38

17. **D.** "Lipids," p. 78

18. **D.** "Proteins," p. 80

19. **C.** "Enzyme Action," pp. 84–85

20. **A.** "Enzyme Action," pp. 84–85

21. **C.** "Enzyme Action," pp. 84–85

22. **C.** "Enzyme Action," pp. 84–85

23. **D.** "Acids and Bases," pp. 82–83

24. **D.** "Acids and Bases," pp. 82–83

25. **A.** "Acids and Bases," pp. 82–83

26. **B.** "Acids and Bases," pp. 82–83

27. **A.** "Acids and Bases," pp. 82–83

28. **A.** "Acids and Bases," pp. 82–83

29. **B.** "Carbohydrates," pp. 75–76

30. **B.** "Factors That Affect the Rate Enzyme Action," p. 85

Chapter 5
Plant Biology

Green plants are amazing organisms. Without green plants, life on earth would not exist as we know it today. Green plants are autotrophs; provide them with carbon dioxide and water, and they are able to make their own food (glucose) by a process called photosynthesis.

$$6CO_2 + 12H_2O \xrightarrow[\substack{\text{chlorophyll,} \\ \text{enzymes}}]{\text{light}} C_6H_{12}O_6 + 6O_2 + 6H_2O$$

The food plants produce also serves as a source of energy for other organisms. Oxygen produced as a by-product of photosynthesis is used in respiration by almost all organisms on the planet.

Plant Structure and Function

Plants have three basic structures (leaves, stems, and roots) that function together as a team, enabling them to be remarkable food- and oxygen-producing machines. The sections that follow review plant structure and function.

Leaf Structure and Function

The leaf is the photosynthetic factory of the plant. Here, carbon dioxide enters through openings called stomates and combines with water brought up from the roots to produce glucose and oxygen. The leaf has specialized cells that contain chloroplasts, which house the green pigment chlorophyll. Chloroplasts are the cellular site for photosynthesis.

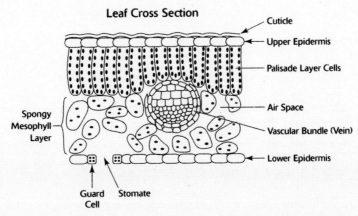

Leaf Cross Section

The following table reviews the structure and function of the leaf and should be studied with the preceding diagram.

Leaf Structure and Function	
Structure	**Function**
Cuticle	A thin, waxy layer that covers the upper epidermis of the leaf, preventing the loss of water.
Upper epidermis layer	A single layer of clear cells that allows light to pass through and prevents the loss of water.
Palisade layer	This layer contains long columnar cells that are packed tightly together. Palisade layer cells contain chloroplasts and are the main cells carrying out photosynthesis.
Vascular bundle:	
xylem	Xylem cells carry water and minerals up from the roots through the stem and into the leaf.
phloem	Phloem cells carry sugar and starch from the leaf down through the stem and into the roots for storage.
Spongy mesophyll layer	The spongy mesophyll layer contains cells with chloroplasts and is a major site of photosynthesis.
Air space	Air spaces are found in the spongy mesophyll layer and function in the exchange of carbon dioxide and oxygen.
Lower epidermis layer	A single layer of clear cells that contains stomates and guard cells.
Stomate	An opening in the lower epidermis that allows carbon dioxide into the leaf and water and oxygen out of the leaf. *Transpiration* is the loss of water by a leaf.
Guard cells	Pairs of guard cells containing chloroplasts are found on the lower epidermis of the leaf and are responsible for forming stomates. During the day the guard cells produce sugar by photosynthesis, causing their walls to curve inward and away from each other and creating a space between the cells called a stomate. At night the process is reversed, and the stomate closes.

Example Problems

These problems review leaf structure and function.

1. What is the function of the cuticle? Why must it be transparent?

   Answer: The cuticle is a thin, waxy layer that covers the upper epidermis of a leaf and prevents the loss of water. The cuticle must be transparent so that light can pass through to the photosynthetic cell layers of the leaf.

2. Which cell layers of the leaf are capable of photosynthesis?

   Answer: The palisade and spongy mesophyll layers are capable of photosynthesis.

3. How is the upper epidermis of the leaf different from the lower epidermis?

   Answer: The upper epidermis of many leaves has a waxy cuticle that the lower epidermis does not. The lower epidermis has guard cells and stomates, which are lacking in the upper epidermis.

4. Why are the vascular bundles of a leaf important?

    Answer: The vascular bundles form the transportation network of the leaf, stem, and root. The phloem cells carry sugar and starch down to the roots for storage. The xylem cells carry water and minerals up from the roots.

5. What is the function of the guard cells?

    Answer: Guard cells control the opening and closing of the stomates. During the day the guard cells produce sugar by photosynthesis, causing their walls to curve inward and away from each other and creating a space between the cells called a stomate. At night the process is reversed, and the stomate closes.

Work Problems

Use these problems on leaf structure and function for additional practice.

1. Why is the leaf an important plant structure?

2. Which cell layer of the leaf performs the most photosynthesis?

3. Why must desert plants have a thick, waxy cuticle?

4. What are the functions of the stomates?

5. Why must plants have both carbon dioxide and oxygen?

Worked Solutions

1. The leaves of plants are important because they are the sites for photosynthesis in plants. During photosynthesis, plants produce glucose that they and other organisms depend on for food. Many organisms use oxygen produced during photosynthesis in the life process of respiration.

2. The palisade layer performs the most photosynthesis. Look at the diagram of the cross section of the leaf:

    The palisade layer cells are closer to the sun.

    Many palisade layer cells are packed tightly together.

    Palisade cells are large cells and contain many chloroplasts.

    Spongy mesophyll layer cells don't perform as much photosynthesis as palisade layer cells. Spongy mesophyll cells are further from the sun, fewer in number, smaller, and don't have as many chloroplasts.

3. Desert plants must have a thick, waxy cuticle to prevent the loss of water.

4. The stomates of a leaf function in photosynthesis, respiration, transpiration, excretion, and evaporation.

Photosynthesis—The stomates allow carbon dioxide into the leaf. Carbon dioxide and water combine to form glucose and oxygen.

Respiration—The stomates allow oxygen into the leaf. Oxygen is used to release the energy stored in glucose.

Transpiration—Water is lost through the stomates.

Excretion—Oxygen is a waste product of photosynthesis that can exit the leaf through the stomate. Carbon dioxide is a waste product of respiration that also exits the leaf through the stomate.

Evaporation—The change of a liquid into a gas.

5. Plants require carbon dioxide for photosynthesis and oxygen for respiration. Plants are capable of both processes. Animals perform only respiration.

The Photosynthetic Process

Look at the chemical equation for photosynthesis shown at the beginning of the chapter. Carbon dioxide and water are the reactants (ingredients) in the equation. Glucose and oxygen are the products (results). Light, chlorophyll, and enzymes are requirements that must be present for photosynthesis to take place.

Light Energy

Plants are actually energy converters rather than food producers. During photosynthesis, light energy is converted into chemical energy, which is stored in the glucose molecule. In cellular respiration, the energy stored in glucose is released, and the cell produces the energy molecule known as ATP (adenosine triphosphate). We will study cellular respiration in Chapter 6. Light can be separated into several different wavelengths: red, orange, yellow, green, blue, indigo, and violet (ROY G BIV). The red wavelength of light is the best for photosynthesis. Green is the worst wavelength for photosynthesis because the leaf reflects this wavelength, making it unavailable to the plant as a source of energy. The colors we see are the wavelengths of light that are reflected back to the eye. The colors we don't see are the wavelengths of light absorbed by the surface they strike.

Chlorophyll

Chlorophyll is a green pigment found inside the chloroplasts of cells in the palisade and spongy mesophyll layers of a leaf. Chlorophyll traps light energy that is used to split water molecules.

Enzymes

The formation of glucose during photosynthesis takes place in two major steps known as the light and dark reactions. Each reaction requires enzymes. Many chemicals used to kill plants work by blocking key enzymes that are needed in photosynthesis.

The Light Reactions

This series of reactions begins the process of photosynthesis. Light energy trapped by chlorophyll is used to split water molecules (*photolysis*). When water is split, oxygen and hydrogen are produced. The oxygen molecules exit the leaf through the stomates and enter the atmosphere. The hydrogen atoms combine with molecules of the coenzyme NADP (<u>n</u>icotinamide <u>a</u>denine <u>d</u>inu-cleotide <u>p</u>hosphate) to form NADPH + H⁺, which is needed in the dark reactions. A *coenzyme* transfers hydrogen and electrons from one reaction to another. (*Reminder:* The hydrogen atom has a proton and an electron.) This movement of electrons is known as the electron transport system.

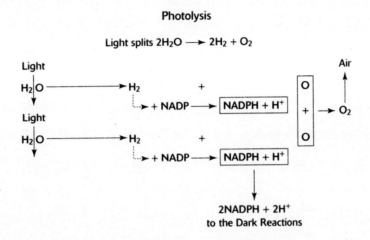

Photolysis

Light splits $2H_2O \longrightarrow 2H_2 + O_2$

2NADPH + 2H⁺
to the Dark Reactions

The Dark Reactions

Light is needed in photosynthesis to split water. All further reactions take place in the absence of light and are called the dark reactions. This series of cyclic reactions was discovered by Melvin Calvin and is often called the Calvin cycle. The Calvin cycle begins when three carbon dioxide mole-cules combine with three molecules of a 5-carbon compound called <u>r</u>ibulose <u>d</u>iphos<u>p</u>hate (RDP); the resulting three 6-carbon compound molecules are unstable and break to form six 3-carbon com-pound molecules called <u>p</u>hospho<u>g</u>lycer<u>a</u>te (PGA). Hydrogen from the light reactions combines with PGA to form six new 3-carbon compounds called <u>p</u>hospho<u>g</u>lycer<u>a</u>ldehyde (PGAL). Three carbon dioxide molecules enter at the start of the cycle to be followed by an additional three molecules of carbon dioxide. Each turn around the cycle can spin off a PGAL to eventually form glucose. An easy way to understand the Calvin cycle is to follow the number of carbon atoms around the cycle.

The Dark Reactions
(The Calvin Cycle)

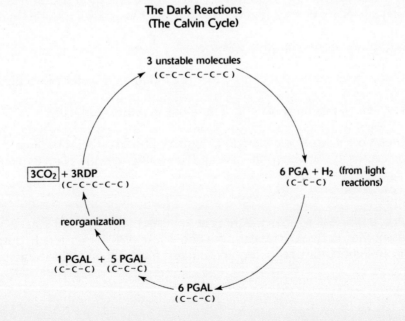

PGAL is the true end product of photosynthesis. Molecules of PGAL can combine and reorganize to form additional RDP molecules, or two PGAL molecules can combine to form glucose. PGAL can also be used to produce proteins and lipids.

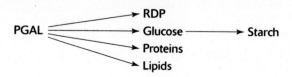

Example Problems

These problems review photosynthesis.

1. Write the chemical equation for photosynthesis. Identify the reactants and the products.

    Answer:

    $$6CO_2 + 12H_2O \xrightarrow[\text{chlorophyll, enzymes}]{\text{light}} C_6H_{12}O_6 + 6O_2 + 6H_2O$$

    The reactants in this chemical equation are carbon dioxide and water. The products in the equation are glucose, oxygen, and water.

2. Identify the process that is the opposite of photosynthesis.

    Answer: Respiration is the process that is the opposite of photosynthesis. Look at the following chemical equation for respiration and compare it to the chemical equation for photosynthesis shown in answer 1.

    $C_6H_{12}O_6 + 6O_2 + 6H_2O \rightarrow$ enzymes $\rightarrow 6CO_2 + 12H_2O +$ energy (ATP)

    In photosynthesis, light energy is converted to chemical energy and stored in the glucose molecule. In respiration, energy stored in the glucose molecule is released in the form of ATP.

3. Why is chlorophyll needed for photosynthesis?

    Answer: Chlorophyll traps light energy that is used to split water molecules.

4. Why is the green wavelength of light of little use in photosynthesis?

    Answer: Green is the worst wavelength of light for photosynthesis because the leaf reflects rather than absorbs this wavelength, making it unavailable to the plant as a source of energy.

5. What is a coenzyme?

    Answer: A coenzyme is a molecule that can transfer hydrogen and electrons from one reaction to another. In photosynthesis, the coenzyme NADP transfers hydrogen and electrons from light to dark reactions. In humans, many vitamins function as coenzymes.

Work Problems

Use these problems on photosynthesis for additional practice.

1. Explain what happens during the light reactions.

2. Where does oxygen produced during photosynthesis come from?

3. Why is PGAL (phosphoglyceraldehyde) a key molecule?

4. How can PGAL (phosphoglyceraldehyde) form RDP (ribulose diphosphate)?

5. At what point do CO_2 molecules enter the Calvin cycle?

Worked Solutions

1. During the light reactions, light energy trapped by chlorophyll is used to split water molecules, producing oxygen and hydrogen. The oxygen molecules leave the leaf through the stomates and enter the atmosphere. The hydrogen atoms combine with molecules of the coenzyme NADP (nicotinamide adenine dinucleotide phosphate) to form NADPH + H^+, which is needed in the dark reactions.

2. Oxygen produced during photosynthesis comes from water molecules. Light energy splits water molecules, producing oxygen and hydrogen. This process is called photolysis.

3. PGAL is a key molecule because it can be used as a starting point for the production of glucose, starch, lipids, proteins, and RDP.

4. The key to correctly answering this question is to keep your eye on the number of carbon atoms. PGAL is a 3-carbon compound. Each turn around the Calvin cycle produces 6 PGAL molecules for a total of 18 carbon atoms. Now subtract 3 carbon atoms for 1 molecule of PGAL that spins off to eventually form glucose. This leaves us with 15 carbon atoms. These atoms are rearranged to form 3 molecules of RDP, which is a 5-carbon compound. The total number of carbon atoms remains the same.

   6 PGAL X (3 C per molecule) = 18 carbon atoms
   −1 PGAL X (3 C per molecule) = −3 carbon atoms (this molecule of PGAL is spun off to later form glucose)
   5 PGAL X (3 C per molecule) = 15 carbon atoms (these atoms now rearrange themselves to form 3 RDP)

   Rearrangement

   3 RDP X (5 C per molecule) = 15 carbon atoms

5. CO_2 molecules enter the Calvin cycle by combining with RDP.

Stem Structure and Function

The stem of a plant is an intermediate structure found between the roots on one end and the leaves on the other. The stem supports the leaves of the plant while serving as a pathway for the transport of water and minerals up from the roots and sugars and starches down from the leaves. Plants contain vascular bundles (veins) that have specialized cells for transport. In dicotyledonous (dicot), or mostly woody plants such as trees and shrubs, the vascular bundles are located directly behind the bark of the stem. In monocotyledonous (monocot), or nonwoody (*herbaceous*) plants, the vascular bundles are scattered throughout the stem. *Lenticels* are small openings in the stem that allow oxygen in and carbon dioxide out.

In dicots the only living cells are those located in the vascular bundles found directly behind the bark. All the other cells in the stem are dead. If a strip of bark is removed all around the stem (*girdling*), the continuity of the vascular bundle is broken. Water is not able to go up, food is not able to go down, and the plant dies. Dicot vascular bundles have three distinct types of cells: phloem, cambium, and xylem.

❑ **Phloem**—Phloem cells are located on the outside of the vascular bundle facing the bark. Phloem cells carry food materials such as sugar and starch through the leaf, *down* the stem, and into the roots. *Translocation* is the movement of food through a plant.

❑ **Cambium**—Cambium cells are located between the phloem and the xylem. The cambium contains *meristematic cells* that are capable of reproduction. Cambium cells produce phloem and xylem. In the winter, cold temperatures kill these cells of the plant. In the spring, the cambium cells reproduce to make new phloem and xylem cells. This allows the plant to live from year to year (*perennial*) and to increase in diameter. Each year dead phloem and xylem produce a ring of cells in the stem. We can often estimate the age of a plant by counting its *annual rings*.

❑ **Xylem**—Xylem cells are located on the inside of the vascular bundle facing the inside of the stem. Xylem cells carry water and minerals *up* from the roots, through the stem, and into the leaves of the plant.

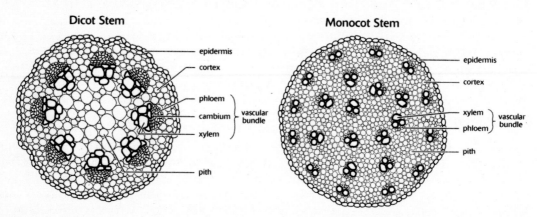

In monocots, the vascular bundles have phloem and xylem but lack cambium cells. In the winter, cold temperatures kill these cells. Without cambium new phloem and xylem cannot form, and the plant dies. Most monocots are *annuals* that live for only one growing season.

Example Problems

These problems review stem structure and function.

1. What is the function of the stem?

 Answer: The stem supports the leaves of the plant while serving as a pathway for the transport of water and minerals up from the roots and sugars and starches down from the leaves.

2. How are monocot stems different from dicot stems?

 Answer: Monocots are nonwoody (herbaceous) plants with vascular bundles that are scattered throughout the stem. Monocot vascular bundles lack cambium cells, and most are annuals, living for only one growing season. Dicots are woody plants with vascular bundles found directly behind the bark. The vascular bundles of dicots contain phloem, cambium, and xylem. Dicots are perennial plants.

3. How are annual rings formed?

    Answer: In the winter, cold temperatures kill phloem and xylem cells, producing a ring of dead cells called an annual ring.

4. What is translocation?

    Answer: Translocation is the movement of food through a plant.

Work Problems

Use these problems on stem structure and function for additional practice.

1. Compare phloem and xylem in terms of the direction in which each transports materials. What materials does each type of cell carry?

2. Why are cambium cells important?

3. Why must a stem have lenticels?

4. How does girdling kill a tree?

Worked Solutions

1. Phloem cells are specialized for downward transport, carrying sugars and starches to the roots. Xylem cells are specialized for upward transport, carrying water and minerals up to the leaves.

2. Cambium cells are important because they are meristematic cells and are capable of producing phloem and xylem.

3. A stem must have lenticels for respiration. These small openings in the stem allow oxygen in and carbon dioxide out. The living cells of the vascular bundle use the oxygen for energy production.

4. Girdling breaks the continuity of the vascular bundles. Water is not able to move up the stem, and food is not able to move down. As a result, the plant eventually dies.

Root Structure and Function

The root of a plant anchors the plant to the ground and serves to support the stem. The root absorbs water and minerals and sends these materials up through the stem to the leaves. The roots also serve as a major storage site of starch.

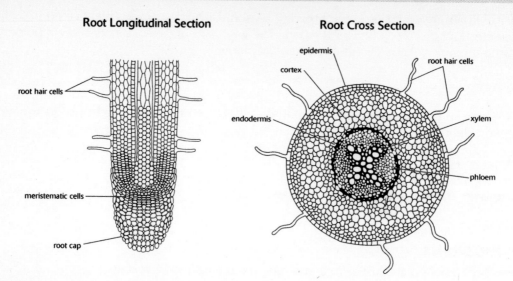

The outer single-cell layer of the root is the epidermis and contains elongated cells called *root hair cells.* These cells have a large surface area and are adapted for the increased absorption of water. In the center of the root is the vascular cylinder, which contains xylem and phloem cells for the transport of materials. Surrounding the vascular cylinder are the cortex cells, which store food. The root cap contains meristematic cells that reproduce to form new cells, causing the root to grow deeper into the ground.

Example Problems

These problems review root structure and function.

1. What are the functions of the root?

    Answer: The root anchors the plant to the ground and supports the stem. Roots absorb water and minerals, sending these materials up through the stem to the leaves. Also, roots serve as a major storage site of starch.

2. Which cells in the root are used for storage?

    Answer: The cortex cells of the root are used for storage.

3. Name the process by which water can enter a root hair cell.

    Answer: Osmosis is the process by which water can enter a root hair cell. In osmosis, water moves from areas of high concentration to areas of low concentration. If you missed this question, review molecular transport in Chapter 2.

Work Problems

Use these problems on root structure and function for additional practice.

1. How are epidermis cells different in structure and function from root hair cells?

2. Why does the plant store food in the roots as starch rather than as glucose?

3. Why don't root cells need chloroplasts?

4. How are meristematic cells in the root important?

Worked Solutions

1. Epidermis cells are the outer layer of cells in a root. Epidermis cells are small and round and function in the absorption of water and minerals from the soil. Root hair cells are elongated cells on the outside of the root with a large surface area that is specialized for the absorption of water and minerals. The large surface area of root hair cells allows them to absorb more water than the epidermis cells.

2. Starch is a polysaccharide formed from many molecules of glucose combined. It is more efficient to store one molecule of starch as opposed to several hundred molecules of glucose.

3. Chloroplasts contain the chlorophyll molecule, which is used to trap light energy needed for photosynthesis. Because roots are underground and not exposed to light, it is pointless for them to have chloroplasts.

4. Meristematic cells are reproductive cells. The meristematic cells of the root are responsible for root growth deeper into the soil.

Chapter Problems and Answers

Problems

The following is a brief paragraph based on plants, their structure and function. For problems 1–10, fill in the missing terms.

Plants are amazing organisms because they are capable of making their own food by the

process of _____. Organisms that can make their own food are known as _____. The
     1   2

reactants needed by a plant to manufacture food are _____ and _____. Specialized green
                                      3   4

structures called _____ contain the green pigment _____ that is essential to start this
                   5   6

process. The plant begins the production of food with a series of chemical reactions

known as the _____. These reactions are followed by another series of reactions that was
                 7

discovered by Calvin known as the _____. At the conclusion of this second set of reactions,
                                8

the plant has produced the food molecule known as _____. These molecules can be
                                        9

combined to make a polysaccharide (called starch) by the process of _____.
                                                      10

For problems 11–16, fill in the name of the plant structure that is associated with each description.

11. Specialized cells for upward transport. _____

12. Reproductive cells. _____

13. Specialized cells in the root for the absorption of water. _____

14. Exchange of gases in the stem. _____

15. Storage of starch. _____

16. Aids in the light reactions of photosynthesis. _____

Problems 17–23 are based on the following cross-section diagram of a leaf. Select the letter of the structure that *best* fits each description.

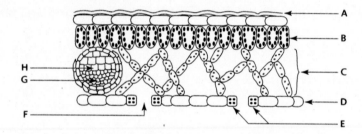

17. Select the structure in the lower epidermis that contains chloroplasts. _____

18. Which structure is the cuticle? _____

19. Identify the cells that control the stomate. _____

20. Which cell layer has air spaces between its cells? _____

21. Select the layer that has the palisade cells. _____

22. Identify the plant structure that disappears in the absence of light. _____

23. Which structure is responsible for the exchange of gases between the leaf and the atmosphere? _____

For problems 24–27, select the molecule from the following list that is associated with each description of a step in the process of photosynthesis. A choice can be used more than once or not at all.

A. Chlorophyll   B. PGAL   C. Coenzyme   D. PGA   E. RDP

24. This molecule combines with carbon dioxide at the start of the dark reactions.

25. This molecule functions in the electron transport system.

26. This molecule traps light energy to begin the process of photosynthesis.

27. This molecule is the true end product of photosynthesis and is capable of forming glucose, lipids, and proteins.

Answers

1. **Photosynthesis.** *Photo* means light, and *synthesis* means to combine. Plants use light energy to combine CO_2 and H_2O to make glucose. The effect of this process is to convert light energy to the chemical-bond energy of the glucose molecule.

2. **Autotrophs.** Any organism that can make its own food is an autotroph.

3. **Carbon dioxide or water.** These two molecules are the reactants (or ingredients) necessary to make glucose.

4. **Water or carbon dioxide.** These two molecules are the reactants (or ingredients) necessary to make glucose.

5. **Chloroplasts.** Only green plants and certain algae have chloroplasts, which contain chlorophyll necessary for photosynthesis.

6. **Chlorophyll.** Chlorophyll traps light energy to split water and begin the photosynthetic process.

7. **Light reactions.** The light reaction (or photolysis) splits water into hydrogen and oxygen to begin the photosynthetic process.

8. **Dark reactions.** The dark reactions are also known as the Calvin cycle and sometimes as *carbon fixation*.

9. **Glucose.** Each turn of the Calvin cycle can spin off one molecule of PGAL. Two PGAL molecules can combine to form one glucose molecule.

10. **Dehydration synthesis.** Glucose molecules can be combined by the removal of water. When hundreds of glucose molecules are combined, a large polysaccharide such as starch can be formed.

11. **Xylem.** Xylem cells are the large cells of a vascular bundle that are specialized for upward transport.

12. **Cambium.** Cambium cells are meristematic (reproductive) cells found between the phloem and xylem.

13. **Root hair cells.** These cells have a large surface area and are adapted for increased absorption of water.

14. **Lenticels.** The small openings, or cracks, in stems of plants are called lenticels. These structures allow oxygen in and carbon dioxide out of the stem.

15. **Cortex cells.** The roots of plants have cortex cells that function in the storage of materials.

16. **B. Chloroplast (chlorophyll).** During the light reactions of photosynthesis, water molecules are split into hydrogen and oxygen. Chlorophyll found in the chloroplasts of green plants captures the light energy needed for this process (photolysis).

17. **E.** The only cells in the lower epidermis that have chloroplasts are the guard cells. When viewed through a microscope, these cells look green.

18. **A.** The cuticle is a layer of wax found on top of the upper epidermis cells of a leaf. The cuticle helps prevent the loss of water from a plant by the process of transpiration.

19. **E.** The guard cells found in the lower epidermis form and close the opening known as the stomate.

20. **C.** The spongy mesophyll layer has air spaces between its cells. At night, oxygen needed for respiration is stored in the air spaces. During the day, CO_2 in the air spaces can be used for photosynthesis.

21. **B.** The cells of the palisade layer are long and packed tightly together.

22. **F.** At night the guard cells close, and the stomate disappears.

23. **F.** The stomates of the lower epidermis allow carbon dioxide to enter a leaf and allow oxygen to exit.

24. **E. RDP.** Ribulose diphosphate is a 5-carbon compound that combines with carbon dioxide (1 carbon) to form an unstable 6-carbon compound that breaks into two molecules of PGA. PGA combines with hydrogen produced during the light reaction to form PGAL. PGAL can form additional molecules of RDP, to maintain the Calvin cycle.

25. **C. Coenzyme.** NADP is the coenzyme that transfers hydrogen and electrons needed in dark reactions.

26. **A. Chlorophyll.** Traps light energy that is used to split water molecules.

27. **B. PGAL.** PGAL (or phosphoglyceraldehyde) can be used as a starting point for the production of glucose, RDP, proteins, and lipids.

Supplemental Chapter Problems

Problems

For problems 1–25, select the *best* answer.

1. The reactants that take part in the process of photosynthesis are:

    A. $H_2O + O_2$ B. $CO_2 + H_2O$ C. $O_2 + CO_2$ D. $C_6H_{12}O_6 + O_2$

2. Photosynthesis takes place in organelles called:

    A. ribosomes   B. the Golgi apparatus   C. mitochondria   D. chloroplasts

3. Which wavelength of light is the least valuable for photosynthesis?

    A. red   B. orange   C. green   D. violet

4. What is the name of the process involved in the movement of water through the stomates of plants?

    A. photolysis   B. osmosis   C. transpiration   D. dehydration synthesis

5. Which molecule produces oxygen that is formed during photosynthesis?

   A. CO_2 B. H_2O C. $C_6H_{12}O_6$ D. NADP

6. Which of the following takes place during the light reactions of photosynthesis?

   A. Hydrogen atoms are combined with carbon dioxide.

   B. Sugar molecules are combined to form starch.

   C. Sugar is split into carbon and water.

   D. Water is split to form hydrogen and oxygen.

7. The hydrogen atoms that a plant uses during the dark reactions of photosynthesis are obtained from:

   A. glucose   B. carbon dioxide   C. water   D. starch

8. Which molecule functions as a coenzyme capable of transporting hydrogen and electrons?

   A. H_2O B. NADP   C. RDP   D. PGAL

9. Which of the following molecules has no effect on the photosynthesis process?

   A. chlorophyll   B. enzymes   C. carbon dioxide   D. nitrogen

10. Which of the following organic molecules is formed during the process of photosynthesis?

    A. oxygen   B. PGAL   C. CO_2 D. chlorophyll

11. In a dicot (or woody) stem, xylem and phloem cells are separated from each other by the:

    A. cambium   B. cortex   C. phloem   D. xylem

12. Meristematic cells can be found in all the following parts of a plant *except* the:

    A. leaves   B. dicot roots   C. dicot stems   D. monocot roots

13. The lenticels found in many plant stems function in the life process of:

    A. photosynthesis   B. transpiration   C. cellular reproduction   D. respiration

14. Which vascular tissue is responsible for the upward transport of materials in the stem of a plant?

    A. cambium   B. cortex   C. phloem   D. xylem

15. Which cells in the stem are responsible for cell division and reproduction resulting in the increased diameter of the stem?

    A. cambium   B. cortex   C. phloem   D. xylem

16. Where are the vascular bundles of nonwoody plants found?

    A. spread throughout the stem   B. anywhere in the root

    C. in the stem just below the bark   D. in the lower epidermis of the leaves

17. The destruction of xylem cells in a plant negatively affects the plant's ability to transport:

    A. water and minerals up to the leaves   B. water out of the leaves through the stomate

    C. sugar from the leaves to the roots   D. oxygen into the leaves through the stomate

18. Girdling a tree results in:

    A. a break in the continuity of the vascular bundles, which results in the death of the tree
    B. stimulated cell growth
    C. decreased respiration due to lenticel damage
    D. a decrease in transpiration and the loss of water

19. Root hair cells are specialized for:

    A. food storage   B. anchorage of the plant
    C. upward water transport   D. increased water absorption

20. Identify the cells in the root that have the ability to store materials produced in the leaf.

    A. meristem   B. cortex   C. phloem   D. xylem

21. Which cells in the root are specialized for carrying out the life function of reproduction?

    A. meristem   B. cortex   C. phloem   D. xylem

22. Which of the following is *not* a function of the root?

    A. absorption of oxygen and carbon dioxide   B. absorption of water and minerals
    C. storage of sugar and starch   D. anchorage and support

23. From the following choices, select the plant structure that is found in all roots, stems, and leaves.

    A. guard cells   B. vascular tissue   C. lenticels   D. spongy mesophyll cells

24. All the following are characteristics of green plants *except:*

    A. respiration   B. transpiration   C. transport   D. heterotrophic nutrition

25. Two end products of photosynthesis are:

    A. $H_2O + O_2$ B. $H_2O + CO_2$ C. $O_2 + CO_2$ D. $C_6H_{12}O_6 + O_2$

Answers

1. **B.** "Chapter Introduction," p. 95

2. **D.** "Leaf Structure and Function," p. 95

3. **C.** "Light Energy," p. 98

4. **C.** "Leaf Structure and Function," p. 95

5. **B.** "The Light Reactions," p. 98

6. **D.** "The Light Reactions," p. 98

7. **C.** "The Light Reactions," p. 98

8. **B.** "The Light Reactions," p. 98

9.   **D.** "The Photosynthesic Process," p. 98

10.   **B.** "The Dark Reactions," p. 99

11.   **A.** "Stem Structure and Function," p. 101

12.   **A.** "Stem Structure and Function," p. 101

13.   **D.** "Stem Structure and Function," p. X101

14.   **D.** "Stem Structure and Function," p. 101

15.   **A.** "Stem Structure and Function," p. 101

16.   **A.** "Stem Structure and Function," p. 101

17.   **A.** "Stem Structure and Function," p. 101

18.   **A.** "Stem Structure and Function," p. 101

19.   **D.** "Root Structure and Function," p. 103

20.   **B.** "Root Structure and Function," p. 103

21.   **A.** "Root Structure and Function," p. 103

22.   **A.** "Root Structure and Function," p. 103

23.   **B.** "Leaf Structure and Function," p. 95

24.   **D.** "Chapter Introduction," p. 95

25.   **D.** "Chapter Introduction," p. 95

Chapter 6
Human Biology

In this chapter on human biology we will study the life functions of nutrition, circulation, respiration, excretion, locomotion, and regulation.

Nutrition and Digestion

Humans are heterotrophic organisms and must ingest (eat) food to obtain energy and the materials necessary to maintain life. Humans have a one-way digestive system. Food enters the mouth, and wastes leave from the large intestines. When food enters the body, it must be broken down by the *digestive system* into molecules small enough to enter the cells. When in the cell, some end products of digestion can be used for energy production, while others are used to build new cells for growth, repair, and reproduction.

Digestive System

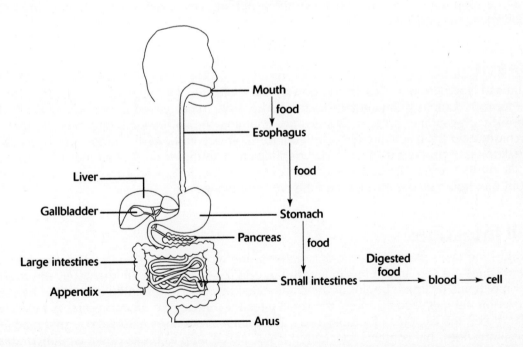

Two kinds of digestion break down food: mechanical (physical) digestion and chemical digestion. *Mechanical or physical digestion* breaks down food into smaller particles, but the type of food is the same. Large carbohydrates are now many small carbohydrates, large proteins are now many small proteins, and so on. *Chemical digestion* breaks down carbohydrates, proteins, and lipids into their basic building blocks. Hydrolysis is the chemical reaction used to break down food during the digestive process. The presence of enzymes speeds up each of the chemical reactions.

$$Starch \xrightarrow{\text{amylase}} Glucose$$

$$Proteins \xrightarrow{\text{protease}} Amino\ acids$$

$$Lipids \xrightarrow{\text{lipase}} Fatty\ acids\ and\ glycerol$$

Mouth

The process of digestion begins in the mouth. Mechanical digestion (chewing) breaks down food into many small pieces, increasing the surface area of the food. This makes chemical digestion more efficient. The salivary glands produce salivary *amylase,* an enzyme that begins the chemical breakdown of starch into sugar. When a person swallows, food enters the esophagus.

Esophagus

The esophagus (food pipe) is a tube that connects the mouth to the stomach. No digestion takes place in the esophagus. Food moves down the esophagus by a series of involuntary muscle contractions known as *peristalsis.*

Stomach

The stomach is a muscle that can grind down food molecules by mechanical digestion. The walls of the stomach contain gastric glands that produce a substance called gastric juice, which begins the chemical digestion of protein. Gastric juice contains *pepsinogen* and hydrochloric acid. Hydrochloric acid (HCl) activates pepsinogen, forming *pepsin* (a gastric protease). Pepsin begins the breakdown of proteins into small polypeptides and amino acids. The stomach has a special mucus lining that resists the corrosive action of the hydrochloric acid. Food stays in the stomach for about four hours before moving into the small intestines.

Small Intestines

Most digestion takes place in the small intestines, and all digestion is completed there. The small intestines are about 6 meters in length but only 2.5 centimeters in diameter. The chemical digestion of lipids by lipase begins in the small intestines. *Bile,* a green fluid secreted by the *liver* and stored in the *gallbladder,* enters the small intestines through a tube to *emulsify* fats (break large fat molecules into small fat molecules). The *pancreas* secretes a fluid through a tube into the small intestines that contains several enzymes, which help to break down carbohydrates, proteins, and lipids. The pancreatic fluid also contains bicarbonate, which neutralizes the acid from the stomach. Digested *food* (sugars, amino acids, glycerol, fatty acids) enters the blood from the small intestines through finger-like projections from the intestinal wall called villi. The circulatory system transports the food to all cells. *Waste* products enter the large intestines.

Large Intestines

The large intestines are about 1.5 meters in length but 6.5 centimeters in diameter. Here, excess water is reabsorbed into the blood, and undigested food is eliminated from the body. The *appendix* off the right side of the large intestines is a small *vestigial* structure (one that no longer has a function).

Example Problems

These problems review nutrition and digestion.

1. Where does the digestion of starch, protein, and lipids begin and end?

    Answer: Starch digestion begins in the mouth, protein digestion begins in the stomach, and lipid digestion begins in the small intestines. All digestion is completed in the small intestines.

2. What is peristalsis?

    Answer: Peristalsis is a series of involuntary muscle contractions of the esophagus that forces food into the stomach.

3. Where does most digestion take place?

    Answer: Most digestion takes place in the small intestines, and all digestion is completed there.

4. What is the role of the appendix in the digestive process?

    Answer: The appendix is a vestigial organ and does nothing for human digestion. At some point in human evolutionary history, it probably did have a digestive function.

5. How is mechanical digestion different from chemical digestion?

    Answer: Mechanical digestion is a physical process that breaks food down into smaller particles, but the type of food remains the same. Chemical digestion breaks down carbohydrates into sugars, proteins into amino acids, and lipids into glycerol and fatty acids.

Work Problems

Use these problems on nutrition and digestion for additional practice.

1. How do the liver and pancreas aid the process of digestion?

2. Why is mechanical digestion important?

3. Why is hydrochloric acid important for digestion?

4. What are the end products of digestion?

5. How do the end products of digestion get to the cells of the body?

Worked Solutions

1. The liver produces bile, which emulsifies fats. The pancreas produces pancreatic fluid that contains several enzymes, which help break down carbohydrates, proteins, and lipids. The pancreatic fluid also contains bicarbonate, which neutralizes the acid from the stomach.

2. Mechanical digestion is important because it increases the surface area of food, making chemical digestion more efficient.

3. Hydrochloric acid is important because it activates pepsinogen, forming *pepsin* (a gastric protease), which begins the breakdown of proteins into small polypeptides and amino acids. Pepsin requires acidic conditions to function.

4. The end products of digestion are sugars, amino acids, glycerol, and fatty acids.

5. The end products of digestion enter the blood from the villi of the small intestines. The circulatory system transports these food molecules to all cells of the body.

Circulation

In humans the blood circulatory system transports materials such as food, water, and oxygen from the environment to all cells of the organism. The circulatory system also carries waste products away from the cells. Humans have a *closed circulatory system* (blood is always in the blood vessels) with a heart that pumps the blood to the organs. Blood vessels called *arteries* have thick muscular walls that pulse and carry blood away from the heart to the organs. *Veins* have thin muscular walls and carry blood back to the heart. Veins have valves that prevent the backward flow of blood. Small blood vessels in the organs called *capillaries* are one cell in diameter and connect arteries to veins.

Human Circulation

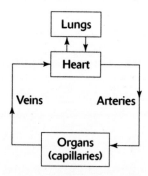

The Structure and Function of the Heart

The function of the heart is to pump blood. The heart is a muscular organ about the size of a fist containing four chambers. The two upper chambers of the heart are the receiving chambers and are called the *atria*. The two lower chambers of the heart are the pumping chambers and are called the *ventricles*. The right atrium receives blood from the *superior vena cava* (upper part of the body) and *inferior vena cava* (lower part of the body). From the right atrium, blood flows into the right ventricle and is pumped into the pulmonary arteries. These arteries carry *deoxygenated blood* (low in O_2, high in CO_2) to the lungs. In the lungs, an exchange of gases takes place; the blood loses CO_2 and picks up O_2. Two pulmonary veins from each lung bring the *oxygenated blood* (high in O_2, low in CO_2) to the left atrium of the heart. From the left atrium, blood flows into the left ventricle and is pumped into the aorta. The heart has valves that prevent the backward flow of blood. The left ventricle has a very thick wall because it must pump blood throughout the body. The aorta is an artery and the largest blood vessel in the body. It branches off to deliver blood to all the organs.

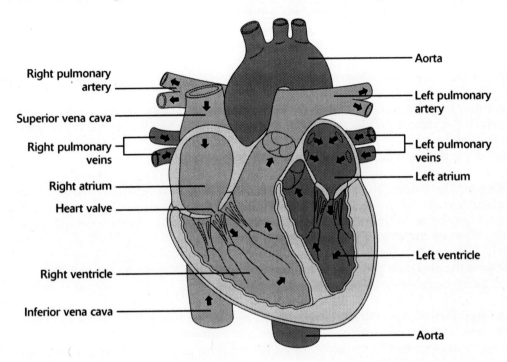

Components (Parts) of the Blood and Their Function

The blood is the river of life in the human body. Humans have about 5.5 liters of blood composed of two major parts: plasma and cells.

Plasma is the straw-colored liquid part of the blood and makes up about 55% of the blood. Plasma consists mostly of water and proteins. One important protein found in plasma is *fibrinogen,* which helps the blood to clot. Plasma carries the end products of digestion to the cells, transports urea wastes from the cells and transports antibodies and hormones.

Red blood cells (RBC) contain the hemoglobin molecule. *Hemoglobin* aids the red blood cell in transporting oxygen to the cells and carbon dioxide from the cells. The mature human red blood cell is an exception to the cell theory because it does not have a nucleus. New red blood cells are made in the bone marrow.

White blood cells (WBC) help the body fight disease. Some white blood cells can engulf and destroy bacteria. Others are capable of producing *antibodies,* which are proteins that can destroy *antigens* (foreign substances in the body). *Antibiotics* are substances produced by bacteria or fungi that can destroy bacteria but not viruses.

Platelets are cell fragments that lack a nucleus. Platelets interact with fibrinogen and help the blood to clot.

Example Problems

These problems review circulation.

1. What is the structure and function of the heart?

    Answer: The heart pumps blood and helps maintain blood pressure. Actually, the heart is a dual pump with two sides. The right atrium receives blood from the body, and the right ventricle pumps blood to the lungs. The left atrium receives blood from the lungs, and the left ventricle pumps blood out to the organs of the body.

2. Why is the human transport system called a *circulatory* system?

    Answer: In humans, blood flows clockwise in a circle beginning and ending with the heart. It is a closed circulatory system because blood never leaves the blood vessels.

3. Name the different blood vessels and state their structure and function.

    Answer: The human blood vessels are arteries, veins, and capillaries. Arteries have thick muscular walls that pulse and carry blood away from the heart to the organs. Veins have thin muscular walls that contain valves and carry blood back to the heart. Capillaries are one cell in diameter and connect arteries to veins.

4. What is the structure and function of each of the following blood components: plasma, red blood cells, white blood cells, and platelets?

    Answer: Plasma is the straw-colored liquid part of the blood and makes up about 55% of the blood. Plasma consists mostly of water and proteins. Plasma carries the end products of digestion to the cells, transports urea wastes from the cells, and transports antibodies, hormones, and other proteins. Red blood cells have the iron-containing hemoglobin molecule that helps transport oxygen to the cells and carbon dioxide from the cells. White blood cells help the body fight disease by destroying bacteria and other foreign substances. Platelets are cell fragments that interact with fibrinogen and help the blood to clot.

Work Problems

Use these problems on circulation for additional practice.

1. Which blood vessels carry the most deoxygenated blood?

2. How does oxygen enter red blood cells?

3. How is the backward flow of blood in the circulatory system prevented?

4. Why must the left ventricle have the thickest walls of the heart?

5. Outline the pathway of blood flow from the right toe to the heart and back to the right toe.

Worked Solutions

1. The two pulmonary arteries carry the most deoxygenated blood from the right ventricle to the lungs. Deoxygenated blood is low in oxygen but high in carbon dioxide. In the lungs, an exchange of gases takes place; the red blood cells lose carbon dioxide and pick up oxygen. Oxygenated blood returns to the heart via the four pulmonary veins. The left ventricle pumps the oxygenated blood to all parts of the body. In general, arteries carry oxygenated blood, and veins carry deoxygenated blood. The pulmonary arteries and pulmonary veins are the exceptions.

2. In the lungs, oxygen enters the red blood cells by diffusion. The concentration of oxygen in the lungs is much higher than in the red blood cells, and the oxygen molecules move from areas of high concentration to areas of low concentration.

3. The backward flow of blood in the circulatory system is prevented by the pumping action of the heart, which helps to keep the blood flowing in a forward direction. Valves in the heart prevent the backward flow of blood in the heart, and valves in the veins prevent the backward flow of blood in those blood vessels.

4. The left ventricle has the thickest walls of the heart because it must pump blood throughout the entire body.

5. Capillaries in the right toe → small veins in the toe → large vein in the right leg → inferior vena cava → right atrium → right ventricle → pulmonary arteries → lungs → left atrium → left ventricle → aorta → large artery in the right leg → small arteries in the right toe → capillaries in the toe.

Respiration

Respiration in humans takes place on an organism level and on a cellular level. On the organism level, oxygen is taken *into* the body (*inhalation*), and carbon dioxide *exits* the body (*exhalation*). On the cellular level, oxygen is used to release energy stored in certain food molecules such as glucose.

Organism Respiration

Air containing oxygen enters the body through the nose or mouth and goes into the *trachea*, which is the windpipe. A small flap called the *epiglottis* covers the trachea when a person swallows, preventing food from entering and keeping a person from choking to death. The trachea has rings of cartilage that prevent it from collapsing. From the trachea, oxygen enters the *bronchi*, which lead into the lungs. The bronchi branch out into thousands of smaller tubes known as *bronchioles* that end in tiny air sacs called *alveoli*. Capillaries surround the alveoli, and the exchange of oxygen for carbon dioxide in the blood takes place there. The *diaphragm* is a muscle located under the lungs. When the diaphragm moves down, the pressure of air outside the lungs is greater than inside, and air flows in (inhalation). When the diaphragm moves up, the pressure of air in the lungs is greater than outside, and air is pushed out (exhalation). The level of carbon dioxide in the blood determines the rate of respiration. The higher the CO_2 level in the blood, the faster the rate of respiration.

Respiratory System

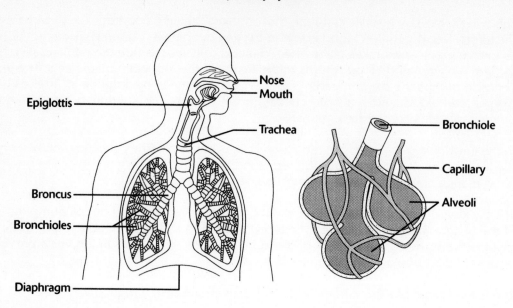

Example Problems

These problems review organism respiration.

1. How does oxygen enter the body?

    Answer: Oxygen enters the body through the nose or the mouth.

2. Why does the trachea have rings of cartilage?

    Answer: The rings of cartilage in the trachea are needed to keep the trachea constantly open. If the trachea collapses, air containing oxygen is not able to enter the body, and this could cause death.

3. How is food kept out of the trachea?

    Answer: When a person swallows, a small flap called the epiglottis covers the trachea and prevents food from entering, keeping the person from choking to death.

Work Problems

Use these problems on organism respiration for additional practice.

1. Outline the pathway of oxygen from the air to a body cell.

2. Why are the alveoli important?

3. How does oxygen enter a red blood cell?

4. What is the role of the diaphragm in respiration?

5. What factor determines how quickly humans breathe?

Worked Solutions

1. Nose/mouth → trachea → bronchi → bronchioles → alveoli → capillaries → red blood cell.

2. The alveoli are air sacs found in the lungs and surrounded by capillaries. Oxygen from the alveoli diffuses into the capillaries.

3. Oxygen enters a red blood cell by the process of diffusion. Diffusion is the movement of molecules from areas of high concentration to areas of low concentration. In the lungs, capillaries that surround the alveoli have a high concentration of oxygen, and red blood cells have a low concentration of oxygen. As a result, oxygen diffuses across the moist membranes of red blood cells to enter these cells.

4. The diaphragm forces air into and out of the lungs. The diaphragm is a muscle located under the lungs. As the diaphragm moves up, air is pushed out of the lungs. As the diaphragm moves down, air is pulled into the lungs.

5. The level of CO_2 in the blood determines the rate of respiration. The higher the CO_2 level in the blood, the faster the rate of respiration.

Cellular Respiration

Cellular respiration is the process by which a cell converts food to energy. The circulatory system brings digested food molecules (glucose, glycerol, fatty acids) and oxygen to the cells. These molecules enter the cell by diffusion and can be used for energy production at the mitochondria. The energy molecule produced by a cell is called ATP, adenosine triphosphate (adenosine with three phosphates attached to it). The greater the number of ATP molecules in a cell, the greater the available energy. ATP can be built from ADP (adenosine diphosphate). The following reaction is reversible:

$$ADP + P \xrightleftharpoons{ATP\text{-}ase} ATP$$

The breakdown of glucose during cell respiration requires two series of reactions, known as anaerobic and aerobic respiration. Each of these reactions needs enzymes, without which they cannot take place. Anaerobic respiration does not require oxygen, but aerobic respiration does.

Anaerobic respiration: $C_6H_{12}O_6 + 2\ ATP \xrightarrow{enzymes} 2\ Pyruvate + 4\ ATP$ (gain of 2 ATP)

Aerobic respiration: $2\ Pyruvate + O_2 \xrightarrow{enzymes} 6CO2 + 6H_2O + \underline{34\ ATP}$ (gain of $\underline{34}$ ATP)

If the two preceding chemical equations are combined, the result is the overall reaction for cellular respiration (as shown):

$$C_6H_{12}O_6 + 6O_2 + 2\ ATP \xrightarrow{enzymes} 6CO_2 + 6H_2O + 38\ ATP\ \text{(gain of 36 ATP)}$$

Anaerobic Respiration

The process of cell respiration begins in the cytoplasm outside the mitochondria. Two molecules of ATP are used to break glucose (*glycolysis*) into two 3-carbon compounds called PGAL (phosphoglyceraldehyde, made during photosynthesis). To break a molecule, energy must be used. Each PGAL molecule loses two hydrogen atoms to become pyruvate, and in the process two molecules of ATP are produced for each pyruvate (for a total of four ATP molecules). During anaerobic respiration, two ATP (four ATP produced minus the two ATP used to break glucose) are gained. The hydrogen atoms lost by PGAL combine with molecules of the coenzyme NAD (nicotinamide adenine dinucleotide) to form NADH + H⁺, where they are needed for aerobic respiration.

Anaerobic Respiration

Fermentation is a form of anaerobic respiration performed by yeast cells. Glucose is broken down to form two pyruvate molecules, each of which is further broken down into the alcohol ethanol, C_2H_5OH and CO_2. The brewing industry uses ethanol for a variety of alcoholic beverages, and the baking industry uses CO_2 to make dough rise.

$$\text{Glucose} + 2\text{ ATP} \xrightarrow{\text{enzymes}} 2\text{ pyruvate} + 4\text{ ATP (2 ATP gain)} \xrightarrow{\text{enzymes}} 2\text{ ethanol} + 2CO_2$$
(C–C–C–C–C–C)   (C–C–C)   (C–C)

Aerobic Respiration

This series of cyclic reactions that takes place inside the mitochondria was discovered by Hans Krebs and is often called the Krebs cycle. Each pyruvate loses a carbon atom to form carbon dioxide and acetate (a 2-carbon compound) that combines with coenzyme A to form acetyl CoA. The Krebs cycle begins when acetyl CoA combines with a 4-carbon compound to form citrate. As a result of going around the Krebs cycle twice (once for each acetyl CoA) and the accompanying electron transport system, 34 ATP are produced. The hydrogen atoms from anaerobic respiration combine with oxygen, the final hydrogen acceptor to form water. If oxygen is not available, pyruvate forms lactic acid. This takes place in muscles and is reversible when oxygen becomes available.

Krebs Cycle

Comparing Cellular Respiration to Photosynthesis

Cellular respiration is a process that produces energy for a cell, and it is the opposite of photosynthesis. In cellular respiration, glucose is broken down to produce energy (ATP), carbon dioxide, and water. In photosynthesis, glucose and oxygen are produced using light energy, carbon dioxide, and water. The following chart summarizes cellular respiration and compares this process to photosynthesis:

Respiration and Photosynthesis Compared		
	Respiration	**Photosynthesis**
Reactants	$C_6H_{12}O_6 + O_2$	$CO_2 + H_2O$
Products	$CO_2 + H_2O$	$C_6H_{12}O_6 + O_2$
Energy considerations: Start → Finish →	2 ATP needed to start. 38 ATP produced (36 ATP gained).	Light energy is needed to start. Glucose produced (has stored chemical energy).
Steps involved	Anaerobic and aerobic respiration	Light and dark reactions
Time of day	24 hours per day	Light hours only
Site in cell	Mitochondria	Chloroplasts
Type of organism	All organisms	Green plants

Example Problems

These problems review cell respiration.

1. Write the overall equation for cellular respiration.

   Answer:

   $$C_6H_{12}O_6 + 6O_2 + 2 \text{ ATP} \xrightarrow{\text{enzymes}} 6CO_2 + 6H_2O + 38\text{ATP}$$

2. How does anaerobic respiration differ from aerobic respiration?

   Answer: Anaerobic respiration takes place in the cytoplasm outside the mitochondria, and oxygen is not needed. In anaerobic respiration, glucose is broken down into two molecules of pyruvate, and 4 ATP are produced. Aerobic respiration takes place in the mitochondria, and oxygen is needed. In aerobic respiration, two pyruvate molecules are converted into CO_2, H_2O, and 34 ATP.

3. Why are two molecules of ATP used to start the process of anaerobic respiration?

   Answer: Two molecules of ATP are needed to break the glucose molecule and start the process of anaerobic respiration. To break a molecule, energy must be used. An investment is made. When glucose is broken down into pyruvate, 4 ATP are produced. When pyruvate is broken down into CO_2 and H_2O, an additional 34 ATP are produced.

4. What is fermentation?

   Answer: Fermentation is a form of anaerobic respiration performed by yeast. Glucose is broken down to yield ethanol, carbon dioxide, and ATP.

$$\text{Glucose} + 2 \text{ ATP} \xrightarrow{\text{enzymes}} 2 \text{ pyruvate} + 4 \text{ ATP (2 ATP gain)} \xrightarrow{\text{enzymes}} 2 \text{ ethanol} + 2CO_2$$

   (C–C–C–C–C–C)   (C–C–C)   (C–C)

5. How can a yeast cell survive with only a gain of two ATP from anaerobic respiration?

   Answer: Yeast cells are small microscopic organisms and do not require more than two ATP. Humans are large organisms and require the additional ATP that can be released from pyruvate by the addition of oxygen during aerobic respiration.

Work Problems

Use these problems on cell respiration for additional practice.

1. How much ATP is gained during anaerobic respiration?

2. How much ATP is gained during aerobic respiration?

3. Which process is more efficient, anaerobic or aerobic respiration? Explain your answer.

4. What are the end products of respiration?

5. How is cellular respiration different from photosynthesis?

Worked Solutions

1. Two ATP are gained during anaerobic respiration. Two ATP are used (invested) to break glucose into pyruvate, producing four ATP.

4 ATP	produced
- 2 ATP	invested
2 ATP	gained

2. Thirty-four ATP are gained during aerobic respiration.

2 ATP	from Krebs cycle
+ 32 ATP	from the electron transport system
34 ATP	gained

3. Aerobic respiration is more efficient than anaerobic respiration. Worked solutions 1 and 2 demonstrate this. Anaerobic respiration produces a gain of only 2 ATP. A great deal of energy remains locked up in the pyruvate molecule. Aerobic respiration produces a gain of 34 ATP by completely breaking down pyruvate. If the ATP gained by the two processes is added, the total gain is 36 ATP for all cellular respiration.

2 ATP	gained from anaerobic respiration
+ 34 ATP	gained from aerobic respiration
36 ATP	gained

4. The end products of respiration are $6CO_2 + 6H_2O + 38$ ATP.

5. In cellular respiration, glucose is broken down into carbon dioxide and water to release the energy (ATP) that is stored in this molecule during its formation by photosynthesis. Cellular respiration takes place at and in the mitochondria in two steps, called anaerobic and aerobic respiration. All organisms perform respiration all the time. Photosynthesis takes place in green plants and green algae but only during light hours. Photosynthesis requires carbon dioxide, water, and light energy to be converted into glucose in two series of reactions called the light and dark reactions.

Excretion

As organisms perform their various life functions, waste products are produced. The major waste products produced in humans are carbon dioxide, water, urea, salt, and heat. These waste products must be removed (*excreted*) because they are harmful or poisonous. The life function of respiration produces the waste products carbon dioxide and water. *Urea* is a poisonous, nitrogenous waste produced from the breakdown of proteins. Excess salt from food is harmful and must be removed. Humans are warm blooded and must maintain a constant body temperature; this requires the removal of excess heat. Several organs in the body work together to carry out the life function of excretion; they are the kidneys, liver, lungs, and skin.

The Kidneys As Organs of Excretion

The human body has two kidneys, located in the back above the waist on each side of the spine. The function of the kidneys is to remove urea, water, and salt from the blood.

Kidney Circulation

Blood leaves the heart through the aorta and travels to the renal arteries, which carry blood into the kidneys. As blood circulates through the kidneys, thousands of little tubes called *nephrons* filter waste products out of the blood. Blood leaves the kidneys through the renal veins, which carry blood to the inferior vena cava and back to the heart. Wastes that have collected in the *renal pelvis* flow down the *ureter* into the *urinary bladder* where they are stored for later removal from the body by the *urethra*.

Nephron Filtration

A small artery from the renal artery carries blood to a ball of capillaries called a *glomerulus,* which is found inside a part of the nephron known as *Bowman's capsule*. Blood circulates through the glomerulus, and a small vein carries blood back to the renal vein. Water, urea, salt, and some food material leave the blood of the glomerulus, enter Bowman's capsule, and progress through the nephron to a *collecting tubule* that gathers wastes from many nephrons. The collecting tubule empties its wastes into the renal pelvis. Blood never leaves the glomerulus or the blood vessels; humans have a closed circulatory system. The *loop of Henle* area of the nephron reabsorbs food materials (glucose, amino acids) and some water into the blood of the surrounding capillaries.

Kidney Excretory System

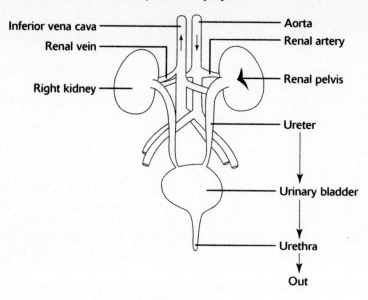

Inferior vena cava — — Aorta
Renal vein — — Renal artery
Right kidney — — Renal pelvis
— Ureter
— Urinary bladder
— Urethra
Out

Nephron

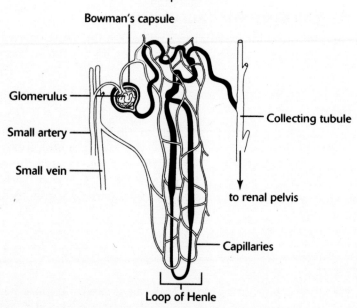

Bowman's capsule
Glomerulus
Small artery — Collecting tubule
Small vein
— to renal pelvis
— Capillaries
Loop of Henle

Example Problems

These problems review the structure and function of the kidneys.

1. Which waste products do the kidneys remove?

    Answer: The kidneys remove urea, water, and salt from the blood. The resulting mixture is known as urine.

2. What is the role of each of the following structures in the urinary system: ureter, urinary bladder, and urethra?

    Answer: The ureter is a tube that carries urine from the renal pelvis to the urinary bladder. The urinary bladder is a temporary storage site for urine. The urethra is a tube that carries urine from the urinary bladder to the outside of the body.

3. Outline the pathway of blood flow from the heart to the kidneys and back to the heart.

    Answer: Heart → aorta → renal artery → small artery → glomerulus → small vein → renal vein → inferior vena cava → heart.

4. What is the glomerulus?

    Answer: The glomerulus is a ball of capillaries inside Bowman's capsule in the nephron. The glomerulus is part of the circulatory system.

Work Problems

Use these problems on the structure and function of the kidneys for additional practice.

1. What is filtration, and where does it occur in the kidneys?

2. Outline the pathway of filtration starting with the glomerulus.

3. Why is the loop of Henle important?

4. Outline the pathway of urine from the kidneys to the outside of the body.

5. How is blood that enters the kidneys different from blood that leaves?

Worked Solutions

1. Filtration in the kidneys takes place when wastes leave the blood of the glomerulus and enter Bowman's capsule in the nephron.

2. Glomerulus → Bowman's capsule in the nephron → nephron tube → collecting tubule → renal pelvis.

3. The loop of Henle is important for the reabsorption of glucose, amino acids and water into the blood of the capillaries that surround the loop.

4. Renal pelvis of the kidney → ureter → urinary bladder → urethra → outside the body.

5. Blood that enters the kidneys is high in urea and salts; blood that leaves the kidneys is low in urea and salts.

The Liver As an Organ of Excretion

One of the many functions of the liver is the removal of dead red blood cells. When removed, the red blood cells are broken down into the nitrogenous waste urea. Excess amino acids have the amino group removed (*deamination*) and then through a series of chemical reactions are converted into ammonia and then into urea. Urea enters the blood and must be removed by the kidneys.

$$\text{Amino acids} \xrightarrow{\text{deamination}} \text{ammonia} \longrightarrow \text{urea}$$

The Lungs As Organs of Excretion

The process of cellular respiration produces CO_2 and H_2O as waste products. These waste products enter the alveoli from the capillaries that surround them. When in the alveoli, the wastes enter the bronchioles and then move into the bronchi to the trachea and then out of the body through the nose or the mouth. This process is called exhalation and is the opposite of the pathway used by oxygen to enter the body.

Example Problems

These problems review excretion by the liver and lungs.

1. Which waste product does the liver produce?

    Answer: The liver produces the waste product urea from dead red blood cells and excess amino acids.

2. Identify the waste products removed by the lungs.

    Answer: The waste products removed by the lungs are carbon dioxide and water.

3. What happens to dead red blood cells?

    Answer: Dead red blood cells are broken down into urea by the liver.

Work Problems

Use these problems on excretion by the liver and lungs for additional practice.

1. Outline the pathway of carbon dioxide from the blood to the outside air.

2. How is urea formed in the liver?

3. How do carbon dioxide and water get from the blood into the alveoli?

4. What process produces the waste products removed by the lungs?

Worked Solutions

1. Blood in capillaries → alveoli → bronchioles → bronchi → trachea → nose/mouth → outside the body.

2. Urea is formed in the liver by the process of deamination. Amino acids lose their amino group to form ammonia, which is converted chemically into urea.

3. Carbon dioxide enters the alveoli by diffusion, and water enters by osmosis. Both of these molecules are moving from an area of high concentration in the capillaries that surround the alveoli to an area of low concentration inside the alveoli. Passive transport is the movement of molecules from areas of high concentration to areas of low concentration. Diffusion is the movement of any molecule except water. Osmosis is the movement of the water molecule.

4. Cellular respiration produces the carbon dioxide and water that is removed by the lungs.

The Skin As an Organ of Excretion

The skin is the largest external organ of the body and is the body's first line of defense against disease. The skin is also an important organ of excretion removing water, urea, and salt in the form of perspiration or sweat. *Sweat glands* in the skin are coiled tubes with a duct that extends to the surface forming an opening called a *pore*. The skin removes heat when water evaporates from its surface. *Evaporation* creates a cooling effect that lowers the body's temperature. Heat is also lost by radiation to the atmosphere when blood flows through dilated (widened) capillaries near the skin's surface. In cold weather, the capillaries constrict (narrow) and less blood flows through the skin, conserving heat. The skin plays a major role in maintaining homeostasis of body temperature.

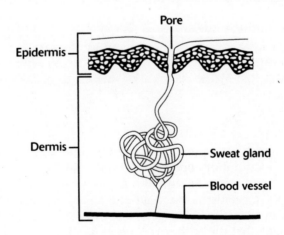

Example Problems

These problems review excretion by the skin.

1. What are the functions of the skin?

    Answer: The skin can function as an organ of excretion by removing sweat (water, urea, and salt) from the body. The skin also removes heat. The skin is the body's first line of defense against disease. As long as the skin is intact, it acts as a barrier against harmful organisms such as bacteria.

2. Describe the structure and function of a sweat gland.

Answer: Sweat glands are coiled tubes with a duct that extends to the surface forming an opening called a pore through which water, urea, and salt wastes can exit.

Work Problems

Use these problems on excretion by the skin for additional practice.

1. How does the skin help maintain temperature homeostasis?

2. What happens if the body cannot remove excess heat?

3. How are wastes removed from the skin different from those removed by the kidneys?

Worked Solutions

1. Heat is lost by radiation to the atmosphere when blood flows through dilated (widened) capillaries near the skin's surface. In cold weather, the capillaries constrict (narrow) and less blood flows through the skin, conserving heat. Humans are warm blooded and must maintain a constant body temperature of approximately 37°C.

2. If excess heat is not removed, body temperature goes up. The active sites of enzymes become distorted, and important chemical reactions in the body slow down and stop. This can cause sickness and possibly death.

3. Both the skin and the kidneys remove urea, water, and salt. However, unlike the kidneys, the skin removes very little urea. Sweat removed by the skin is about 99% water. The skin plays a major role in regulating the body's temperature by either removing or conserving heat. Almost no heat is lost by the removal of wastes from the kidneys.

Locomotion

Humans are heterotrophic and must eat to obtain food for energy and life functions. To get food, humans must be capable of locomotion. *Locomotion* is the ability of an organism to move from one place to another. In addition to obtaining food, animals depend on locomotion for finding a mate and avoiding predators. Organisms that have the ability of locomotion are called *motile;* organisms that don't move are *sessile*. In humans, locomotion is dependent on the bones of the endoskeleton and the muscles that move them. An *endoskeleton* is one where the bones and cartilage are found inside the body. This arrangement is typical of vertebrates. Some invertebrate organisms such as insects have an *exoskeleton* (outside the body) made of a hard, shell-like substance called *chitin*.

The Endoskeleton

The endoskeleton is made up of bones and cartilage. The skeleton acts like a hanger. The muscles of the body are attached to the bones of the skeleton, and many of the organs of the body are surrounded, supported, and protected by the bones. The long bones of the body also produce red blood cells. The bones act as levers for movement. *Joints* are areas in the skeleton where bones meet. *Ligaments* connect bones to each other. *Cartilage* is a connective tissue that is flexible and not as hard as bone. Cartilage can be found in the ears, nose, trachea, ends of the ribs, and between the vertebrae in the back.

The Muscles

Humans have three kinds of muscle tissue: skeletal muscle, smooth muscle, and cardiac muscle.

Skeletal muscle is made of cells that have combined to form fibers called *striations,* which give this muscle tissue strength. Skeletal muscles are *voluntary muscles,* that is to say their use is consciously controlled. Humans decide whether to use these muscles or not. Skeletal muscles are attached to the bones by *tendons.* The function of these muscles is to move the bones during locomotion. Skeletal muscles tend to work in pairs. *Flexors* are skeletal muscles that move bones toward the body, and *extensors* move bones away from the body.

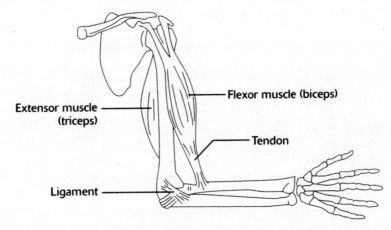

Extensor muscle (triceps)

Flexor muscle (biceps)

Tendon

Ligament

Smooth muscle tissue is composed of individual cells with their own nuclei and no striations. Smooth muscles are found in the esophagus, stomach, diaphragm, and blood vessels. Smooth muscles are *involuntary muscles,* meaning that their activity is not consciously controlled. For example, after swallowing food, humans have no control over the events that follow. Muscles of the esophagus automatically push the food down into the stomach.

Cardiac muscle tissue, composed of striated cells, is involuntary and found in the heart.

Example Problems

These problems review locomotion.

1. Why must some organisms be capable of locomotion?

    Answer: Locomotion is the ability of an organism to move from one place to another to obtain food, find a mate, and avoid predators.

2. How is an endoskeleton different from an exoskeleton?

    Answer: An endoskeleton is one where the bones and cartilage are found inside the body. An exoskeleton is found outside the body and is made of a hard, shell-like substance called chitin.

3.   What is a joint?

Answer: A joint is where two bones meet.

4.   How are bones connected to each other?

Answer: Ligaments connect bones to each other.

Work Problems

Use these problems on locomotion for additional practice.

1.   Describe the structure and function of skeletal muscle.

2.   How are skeletal muscles connected to bones?

3.   How do muscles work together in pairs to achieve locomotion?

4.   How are voluntary muscles different from involuntary muscles?

Worked Solutions

1.   Skeletal muscle is made of cells that have combined to form fibers called striations. The function of these muscles is to move the bones during locomotion.

2.   Skeletal muscles are connected to bones by tendons.

3.   Muscles of the endoskeleton work together in pairs called flexors and extensors. Flexors move bones toward the body, and extensors move bones away from the body. As the bones move, so does the person.

4.   Voluntary muscles are muscles that are consciously controlled. Skeletal muscles are an example of voluntary muscles; decisions can be made about when to use them. Involuntary muscles are muscles that are not consciously controlled. Smooth muscle and cardiac muscle are examples of involuntary muscles.

Regulation

Regulation is the ability of an organism to respond to changes in its environment. Any change in the environment of an organism is called a *stimulus*. The reaction of the organism is the *response*. Regulation is the life function that allows an organism to maintain homeostasis. Two systems in humans help regulate the body; they are the *nervous system* and the *endocrine system*.

The Structure of the Nervous System

The nervous system is composed of nerve cells, a spinal cord, and a brain. The *neuron* is the basic unit of structure and function in the nervous system. *Receptors* are organs in the body capable of detecting a stimulus. *Effectors* are organs that carry out a response.

The Function of the Nervous System

Neurons in a receptor detect a stimulus and transmit an impulse to the brain or spinal cord. In the brain or spinal cord, the information is processed, and other neurons transmit an impulse to an effector, which carries out the response. Nervous system responses are fast, taking only a fraction of a second. Also, nerve responses have a short duration. When a response has taken place, it's over! Nervous system responses are limited in scope.

$$\text{Receptors} \xrightarrow{\text{nerve cells}} \text{spinal cord or brain} \xrightarrow{\text{nerve cells}} \text{effectors}$$

The following tables list the different receptors in the body and the stimuli that they can detect, as well as the different effectors in the body and the response they can produce.

Receptors and Stimuli Detected	
Receptor	**Stimulus Detected**
Eyes	Light
Ears	Sound
Tongue	Taste
Nose	Odor
Skin	Touch, temperature

Effectors and Their Responses	
Effector	**Response**
Muscles	Move bones
Glands	Produce hormones/secretions

Example Problems

These problems review the structure and function of the nervous system.

1. Which systems in the body are responsible for carrying out the life function of regulation?

   Answer: The nervous system and the endocrine system, working together, carry out the life function of regulation.

2. Distinguish between a stimulus and a response.

   Answer: A stimulus is a change in the environment of an organism. A response is the reaction of an organism to a stimulus.

3. How are receptors different from effectors?

   Answer: Receptors are organs in the body capable of detecting a stimulus. The effectors are the organs that carry out the response. The receptors are the five sensory organs of the body: eyes, ears, nose, tongue, and skin. The effectors are the muscles and glands of the body. The skeletal muscles help with locomotion, and the endocrine glands help regulate various organ systems.

Work Problems

Use these problems on the structure and function of the nervous system for additional practice.

1. How does the life function of regulation help a person survive?

2. What are the characteristics of nervous system responses?

3. What is the role of the nerve cells of the body?

4. Describe the role of the brain?

5. What is the effect on the body if the nerve cells from the spinal cord to the effectors are damaged?

Worked Solutions

1. The life function of regulation helps a person adapt to changes in the environment. The reaction of the person is the response. The ability to respond to environmental change enables the individual to maintain homeostasis and survive. For example, a person walking down a street sees a large box and walks around it. The box was the stimulus, and walking around the box was the response. However, if the person didn't see the box and tripped over it, he might be injured or break his neck and die.

2. Nervous system responses are fast, have a short duration and are limited in scope. For example, if a person touches a hot object, his hand is removed in a fraction of a second. The response is over quickly and is limited to just the person's hand.

3. The nerve cells are the basic unit of structure and function in the nervous system. The nerve cells act as lines of communication between the various parts of the body. Nerve cells act as a link between the receptors and the brain or spinal cord. Nerve cells keep the brain and spinal cord in contact with the effectors of the body. Also, the brain is made of millions of nerve cells.

4. The brain is the processing center of the body. The brain receives information from the receptors, interprets this information, and signals an effector to carry out an appropriate response. The brain is not a receptor organ. Without receiving information from a receptor, the brain doesn't know anything.

5. If the nerve cells from the spinal cord to the effectors are damaged, a person might be paralyzed in part of his body. The nerve cells from the spinal cord signal the effectors to carry out an appropriate response. Damaged nerve cells break the line of communication to the effectors, preventing them from responding.

The Structure of a Neuron (Nerve Cell)

The *cyton* is the cell body of a neuron or nerve cell and includes the following structures:

❑ **Nucleus:** Controls the nerve cell and functions in reproduction

❑ **Cytoplasm:** Site of the organelles and chemical reactions

❑ **Dendrites:** Detect stimuli

The *axon* transmits nerve impulses from the cyton to the terminal branches. Some axons have a *myelin sheath,* which is a covering that insulates the axon and speeds up the nerve impulse. Inside the axon, the nerve impulse is electrical and can best be described as a *depolarization wave.* The axon of a resting nerve cell is positively charged on the outside and negatively charged on the inside; as an impulse goes down the axon, the outside becomes positively charged, and the inside becomes negatively charged. A *nerve* is a bundle of axons surrounded by connective tissue.

The *terminal branches* produce a chemical called a *neurotransmitter* that carries the nerve impulse across the *synapse* (a space between two nerve cells) to the next nerve cell. An example of a neurotransmitter is *acetylcholine. Cholinesterase* is an enzyme that can break down acetylcholine.

A nerve impulse travels from the dendrites to the terminal branches. A nerve impulse is electrical and chemical, electrical in the nerve cell and chemical in the synapse between nerve cells.

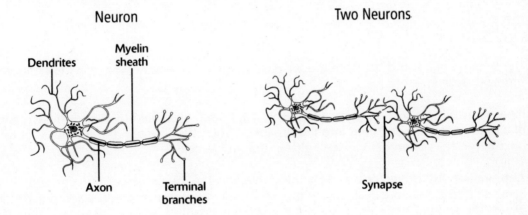

Example Problems

These problems review the structure of neurons.

1. What is the relationship between neurons, nerve cells, and nerves?

    Answer: Neurons and nerve cells are different names for the same structure. A nerve is a bundle of axons surrounded by connective tissue.

2. What are the parts of the cyton and their function?

    Answer: The parts of the cyton are:

    Nucleus: Controls the nerve cell and functions in reproduction

    Cytoplasm: The site of the organelles and chemical reactions

    Dendrites: Detect stimuli

3. Describe the role of the axon.

    Answer: The axon connects the cyton to the terminal branches and carries the nerve impulse from the cyton to the terminal branches.

Work Problems

Use these problems on the structure of neurons for additional practice.

1. Describe the synapse.

2. How does the nerve impulse get from one nerve cell to the next?

3. Describe the *nerve impulse.*

4. In which direction do nerve impulses travel?

5. Why is cholinesterase important?

Worked Solutions

1. The synapse is a space between the terminal branches of one neuron and the dendrites of the next neuron.

2. Neurotransmitters such as acetylcholine are chemicals produced by the terminal branches of a neuron that are secreted into the synapse. When they contact the dendrites of the next nerve cell, a nerve impulse begins in that nerve cell.

3. A nerve impulse is electrical and chemical, electrical in the nerve cell and chemical in the synapse between nerve cells. Inside the nerve cell, the nerve impulse is a depolarization wave. The axon of a resting nerve cell is positively charged on the outside and negatively charged on the inside; as an impulse goes down the axon, the outside becomes positively charged, and the inside becomes negatively charged. Between nerve cells, chemicals called neurotransmitters transmit the nerve impulse to the next nerve cell.

4. Nerve impulses always travel from the cyton through the axon to the terminal branches.

5. Cholinesterase is important because it is the enzyme that breaks down acetylcholine in the synapse so that the same message does not keep repeating itself. If acetylcholine remains in the synapse, the dendrites of the next neuron are continuously stimulated. Cholinesterase breaks down acetylcholine to prevent this from happening. After cholinesterase does its work, it breaks down so that in the future, acetylcholine can again be produced in the synapse.

The Brain

The brain is located inside the skull and is composed of neurons and organized into three parts. The *cerebrum* is the largest part of the human brain. The cerebrum is responsible for interpreting information received from the receptors. This is the conscious part of the brain and controls voluntary activities, as well as judgment and memory. The *cerebellum* controls coordination and balance. The *medulla* controls involuntary body activities such as breathing, heartbeat, digestion, blinking, and sneezing.

The Spinal Cord

The spinal cord begins at the base of the medulla and extends down through the vertebrae in the back. The spinal cord plays a central role in the reflex arc helping to control simple reflex actions such as the knee jerk and removing a hand from a hot or sharp object.

The Reflex Arc

A *reflex* is an automatic, involuntary, unlearned response to a stimulus. For example, when a person touches a hot object, the hot object is the stimulus, and the response (reflex) is to pull the hand away. A similar situation occurs if a person touches a sharp object. This behavior (response) involves three neurons and the spinal cord.

The *sensory neuron* detects a stimulus and transmits the nerve impulse from a receptor to an interneuron. An *interneuron* is a neuron located in the spinal cord that receives the nerve impulse from the sensory neuron and relays the impulse to the motor neuron. The *motor neuron* receives the impulse from the interneuron and transmits the impulse to an effector to carry out the response.

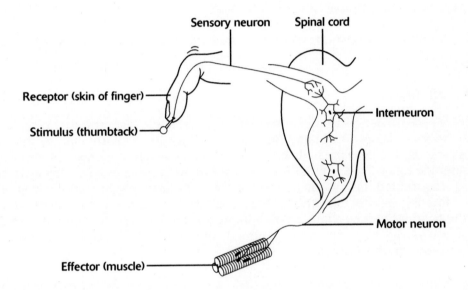

Example Problems

These problems review the brain and spinal cord.

1. What is the role of each of the following neurons: sensory neuron, interneuron, and motor neuron?

   Answer: The sensory neuron in a receptor detects the stimulus and transmits the nerve impulse to the interneuron. The interneuron is located in the spinal cord and receives the nerve impulse from the sensory neuron and relays the impulse to the motor neuron. The motor neuron transmits the impulse to an effector to carry out the response.

2. Why is the reflex arc important?

   Answer: A reflex is an automatic, involuntary, unlearned response to a stimulus. The reflex arc is a great way to respond to a potentially harmful situation such as coming into contact with a sharp object or a flame, where a fast response is necessary to avoid injury.

3. Identify the parts of the brain and their function.

   Answer: The brain has three major parts. The cerebrum is responsible for interpreting information received from the receptors. This is the conscious part of the brain and controls voluntary activities, judgment, and memory. The cerebellum controls coordination and balance. The medulla controls involuntary body activities such as breathing, heartbeat, digestion, blinking, and sneezing.

Work Problems

Use these problems on the brain and spinal cord for additional practice.

1. Why is it important that the brain not be involved in a simple reflex?

2. Which part of the human brain is the largest?

3. Describe the function of the spinal cord.

4. Why is the cerebellum in squirrels large compared to their cerebrum? In humans, the cerebellum is small compared to the cerebrum.

5. Why is it important that the cerebrum not have control over respiratory activities?

Worked Solutions

1. A simple reflex must be a fast, immediate, and automatic response to a stimulus for a person to avoid injury. The brain takes too long to evaluate information by thinking and processing. By the time the brain has finished doing its job, the person is probably injured. Consider the following example: A person accidentally touches a hot object. If the sensory neuron transmitted this information to the brain, the brain might start thinking along these lines, "The object is hot. How hot is it? Is it so hot that the hand must be removed now? Can the hand be removed later? How quickly should the hand be removed?" By this time the person's hand might be severely burned. The simple reflex takes the thinking out of the response and gives a fast result!

2. The cerebrum is the largest part of the brain and controls the five senses, judgment, memory, voluntary activities, and much more.

3. The spinal cord plays a central role in the reflex arc, helping to control simple reflex actions when fast responses are needed.

4. A squirrel has the ability to walk on the branch of a tree and to jump from limb to limb without falling. Squirrels have a terrific sense of balance and coordination as a result of a large cerebellum in comparison to the size of their cerebrum. However, squirrels are not known for their thinking abilities.

5. Respiration is an activity that goes on all the time in humans. If the cerebrum controlled the process of respiration, little time would be left over for thinking. The cerebrum would be too concerned with when to breathe in, when to exhale, and so on. This is why the medulla controls activities that there isn't time to think about, such as respiration and digestion.

The Structure and Function of the Endocrine System

The endocrine system is composed of *ductless glands* (glands that do not have tubes). Endocrine glands secrete *hormones,* which are chemical "messengers" that travel through the blood to regulate the activity of a target organ. Glands that have ducts are called *exocrine glands.* The secretions of exocrine glands reach their target by traveling through a duct (tube). Exocrine glands are not part of the endocrine system. Some examples of exocrine glands are sweat glands and salivary glands. The diagram that follows shows the major endocrine glands and their locations.

Endocrine responses are slow because hormones must travel through the blood to reach their target organ(s). Also, the duration of the response is long because the hormone must be filtered out of the blood by the kidneys. Hormones often have multiple targets; as a result, their effects in the body are widespread.

The Endocrine System

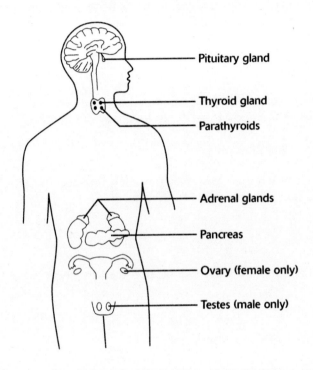

- Pituitary gland
- Thyroid gland
- Parathyroids
- Adrenal glands
- Pancreas
- Ovary (female only)
- Testes (male only)

Hormone Action

The following diagrams explain how hormones function. The pituitary gland produces a *stimulating hormone (SH)* that travels through the blood to a target endocrine gland. When the endocrine gland detects the SH, it produces its own hormone, which enters the blood to reach a target organ. Meanwhile, the pituitary detects the endocrine gland's hormone in the blood and stops the production of SH. This is called *negative feedback* and shuts down this endocrine pathway. In the following example, the pituitary gland produces *thyroid stimulating hormone (TSH)* that travels through the blood to the thyroid gland. The thyroid gland produces thyroxin that targets the body cells, regulating their rate of metabolism. When the pituitary detects thyroxin in the blood, the production of TSH stops, and the system shuts down.

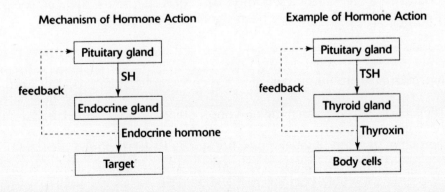

Mechanism of Hormone Action

Pituitary gland → SH → Endocrine gland → Endocrine hormone → Target

feedback

Example of Hormone Action

Pituitary gland → TSH → Thyroid gland → Thyroxin → Body cells

feedback

Example Problems

These problems review the structure and function of the endocrine system.

1. How are endocrine glands different from exocrine glands?

   Answer: Endocrine glands are ductless glands whose secretions (hormones) reach their target by traveling through the blood. Exocrine glands have ducts through which their secretions travel to reach their target.

2. What are hormones?

   Answer: Hormones are chemical messengers produced by endocrine glands. Hormones signal other organs of the body to perform a variety of functions.

3. How do hormones reach their target?

   Answer: Hormones reach their target by traveling through the blood. Endocrine glands do not have ducts that connect them to their target.

Work Problems

Use these problems on the structure and function of the endocrine system for additional practice.

1. Describe the mechanism of hormone action.

2. What is feedback and why is it important?

3. How is the nervous system similar to the endocrine system?

4. How is the nervous system different from the endocrine system?

Worked Solutions

1. The pituitary gland produces a stimulating hormone (SH) that travels through the blood to an endocrine gland. The endocrine gland detects the SH and produces its own hormone to reach a target. When the pituitary gland detects the endocrine gland's hormone, the production of SH is stopped.

2. In the endocrine system feedback is important because it shuts down an endocrine pathway after the target organ has been signaled.

3. The nervous system and the endocrine system are similar in that they both play a major role in maintaining homeostasis, and both secrete chemical messengers. The nervous system produces neurotransmitters, and the endocrine system produces hormones.

4. Three differences exist between the nervous system and the endocrine system:

   Nerve impulses travel through neurons; hormones travel through the blood.

   Nerve responses are faster than endocrine responses.

   Nerve responses are of shorter duration than endocrine responses.

Pituitary Gland

The pituitary gland is attached to a part of the brain called the *hypothalamus*. The hypothalamus controls the pituitary gland through the action of hormones and through nerve fibers that extend into this gland. The pituitary gland is about the size of a green pea, measuring about 1 centimeter in diameter. The pituitary gland produces many hormones that control the other endocrine glands and is often called the *master gland*. Stimulating hormones (SHs) target other endocrine glands to produce their hormones. *Growth hormone (GH)* made by the pituitary gland targets bone and muscle to promote growth.

Thyroid Gland

The thyroid gland is located in the neck and produces the hormone *thyroxin*. Thyroxin controls the rate of metabolism in most cells of the body, along with physical and mental development. Undersecretion of thyroxin results in *hypothyroidism* (hypo means under), which can cause weight gain, tiredness, and cretinism (a form of mental retardation). Thyroxin contains the element *iodine*. People lacking iodine in their diet can develop a *goiter*, which is an enlargement of the thyroid as it attempts to filter more iodine from the blood to produce sufficient amounts of thyroxin. *Hyperthyroidism* (hyper means over) is caused by an oversecretion of thyroxin and results in increased rates of metabolism, blood pressure, energy, and weight loss.

Parathyroid Glands

The four small parathyroid glands are located on the back of the thyroid. The parathyroid glands secrete a hormone called *parathyroid hormone (PTH)* or *parathormone*. PTH regulates calcium and phosphorous levels in the blood and bones.

Adrenal Glands

The adrenal glands are located on top of the kidneys. The adrenal glands produce several important hormones; among them are adrenalin, cortisone, cortisol, and aldosterone. *Adrenalin* (or *epinephrine*) regulates the rate of breathing, heartbeat, blood pressure, and blood clotting. Adrenalin is the body's emergency hormone, preparing the body for fight or flight. *Cortisone* and *cortisol* are *glucocorticoids* (or steroids) produced by the adrenal cortex. The anti-inflammatory action of cortisone and cortisol can be used to treat arthritis in the joints. Their antiallergic effects can be used to treat asthma and allergic reactions. Cortisone can treat various skin diseases and help manage stress. *Aldosterone* is a steroid that regulates the level of salts in the blood and helps maintain the body's electrolyte balance. Aldosterone targets the kidneys to decrease the level of sodium in urine, return water to the body, and increase blood pressure.

Islets of Langerhans (Pancreas)

The Islets of Langerhans are specialized cells found inside the pancreas that produce the hormones insulin and glucagon. *Insulin* targets the liver and instructs it to remove sugar from the blood and to produce glycogen (animal starch). An undersecretion of the hormone insulin results in too much sugar in the blood (*hyperglycemia*) and the disease called diabetes. Diabetics can be treated with insulin or other medications that can lower blood-sugar levels. Too much insulin results in a low blood-sugar level (*hypoglycemia*). *Glucagon* targets the liver, causing it to break down glycogen into sugar, which is released into the blood. The pancreas is also an exocrine gland, producing pancreatic fluid, which enters the small intestines through a duct. Pancreatic fluid contains enzymes needed for digestion.

Ovaries

The ovaries are the female sex glands (*gonads*), and they produce the hormones estrogen and progesterone. Estrogen is responsible for female secondary sex characteristics such as a high voice, breast development, a narrow waist, body hair, and widening of the hips. *Progesterone* is a hormone that helps maintain the lining of the uterus. Estrogen and progesterone function together to help regulate the menstrual cycle.

Testes

The testes are the male sex glands (*gonads*), and they produce the hormone testosterone. Testosterone is responsible for male secondary sex characteristics such as a deep voice, facial hair, body hair, broad shoulders, and narrow hips. Testosterone is also needed for sperm cell production.

Example Problems

These problems review endocrine glands and their function.

1. Why is the pituitary gland called the master gland?

   Answer: The pituitary gland is called the master gland because it produces hormones that control the other endocrine glands. For example, stimulating hormones (SHs) target other endocrine glands to produce their hormones.

2. What organ controls the pituitary gland?

   Answer: The hypothalamus of the brain produces hormones that control the pituitary gland and has nerve fibers that extend into this gland. The hypothalamus is the organ that ties the nervous system to the endocrine system.

3. What is the function of the thyroid gland?

   Answer: The thyroid gland produces the hormone thyroxin. Thyroxin controls the rate of metabolism in most cells of the body, along with physical and mental development.

4. Why is adrenalin called the body's emergency hormone?

   Answer: Adrenalin is called the body's emergency hormone because, in an emergency, adrenalin can signal the liver to release sugar into the blood for use by the muscles to produce energy. Adrenalin can increase the breathing rate, heartbeat, blood pressure, and blood clotting, preparing a person for fight or flight.

5. Identify the hormones produced by the gonads and discuss their function.

   Answer: The ovaries of a female produce the hormones estrogen and progesterone. Estrogen is responsible for female secondary sex characteristics. Progesterone is a hormone that helps maintain the lining of the uterus. Estrogen and progesterone function together to help regulate the menstrual cycle. The testes of the male produce the hormone testosterone. Testosterone is responsible for male secondary sex characteristics and is needed for sperm cell production.

Work Problems

Use these problems on endocrine glands and their function for additional practice.

1. Explain how blood sugar levels in the body are regulated.

2. How do the parathyroid glands regulate the level of calcium in the blood?

3. Which hormones are produced by the adrenal cortex, and what is their function?

4. Why is the pancreas considered both an exocrine and endocrine gland?

5. How should a person with hypoglycemia be treated?

Worked Solutions

1. Insulin and glucagon regulate blood-sugar levels. Insulin targets the liver and instructs it to remove sugar from the blood and to produce glycogen (animal starch). Glucagon targets the liver, causing it to break down glycogen into sugar, which is released into the blood.

2. The parathyroid glands regulate the level of calcium in the blood by producing the parathyroid hormone, which targets the bones to release calcium. Bones are mostly calcium and can act like a bank for this mineral.

3. Cortisone, cortisol, and aldosterone are produced by the adrenal cortex. Cortisone and cortisol are anti-inflammatories and have antiallergic effects. Cortisone can treat skin diseases and help manage stress. Aldosterone regulates the level of salt in the blood and helps maintain the body's electrolyte balance.

4. The pancreas produces pancreatic fluid, which enters the small intestines through a duct, acting as an exocrine gland. The pancreas contains the Islets of Langerhans cells that produce the hormones insulin and glucagon, which reach their targets by traveling through the blood. In this way, the pancreas functions as an endocrine gland (or ductless gland).

5. A person with hypoglycemia has a low blood-sugar level. It might be tempting to treat this condition by simply telling the person to eat foods with sugar. This is a mistake that could result in death. The mere symptoms of a disease should not be treated; the cause of a disease should be. When an individual with hypoglycemia eats foods with sugar, the Islets of Langerhans respond by oversecreting insulin, lowering the sugar level further. A person with hypoglycemia should be put on a low-carbohydrate diet to prevent the oversecretion of insulin and the lowering of sugar levels that follow.

Chapter Problems and Answers

Problems

The following is a brief paragraph based on the life functions of nutrition, circulation, and respiration. For problems 1–10, fill in the missing terms.

During lunch, a high school student bought a hamburger and French fries in the school

cafeteria. The student sat down at a table and began to eat his lunch. The chemical

digestion of starch in the bun began in the student's _____. The breakdown of starch

1

required the presence of the enzyme _____. When the student swallowed, the food

2

went down the esophagus and not the trachea because the _____ (a flap) covered the

3

trachea preventing this from happening. Muscle contractions of the esophagus known

as _____ forced the food into the stomach. In the stomach, the chemical digestion of

4

_____ began. From the stomach, food entered the small intestines where all digestion

5

was _____. From the small intestines, food entered the _____ through finger-like projections

6   7

in the wall known as _____. Digested food was transported to all the cells of the body by

8

the _____. The cells used the food to produce energy at organelles known as _____.

9   10

For problems 11–14, give the name of the structure associated with each description.

11. Blood vessels that connect arteries to veins. _____

12. Structures that prevent the backward flow of blood. _____

13. Blood vessels that surround the alveoli and function in gas exchange.

_____

14. Blood vessels that carry blood to the atria. _____

For problems 15–18, select the molecule from the following list that is associated with each description in the process of respiration. A choice can be used more than once or not at all.

A. ATP   B. coenzyme   C. ethanol   D. PGAL   E. pyruvate

15. Which molecule is a starting point for aerobic respiration?

16. This molecule functions in the electron transport system.

17. Which molecule is an end product of fermentation?

18. Select the molecule that is the true end product of photosynthesis and is an intermediate product in the breakdown of glucose during cellular respiration.

Problems 19–22 are based on the following four chemical equations.

A. Glucose + 2 ATP $\xrightarrow{\text{enzymes}}$ 2 pyruvate + 4 ATP (2 ATP gain) $\xrightarrow{\text{enzymes}}$ 2 ethanol + 2CO$_2$

B. $C_6H_{12}O_6$ + 6O$_2$ + 2 ATP $\xrightarrow{\text{enzymes}}$ 6CO$_2$ + 6H$_2$O + 38 ATP (gain of 36 ATP)

C. $C_6H_{12}O_6$ + 2 ATP $\xrightarrow{\text{enzymes}}$ 2 pyruvate + 4 ATP (gain of 2 ATP)

D. 2 pyruvate + O$_2$ $\xrightarrow{\text{enzymes}}$ 6CO$_2$ + 6H$_2$O + 34 ATP (gain of 34 ATP)

19. Which equation represents both anaerobic and aerobic respiration? _____

20. Select the equation that best represents fermentation. _____

21. Identify the equation the produces the greatest amount of energy. _____

22. Select the equation that represents the aerobic stage of cellular respiration. _____

For problems 23–26, give the name of the endocrine gland associated with each description.

23. This endocrine gland produces stimulating hormones that control other endocrine glands.
_____

24. This gland secretes the hormones estrogen and progesterone. _____

25. This endocrine gland produces glucagon, which causes the liver to break glycogen into glucose. _____

26. This gland secretes a hormone that regulates the levels of calcium and phosphorous in the blood. _____

Problems 27–30 are based on the following diagram. Select the letter of the structure that *best* fits the description.

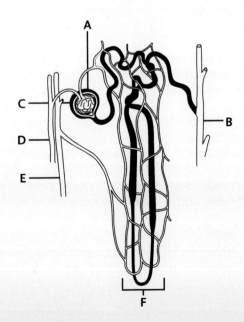

27. Where does reabsorption take place? _____

28. Which structure is the first to receive wastes from the blood? _____

29. Select the structure that brings wastes to the renal pelvis. _____

30. Which structure is the glomerulus? _____

For problems 31–35, select the item from the following list that *best* answers each statement. A choice can be used more than once or not at all.

   A. neurotransmitter   B. cholinesterase   C. depolarization wave   D. synapse   E. NADP

31. Which choice carries a nerve impulse from one neuron to the next? _____

32. Identify the substance that breaks down chemicals produced by the terminal branches of a neuron. _____

33. Which choice has no association with nerve impulses? _____

34. Identify the term that describes a nerve impulse inside an axon. _____

35. Select the term that indicates the space between two neurons. _____

Answers

1. **Mouth.** Both chemical and physical digestion take place in the mouth. The chemical digestion of starch begins in the mouth. Protein digestion begins in the stomach, and lipid digestion begins in the small intestines.

2. **Amylase.** Salivary amylase is the enzyme in the mouth that breaks down starch. The small intestines and the pancreas also produce amylase. Chemical digestion of starch is completed in the small intestines.

3. **Epiglottis.** The esophagus and trachea are next to each other. To prevent food from entering the trachea, its opening must be covered up when a person swallows. This is the function of the epiglottis.

4. **Peristalsis.** The esophagus is a muscular tube that connects the mouth to the stomach. The muscle walls of the esophagus contract and squeeze food into the stomach. These muscular contractions are involuntary and are known as peristalsis.

5. **Proteins.** Each major organ of the digestive system begins the chemical breakdown of a major food group. The chemical digestion of starch begins in the mouth. Protein digestion begins in the stomach, and lipid digestion begins in the small intestines.

6. **Completed.** The small intestines are the main organ of the digestive system; all nutrients are chemically digested here, and digestion is completed here.

7. **Blood** or **circulatory system.** Digested food enters the blood through the wall of the small intestines. Undigested food goes into the large intestines and is eventually removed from the body.

8. **Villi.** The villi are finger-like projections from the wall of the small intestines that increase the surface area for the absorption of food into the blood.

9. **Circulatory system.** The circulatory system is the transport system of the body and carries digested food to the cells.

10. **Mitochondria.** The mitochondria are the energy-producing factories of a cell. Anaerobic respiration takes place in the cytoplasm just outside the mitochondria. Aerobic respiration takes place inside the mitochondria.

11. **Capillaries.** Capillaries are one cell in diameter and connect arteries to veins; they are generally found in the organs of the body.

12. **Valves.** The heart and the veins have valves that prevent blood from flowing backward.

13. **Capillaries.** Capillaries surround the alveoli in the lungs. CO_2 in the alveoli is exchanged for O_2 found in the blood of the capillaries.

14. **Veins.** Veins carry blood to the heart. The atria are the chambers of the heart that receive blood from veins. The superior and inferior vena cavas are veins that carry blood to the right atrium. The pulmonary veins carry blood to the left atrium.

15. **E. Pyruvate.** During anaerobic respiration, glucose is broken down into two pyruvate molecules, and ATP is produced. Pyruvate is the starting point for aerobic respiration and can be broken down further if oxygen is available.

16. **B. Coenzyme.** In cellular respiration, the molecule NADP functions as a coenzyme in the electron transport system to bring hydrogen into aerobic respiration.

17. **C. Ethanol.** Fermentation is an anaerobic form of cell respiration in yeast that results in the production of ethanol and CO_2.

18. **D. PGAL.** When the glucose molecule is split, two molecules of PGAL form, which are later converted to pyruvate. The dark reactions result in the formation of PGAL. When two molecules of PGAL are combined, glucose is formed.

19. **B.** If chemical equations **C** (anaerobic respiration) and **D** (aerobic respiration) are combined, the result is the overall reaction for photosynthesis (as seen in **B**).

20. **A.** Fermentation is an anaerobic process that results in the production of ethanol and carbon dioxide.

21. **B.** Chemical equation **B** shows the overall reaction for photosynthesis. Thirty-eight ATP are produced for a net gain of 36 ATP.

22. **D.** Aerobic respiration requires oxygen. Equations **B** and **D** both have oxygen. However, **B** is incorrect because it is the summary equation for both anaerobic and aerobic respiration. Equation **D** shows only aerobic respiration.

23. **Pituitary.** Many hormones produced by the pituitary gland signal other endocrine glands to begin producing their hormone. This is why the pituitary gland is often called the master gland.

24. **Ovaries.** The ovaries are the female gonads (sex glands) that produce the sex hormones estrogen and progesterone.

25. **Islets of Langerhans.** The Islets of Langerhans are specialized endocrine cells found inside the pancreas that produce the hormones glucogon and insulin.

26. **Parathyroids.** The parathyroids produce the hormone parathormone, which regulates the levels of calcium and phosphorus in the blood.

27. **F. Loop of Henle.** Excess water and digested foods that are filtered out of the blood can be reabsorbed into the blood in this area of the nephron.

28. **A. Bowman's capsule.** Choice **C** is incorrect because that structure is the ball of capillaries from which the wastes exit before entering Bowman's capsule. Bowman's capsule is the part of the nephron that looks like the letter *C* written backward.

29. **B. Collecting tubule.** Each collecting tubule gathers wastes from several nephrons before emptying into an area of the kidney called the renal pelvis.

30. **C.** The glomerulus is a ball of capillaries located within Bowman's capsule in the nephron.

31. **A. Neurotransmitter.** Neurotransmitters are chemicals produced by the terminal branches of a neuron that bridge the synapse and transmit the nerve impulse to the next neuron. Acetylcholine is an example of a neurotransmitter.

32. **B. Cholinesterase.** Cholinesterase is an enzyme (with the suffix *-ase*) that can break down acetylcholine, which is a neurotransmitter.

33. **E. NADP.** This compound is a coenzyme involved with the electron transport system in the dark reactions.

34. **C. Depolarization wave.** The axon of a resting nerve cell is positively charged on the outside and negatively charged on the inside; as an impulse goes down the axon, the outside becomes positively charged, and the inside becomes negatively charged.

35. **D. Synapse.** The synapse is the space between the terminal branches of one neuron and the dendrites of the next neuron.

Supplemental Chapter Problems

Problems

For problems 1–23, select the *best* answer.

1. The chemical reaction that breaks down food during digestion is:

   A. fermentation   B. dehydration synthesis   C. hydrolysis   D. photosynthesis

2. Chemical digestion of food becomes more efficient when mechanical digestion:

   A. increases the pH of the food   B. decreases the volume of the food
   C. increases the surface area of the food   D. decreases the surface area of the food

3. Which of the following represents the correct order of digestion in humans?

    A. mouth → stomach → esophagus → small intestines

    B. mouth → esophagus → stomach → large intestines

    C. mouth → stomach → esophagus → small intestines

    D. mouth → esophagus → stomach → small intestines

4. Which of the following fluids is produced by the liver?

    A. hydrochloric acid   B. bicarbonate   C. bile   D. gastric juice

5. Where does the reabsorption of water into the blood take place?

    A. small intestines   B. large intestines   C. villi   D. appendix

6. Select the chamber of the heart that pumps blood to the lungs.

    A. right atrium   B. right ventricle   C. left atrium   D. left ventricle

7. Identify the chamber of the heart that receives blood from the lower part of the body.

    A. right atrium   B. right ventricle   C. left atrium   D. left ventricle

8. Which of the following represents the correct order of circulation in humans?

    A. right atrium → right ventricle → pulmonary arteries → lungs → pulmonary veins → left atrium → left ventricle

    B. right atrium → right ventricle → pulmonary veins → lungs → pulmonary arteries → left atrium → left ventricle

    C. left atrium → left ventricle → pulmonary veins → lungs → pulmonary arteries → right atrium → right ventricle

    D. right atrium → left atrium → pulmonary veins → lungs → pulmonary arteries → right ventricle → left ventricle

9. Select the part of the blood that carries the iron-containing hemoglobin molecule.

    A. plasma   B. red blood cell   C. white blood cell   D. platelet

10. Which two blood components are most closely associated with the clotting process?

    A. white blood cells and antibiotics   B. platelets and plasma

    C. antibodies and antigens   D. platelets and fibrinogen

11. Identify the proteins that can destroy bacteria.

    A. antibodies   B. antigens   C. hormones   D. hemoglobin

12. Which part of the blood carries the end products of digestion?

    A. plasma   B. red blood cells   C. white blood cells   D. platelets

13. What happens to the diaphragm when humans inhale?

    A. The diaphragm moves up.

    B. The diaphragm moves down.

    C. The diaphragm does not change its position at all.

    D. The diaphragm moves deeper into the chest cavity.

14. As the level of carbon dioxide in the blood increases, a person's rate of respiration:

   A. increases   B. decreases   C. remains the same   D. varies

15. Where in the human respiratory system does an exchange of gases take place?

   A. alveoli   B. bronchioles   C. bronchi   D. stomates

16. Which of the following biological processes is the reverse of respiration?

   A. dehydration synthesis   B. photosynthesis   C. hydrolysis   D. peristalsis

17. How many ATP molecules are gained from the complete breakdown of glucose during cellular respiration?

   A. 2   B. 4   C. 36   D. 38

18. Which of the following statements best describes cellular respiration?

   A. Cellular respiration converts light energy into chemical energy.

   B. Cellular respiration converts chemical energy stored in the glucose molecule into molecules of ATP.

   C. Cellular respiration produces glucose from water and carbon dioxide.

   D. Cellular respiration converts ATP into chemical bond energy stored in the glucose molecule.

19. Which of the following is *not* an organ of excretion?

   A. small intestines   B. liver   C. skin   D. lungs

20. Which organ contains nephrons?

   A. lungs   B. skin   C. kidneys   D. liver

21. Sweating cools the body as a result of the physical process of:

   A. hydrolysis   B. condensation   C. precipitation   D. evaporation

22. Which of the following best describes sweat glands?

   A. Sweat glands are exocrine glands.   B. Sweat glands are endocrine glands.
   C. Sweat glands produce hormones.   D. Sweat glands are ductless glands.

23. Select the structure that can connect two bones together.

   A. ligaments   B. tendons   C. striations   D. muscles

Problems 24–26 are based on the following diagram:

24. Nerve impulses travel from:

   A. 3 to 2   B. 3 to 1   C. 2 to 1   D. 1 to 3

25. Which structure can produce the neurotransmitter acetylcholine?

    A. 1   B. 2   C. 3

26. Select the structure that detects stimuli.

    A. 1   B. 2   C. 3

27. Which of the following is an example of an effector?

    A. tongue   B. kidneys   C. salivary glands   D. brain

28. In humans, which gland functions as both an exocrine and endocrine gland?

    A. adrenal gland   B. pituitary gland   C. thyroid gland   D. pancreas

29. When a high concentration of sugar is in the blood, the body produces:

    A. insulin   B. glucagon   C. adrenalin   D. thyroxin

30. Select the hormone that regulates the rate of metabolism in body cells.

    A. thyroxin   B. adrenalin   C. parathormone   D. estrogen

Answers

1. **C.** "Nutrition and Digestion," p. 113

2. **C.** "Mouth," p. 114

3. **D.** "Nutrition and Digestion," p. 113

4. **C.** "Small Intestines," p. 114

5. **B.** "Large Intestines," p. 115

6. **B.** "The Structure and Function of the Heart," p. 116

7. **A.** "The Structure and Function of the Heart," p. 116

8. **A.** "The Structure and Function of the Heart," p. 116

9. **B.** "Components (Parts) of the Blood and Their Function," p. 117

10. **D.** "Components (Parts) of the Blood and Their Function," p. 117

11. **A.** "Components (Parts) of the Blood and Their Function," p. 117

12. **A.** "Components (Parts) of the Blood and Their Function," p. 117

13. **B.** "Organism Respiration," p. 119

14. **A.** "Organism Respiration," p. 119

15. **A.** "Organism Respiration," p. 119

16. **B.** "Table: Respiration and Photosynthesis Compared," p. 123

17. **C.** "Cellular Respiration," p. 121

18. **B.** "Cellular Respiration," p. 121

19. **A.** "Excretion," p. 125

20. **C.** "Kidney Circulation," p. 125

21. **D.** "The Skin As an Organ of Excretion," p. 129

22. **A.** "The Skin As an Organ of Excretion," p. 129

23. **A.** "The Endoskeleton," p. 130

24. **D.** "The Structure of a Neuron (Nerve Cell)," p. 134

25. **C.** "The Structure of a Neuron (Nerve Cell)," p. 134

26. **A.** "The Structure of a Neuron (Nerve Cell)," p. 134

27. **C.** "The Structure of the Nervous System," p. 132

28. **D.** "Islets of Langerhans (Pancreas)," p. 141

29. **A.** "Islets of Langerhans (Pancreas)," p. 141

30. **A.** "Thyroid Gland," p. 141

Chapter 7
Animal Biology

The amoeba, paramecium, hydra, earthworm and grasshopper are often referred to as the *fabulous five*. These organisms are representative of key phyla in the kingdom Protista and Animalia. The amoeba and paramecium are one-celled animals. The way they carry out their life functions is typical of individual cells. The hydra is interesting because it demonstrates what happens when cells work together in a simple multicellular organism. The earthworm is amazing in that, on a basic level, its life functions resemble those in humans. Grasshoppers are members of class Insecta. Insects are the largest and most successful class in the animal kingdom, but they conduct their life functions very differently from humans. The fabulous five are presented here in order of increasing complexity and evolutionary development. Compare how these organisms perform their life functions with those of humans (presented in Chapter 6).

Animal-Like Protists

Amoeba and paramecia are animal-like protists in the kingdom Protista. The amoeba is in the phylum Rhizopoda, and the paramecium is in the phylum Ciliophora. These unicellular organisms live in water.

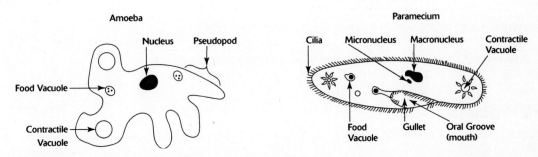

- **Nutrition:** Amoeba and paramecia eat small one-celled organisms found in water. Their food is digested in food vacuoles, and digestion is intracellular. All other organisms have both *extracellular* (outside the cell) and *intracellular* (inside the cell) digestion.

- **Circulation:** Amoeba and paramecia do not need a circulatory system because they are unicellular and in direct contact with their environment. However, materials circulate throughout the cell by the movement of the cytoplasm.

- **Respiration:** Animal-like protists obtain oxygen by diffusion from the surrounding water.

- **Excretion:** Carbon dioxide, ammonia, and water are waste products excreted by animal-like protists. Carbon dioxide and ammonia (a highly poisonous, nitrogenous waste) are removed by diffusion into the surrounding water. Animal-like protists have contractile vacuoles that regulate water. Water enters the contractile vacuole by osmosis and is forced out by active transport when the vacuole is filled.

❏ **Locomotion:** Amoeba have *pseudopods* (or false feet), which are cellular extensions that constantly change shape. Pseudopods are used for locomotion and to capture or engulf food by a process called *phagocytosis*. Paramecia have *cilia* (or hair-like projections) that beat back and forth in the water. Cilia are also used to sweep food into the oral groove (mouth) for ingestion.

❏ **Regulation:** Because these are one-celled organisms, they do not have endocrine or nervous systems. However, they are able to recognize and respond to stimuli such as food.

❏ **Reproduction**: Amoeba and paramecia reproduce by binary fission (splitting in two). Under stressful conditions paramecia and other ciliates reproduce by conjugation (exchanging genetic material).

Example Problems

These problems review the life functions of the animal-like protists.

1. How is extracellular digestion different from intracellular digestion?

   Answer: Extracellular digestion takes place outside the cells of an organism in an organ or organ system specialized for this purpose. Intracellular digestion takes place inside cells in organelles called food vacuoles. Most organisms have both extra- and intracellular digestion. The exception is unicellular organisms, which have only intracellular digestion.

2. How do animal-like protists excrete carbon dioxide and ammonia?

   Answer: Carbon dioxide and ammonia are excreted by diffusion through the cell membrane into the surrounding water.

3. Where does digestion in animal-like protists take place?

   Answer: Animal-like protists have food vacuoles for digestion.

Work Problems

Use these problems on the life functions of the animal-like protists for additional practice.

1. Why do we study animal-like protists?

2. How do the pseudopods of amoeba help with locomotion and nutrition?

3. How do animal-like protists maintain water balance?

4. Which food nutrient results in the production of ammonia as a waste product?

5. Why don't the amoeba and paramecium need a transport system?

Worked Solutions

1. Animal-like protists are one-celled organisms that carry out their life functions in a manner typical of individual cells. They give us a better understanding of the functioning of individual cells in humans. For example, the life functions of human white blood cells called *phagocytes* are similar to those of amoeba.

2. The pseudopods of amoeba are false feet, which are used for locomotion and to capture food. The amoeba is like a blob that is constantly changing shape as its pseudopods move about. The amoeba uses its pseudopods to engulf food by a process called phagocytosis.

3. Animal-like protists live in water. As a result, water is constantly entering by osmosis. If water is not removed, the organism would swell and eventually explode. The contractile vacuoles of animal-like protists are organelles that squeeze excess water out by active transport.

4. Proteins are the only nutrients in food that contain the element nitrogen. Ammonia is a nitrogenous waste that is produced as a result of protein digestion.

5. The amoeba and paramecium do not need a transport system because they are unicellular and in direct contact with their environment. Materials circulate within these organisms by the movement of the cytoplasm.

Phylum Cnidaria

The phylum Cnidaria is in the kingdom Animalia. Cnidaria include: hydra, jellyfish, coral, and sea anemone. These organisms live in fresh or marine water.

The hydra has two cell layers; the outer layer is called the ectoderm, and the inner layer is the endoderm. A hydra has an opening (mouth) for ingestion, a hollow body cavity for digestion, and tentacles for the capture of food.

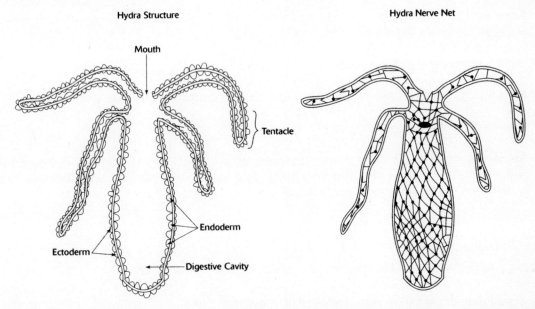

Hydra Structure

Hydra Nerve Net

Mouth

Tentacle

Endoderm

Ectoderm

Digestive Cavity

❑ **Nutrition:** A hydra eats small animals, which it captures with its tentacles. *Nematocysts* are harpoon-like structures that shoot out from the tentacles. They capture and poison the prey of the hydra. Food is put into the mouth and digested in the digestive cavity by specialized cells in the endoderm. Digestion in a hydra is extracellular in the digestive cavity and intracellular in the food vacuoles of its cells. The hydra has a two-way digestive system. Food enters through the mouth, and undigested food can exit from the mouth.

❑ **Circulation:** A hydra does not have a circulatory system because its endoderm cells are in direct contact with water in the digestive cavity, and the ectoderm cells are in direct contact with water from their surrounding environment. Digested food from the digestive cavity can diffuse into the endoderm cells and then to the ectoderm cells. Within the cells, materials are transported by the movement of the cytoplasm.

❑ **Respiration:** The ectoderm cells of a hydra obtain oxygen by diffusion from the surrounding water, while endoderm cells get their oxygen from the digestive cavity.

❑ **Excretion:** The hydra excretes carbon dioxide, ammonia, and water. Carbon dioxide and ammonia from the endoderm cells are excreted into the digestive cavity. The ectoderm cells excrete carbon dioxide and ammonia into the surrounding water. Excess water is excreted by the contractile vacuoles of the cells.

❑ **Locomotion:** For the most part, the hydra is sessile. However, it can glide on its base and somersault. The hydra can use water currents to transport itself to a new location.

❑ **Regulation:** A hydra has a nerve net between the ectoderm and the endoderm cells but no brain or nerve cord to coordinate responses. Touch any part of a hydra, and the entire organism responds.

❑ **Reproduction:** Cnidarians reproduce asexually and sexually. The hydra reproduces asexually by budding (a small new hydra forms off of the parent) or sexually (by the combination of sperm and egg).

Example Problems

These problems review the life functions of a hydra.

1. Describe the digestive system of a hydra.

   Answer: The hydra has a two-way digestive system. Food enters through the mouth, and undigested food can exit from the mouth.

2. What wastes products are produced by a hydra?

   Answer: A hydra can produce carbon dioxide and ammonia as waste products.

3. How does a hydra excrete wastes?

   Answer: Carbon dioxide and ammonia are excreted by diffusion through the cell membrane of the hydra's ectoderm cells into the surrounding water. The endoderm cells excrete these wastes into the water of the digestive cavity.

Work Problems

Use these problems on the life functions of a hydra for additional practice.

1. What kinds of digestion take place inside a hydra?

2. Why doesn't a hydra need a transport system?

3. How can a hydra move from one place to another?

4. Why do all the tentacles of a hydra move even though only one is touched?

Worked Solutions

1. The hydra has extracellular digestion that takes place in the digestive cavity and intracellular digestion that takes place in the food vacuoles of the ectoderm and endoderm cells.

2. A hydra does not need a transport system because all its cells are in contact with its watery environment. The endoderm cells are in contact with water in the digestive cavity, and the ectoderm cells are in contact with the surrounding water.

3. A hydra is basically sessile. However, it can glide on its base and somersault. A hydra can move long distances by allowing water currents to carry it to a new location.

4. The hydra has a nerve net but does not have a brain or nerve cord to control its responses.

Phylum Annelida

The phylum Annelida is in the kingdom Animalia. Annelida include: earthworms, sandworms, and leeches.

Earthworm Anatomy

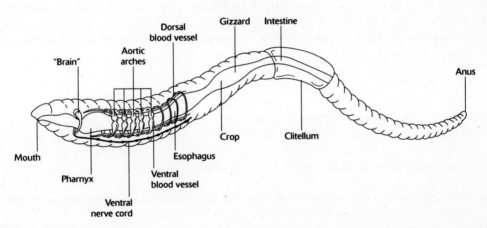

- **Nutrition:** The earthworm has a one-way digestive system, a tube-within-a-tube body, and extra- and intracellular digestion. Food enters the mouth, and waste exits the anus. The muscular pharynx sucks in food from the mouth to the esophagus, where peristalsis forces the food into the crop. Food is stored in the crop and is eventually pushed into the gizzard, where mechanical digestion takes place. From the gizzard, food enters the intestines where chemical digestion occurs. From the intestines, digested food enters the blood, and waste leaves through the anus.

- **Circulation:** The earthworm has a closed circulatory system, dorsal and ventral blood vessels surrounded by five aortic arches (hearts) that pump blood. Capillaries from the major blood vessels lead to the organs, supplying food and oxygen.

- **Respiration:** Oxygen in the air diffuses through the moist skin of the earthworm into capillaries, which carry the oxygen to the two major blood vessels.

- **Excretion:** Most segments of the earthworm have a pair of tubes called *nephridia* that are surrounded by capillaries. Ammonia and urea are filtered out of the blood by the nephridia and

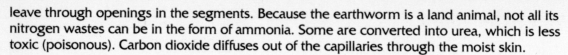

leave through openings in the segments. Because the earthworm is a land animal, not all its nitrogen wastes can be in the form of ammonia. Some are converted into urea, which is less toxic (poisonous). Carbon dioxide diffuses out of the capillaries through the moist skin.

❑ **Locomotion:** The segments of the earthworm have muscles and four pairs of bristles (called *setae*) that grab the soil and allow the worm to dig.

❑ **Regulation:** At the anterior (head) end, the earthworm has a primitive brain that is composed of two *ganglia* (a group of nerve cells and cytons). Connected to the ganglia is a ventral nerve cord. Together, the ganglia and the nerve cord coordinate the activities of the worm.

❑ **Reproduction:** Annelids reproduce sexually. Earthworms produce both sperm and eggs within the same individual, thus they are *hermaphrodites*.

Example Problems

These problems review the life functions of the earthworm.

1. Describe the digestive system of an earthworm.

    Answer: The earthworm has a one-way digestive system. The pharynx sucks in food from the mouth to the esophagus, where peristalsis forces the food into the crop. Food is stored in the crop and pushed into the gizzard for mechanical digestion. From the gizzard, food enters the intestines where chemical digestion occurs. From the intestines, digested food enters the blood, and waste leaves through the anus.

2. How does digested food get into the cells of the earthworm?

    Answer: In earthworms, digested food enters the blood from the intestines. The capillaries from dorsal and ventral blood vessels distribute the digested food to the cells.

3. What waste products does the earthworm produce?

    Answer: Earthworms produce carbon dioxide and water as waste products from respiration. They produce ammonia and urea as nitrogen wastes.

Work Problems

Use these problems on the life functions of the earthworm for additional practice.

1. What kinds of digestion take place inside an earthworm?

2. Describe the circulatory system of an earthworm.

3. How can the earthworm move from one place to another?

4. Describe the nervous system of an earthworm.

5. How does oxygen get into the cells of the earthworm?

Worked Solutions

1. The earthworm has extracellular digestion that takes place in the digestive system and intracellular digestion that takes place in the cells of the organs.

2. The earthworm has a closed circulatory system, dorsal and ventral blood vessels surrounded by five aortic arches (hearts) that pump blood. Capillaries from the major blood vessels to the organs supply food and oxygen.

3. The segments of the earthworm have muscles and four pairs of bristles (called setae) that grab the soil and allow the worm to dig.

4. The earthworm has a primitive brain that is composed of two ganglia connected to a ventral nerve cord. Together, the ganglia and the nerve cord coordinate the activities of the worm.

5. Oxygen in the air diffuses through the moist skin of the earthworm into capillaries, which carry the oxygen to the two major blood vessels.

Phylum Arthropoda

The phylum Arthropoda is in the kingdom Animalia. Three of the classes that make up the phylum Arthropoda are Crustacea, Arachnida and Insecta. The grasshopper is in the class Insecta.

Grasshopper Anatomy

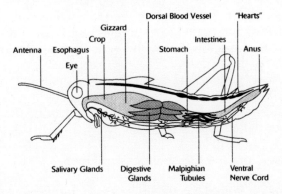

❏ **Nutrition:** The grasshopper has a one-way digestive system, a tube-within-a-tube body, and extra- and intracellular digestion. Food enters the mouth, and waste exits through the anus. The grasshopper has a mouth that chews, crushes, and grinds food. From the mouth, food enters the esophagus and moves into the crop for storage. Food in the crop eventually enters the gizzard (where mechanical digestion takes place) and then moves into the stomach (where chemical digestion occurs). From the stomach, digested food enters the blood, and waste enters the large intestines, moves to the small intestines, and exits through the anus.

❏ **Circulation:** The grasshopper has an open circulatory system, with a dorsal blood vessel that has bulges (hearts) that pump blood. The blood sprays out of the dorsal blood vessel to bathe the organs and supply them with food. Insect blood is colorless and does not carry oxygen.

❏ **Respiration:** Openings in the grasshopper's abdomen (called *spiracles*) lead to a network of tubes (called *tracheal tubes*) that carry oxygen to the cells.

❏ **Excretion:** The grasshopper has a network of tubes called *Malpighian tubules* that filter the blood, removing nitrogen waste in the form of uric acid (a solid that exits through the anus). Water is conserved and not excreted. Carbon dioxide leaves through the spiracles.

❏ **Locomotion:** Most insects have three pairs of legs and two pairs of wings that they use for movement.

❏ **Regulation:** Insects have two large, combined ganglia in their heads that act as a brain. Attached to the ganglia are two ventral nerve cords. The brain and nerve cords coordinate the activities of the insect.

❏ **Reproduction:** Insects reproduce sexually. The female grasshopper deposits its eggs in the soil. The grasshopper exhibits complete metamorphosis (changing its body form from a larva to an adult). Some insects (bees) can reproduce by parthenogenesis (unfertilized eggs).

Example Problems

These problems review the life functions of the grasshopper.

1. Describe the digestive system of a grasshopper.

    Answer: The grasshopper has a one-way digestive system. The grasshopper has a mouth for the mechanical digestion of food. From the mouth, food enters the esophagus and moves into the crop for storage. Food in the crop enters the gizzard (where mechanical digestion takes place) and then moves into the stomach (where chemical digestion occurs). From the stomach, digested food enters the blood, and waste enters the large intestines, moves to the small intestines, and exits through the anus.

2. How does digested food get into the cells of the grasshopper?

    Answer: From the stomach, digested food enters the blood, and the dorsal blood vessel sprays the blood over the organs.

3. What wastes products does the grasshopper produce?

    Answer: Carbon dioxide from the cells diffuses into the tracheal tubes and exits through the spiracles. Uric acid is a solid, nitrogenous waste that exits through the anus.

Work Problems

Use these problems on the life functions of the grasshopper for additional practice.

1. What kinds of digestion take place inside grasshoppers?

2. Describe the circulatory system of a grasshopper.

3. How can the grasshopper move from one place to another?

4. Describe the nervous system of a grasshopper.

5. Which digestive organs are found in insects but are not found in humans?

Worked Solutions

1. The grasshopper has extracellular digestion that takes place in the digestive system and intracellular digestion that takes place in the cells of the organs.

2. The grasshopper has an open circulatory system with a dorsal blood vessel that can pump blood. The blood sprays out of the dorsal blood vessel to bathe the organs, supplying food to the cells.

3. Grasshoppers can walk and jump because they have three pairs of legs. Grasshoppers can fly because they have two pairs of wings.

4. The grasshopper has a brain that consists of two large, combined ganglia that are attached to two ventral nerve cords. The brain and nerve cords coordinate the activities of the grasshopper.

5. The crop and gizzard are digestive organs found in insects but not in humans.

Chapter Problems and Answers

Problems

For problems 1–10, fill in the term that *best* completes the sentence.

1. Digestion within a cell takes place in organelles called _____.

2. Homeostasis of water in a paramecium is controlled by organelles called
_____.

3. To surround and engulf food with its pseudopods, the amoeba must use energy. This is an example of _____.

4. In a hydra, digestion is both _____ and
_____.

5. Oxygen enters the cells of a hydra by the process of _____
_____.

6. An earthworm gets rid of carbon dioxide through its _____
_____.

7. Earthworms have four pairs of bristles called setae on each segment that are used for_____.

8. Air enters the respiratory system of a grasshopper through structures called
_____.

9. The solid, nitrogenous waste product produced by insects is _____
_____.

10. Earthworms and insects have a _____ on their ventral side that helps to coordinate nervous system responses.

For problems 11–15, select the structure from the following list that is associated with each description. A choice can be used more than once or not at all.

A. ganglia   B. Malpighian tubules   C. moist skin   D. tracheal tubes   E. nephridia

11. This structure exchanges oxygen with the cells of an insect. _____

12. This structure removes nitrogen waste from an insect's blood. _____

13. This structure removes ammonia from an earthworm's blood. _____

14. Select the respiratory structure found in earthworms. _____

15. This structure does not have an excretory function. _____

For problems 16–20, select the structure from the following list that is associated with each step in the process of regulation. A choice can be used more than once or not at all.

A. ganglia   B. nerve cord   C. nematocysts   D. nerve net   E. cilia

16. Select the structure in a paramecium that can detect food and sweep it toward the mouth. _____

17. Identify the structure associated with nervous system responses in a hydra. _____

18. Select the structure that can be found on the ventral side of an earthworm. _____

19. Which structure forms part of the primitive brain of an earthworm? _____

20. Select the structure in a hydra that can be used to capture food. _____

Answers

1. **Food vacuoles.** One-celled organisms do not have a digestive system. All digestion takes place within food vacuoles.

2. **Contractile vacuoles.** Paramecia have contractile vacuoles that regulate water. Water enters the contractile vacuole by osmosis and is forced out by active transport (meaning that energy is needed).

3. **Active transport.** Active transport includes all processes within a cell that involve the use of energy to move molecules.

4. **Extracellular and intracellular.** In a hydra, digestion that takes place in the digestive cavity is extracellular. Digestion inside the food vacuoles of endoderm and ectoderm cells is intracellular.

5. **Osmosis.** Oxygen dissolved in water enters the cells of a hydra by osmosis (the movement of water molecules from areas of high concentration to areas of low concentration).

6. **Skin.** Earthworms do not have lungs for respiration. Their moist skin serves as a respiratory structure for the exchange of gases between the organism and its environment.

7. **Locomotion.** The interaction between the setae and muscles in the earthworm enables the organism to move.

8. **Spiracles.** The abdomen of a grasshopper has several openings called spiracles that allow air in and out. The spiracles lead to a network of tracheal tubes that carry oxygen to each cell.

9. **Uric acid.** Uric acid is solid, allowing insects to conserve water.

10. **Nerve cord(s).** The earthworm has one ventral nerve cord, and insects have two ventral nerve cords. In both of these organisms, the nerve cord(s) is attached to combine ganglia in the head.

11. **D. Tracheal tubes.** The spiracles of insects lead to a network of tracheal tubes. Oxygen in the tracheal tubes diffuses into the cells, and carbon dioxide from the cells diffuses into the tubes.

12. **B. Malpighian tubules.** These tubules remove nitrogen wastes from the blood and pass this waste into the large intestines. Uric acid forms that is eventually removed through the anus.

13. **E. Nephridia.** The segments of earthworms have pairs of tubes called nephridia that are surrounded by capillaries. Ammonia and urea are filtered out of the blood by the nephridia and leave through openings in the segments.

14. **C. Moist skin.** The moist skin of an earthworm serves as a respiratory structure for the exchange of gases between the organism and its environment.

15. **A. Ganglia.** Ganglia are groups of nerve cells and cytons that function in nervous system regulation.

16. **E. Cilia.** The cilia are hair-like projections that cover the paramecium. When they make contact with food, the cilia sweep the food toward the oral groove (mouth).

17. **D. Nerve net.** A hydra has a network of nerve cells located between the ectoderm and endoderm cells that enables it to respond to stimuli.

18. **B. Nerve cord.** The earthworm has one ventral nerve cord and insects have two. The nerve cord(s) and the ganglia make up the nervous systems of these organisms.

19. **A. Ganglia.** Ganglia are groups of nerve cells and cytons that form the brain of earthworms and insects.

20. **C. Nematocysts.** Nematocysts are harpoon-like structures that shoot out from the tentacles of a hydra to poison and capture prey.

Supplemental Chapter Problems

Problems

For problems 1–21, select the *best* answer.

1. An amoeba can engulf a small organism by the process of:

    A. phagocytosis   B. cyclosis   C. osmosis   D. diffusion

2. A tube-within-a-tube body organization is found in:

    A. amoeba and paramecia   B. earthworms and humans
    C. hydra and grasshoppers   D. animal-like protists and humans

3. Which of the following organisms has only intracellular digestion?

    A. humans   B. grasshoppers   C. earthworms   D. animal-like protists

4. Select the organism that does *not* have a one-way digestive system.

    A. hydra   B. earthworm   C. grasshopper   D. human

5. Where does mechanical digestion in earthworms and grasshoppers take place?

    A. crop   B. esophagus   C. gizzard   D. intestines

6. The transport of materials within animal-like protists takes place by a process called:

    A. diffusion   B. osmosis   C. movement of the cytoplasm   D. passive transport

7. Water enters the amoeba from its surrounding environment by:

    A. diffusion   B. osmosis   C. cyclosis   D. active transport

8. Which of the following best describes transport in animal-like protists?

    A. a system of tracheal tubes   B. an open circulatory system
    C. a closed circulatory system   D. osmosis, diffusion, movement of the cytoplasm

9. In a hydra, digested food found in the digestive cavity enters the endoderm cells by the process of:

    A. diffusion   B. osmosis   C. cyclosis   D. active transport

10. Which of the following best describes transport in earthworms?

    A. an open circulatory system with tubular hearts that pump blood

    B. an open circulatory system with five aortic arches that pump blood

    C. a closed circulatory system with five aortic arches that pump blood

    D. a closed circulatory system with a four-chambered heart that pumps blood

11. Which of the following substances is never found in the blood of a grasshopper?

    A. digested food   B. oxygen   C. water   D. cells

12. Select the process by which animal-like protists excrete water into their surrounding environment.

   A. diffusion   B. osmosis   C. cyclosis   D. active transport

13. Insects use structures called spiracles to:

   A. digest food   B. circulate water   C. reproduce   D. breathe

14. Which organism has ammonia as its main nitrogenous waste?

   A. hydra   B. earthworm   C. grasshopper   D. human

15. Which organism depends on Malpighian tubules for excretion?

   A. hydra   B. earthworm   C. grasshopper   D. human

16. Select the organism that has paired nephridia for the removal of nitrogenous wastes.

   A. hydra   B. earthworm   C. grasshopper   D. human

17. Identify the organism that produces uric acid as a waste to conserve water.

   A. hydra   B. earthworm   C. grasshopper   D. human

18. Which of the following organisms does *not* depend on neurotransmitters for nervous system control?

   A. animal-like protists   B. earthworm   C. grasshopper   D. human

19. Which organism has only one ventral nerve cord?

   A. animal-like protists   B. earthworm   C. grasshopper   D. human

20. Which of the following organisms does not depend on hormones for endocrine system control?

   A. animal-like protists   B. earthworm   C. grasshopper   D. human

21. Which of the following shows the correct order of organisms, with respect to increasing complexity in the handling of life functions?

   A. human → grasshopper → earthworm → hydra → animal-like protists

   B. hydra → animal-like protists→ earthworm → grasshopper → human

   C. animal-like protists→ hydra → grasshopper → earthworm → human

   D. animal-like protists→ hydra → earthworm → grasshopper → human

Answers

   1. **A.** "Animal-Like Protists," p. 154

   2. **B.** "Phylum Annelida," p. 157

   3. **D.** "Animal-Like Protists," p. 153

   4. **A.** "Animal-Like Protists," p. 153

5.   **C.** "Phylum Annelida," p. 157; "Phylum Arthropoda," p. 159

6.   **C.** "Animal-Like Protists," p. 153

7.   **B.** "Animal-Like Protists," p. 153

8.   **D.** "Animal-Like Protists," p. 153

9.   **A.** "Phylum Cnidaria," p. 155

10.   **C.** "Phylum Annelida," p. 157

11.   **B.** "Phylum Arthropoda," p. 159

12.   **D.** "Animal-Like Protists," p. 153

13.   **D.** "Phylum Arthropoda," p. 160

14.   **A.** "Phylum Cnidaria," p. 156

15.   **C.** "Phylum Arthropoda," p. 160

16.   **B.** "Phylum Annelida," pp. 157–158

17.   **C.** "Phylum Arthropoda," p. 160

18.   **A.** "Animal-Like Protists," p. 154

19.   **B.** "Phylum Annelida," p. 158

20.   **A.** "Animal-Like Protists," p. 154

21.   **D.** "Chapter Introduction," p. 153

Chapter 8
Patterns of Reproduction

Reproduction is the life function by which organisms produce new individuals of the same kind (species). Two patterns of reproduction exist: asexual reproduction and sexual reproduction. *Asexual reproduction* is the formation of a new individual from one parent. *Sexual reproduction* is the formation of a new individual from two parents (fusion of gametes).

Asexual Reproduction

Unicellular organisms such as amoeba, paramecia and yeast reproduce asexually. Some multicellular organisms such as molds and green plants can reproduce asexually to form new individuals. In all multicellular organisms, asexual reproduction produces new cells that are necessary for growth and repair. In the most basic form of asexual reproduction, one cell (the parent cell) divides to form two cells called daughter cells. The daughter cells are identical to each other and to the parent cell that formed them. During cell division (*mitosis*), two events occur that are necessary to end up with identical daughter cells:

❑ **Replication (duplication) of the chromosomes** and their equal distribution into each cell so that the daughter cells have the same number of chromosomes as the parent. Thus, the chromosome number for the species remains the same from one generation to the next.

❑ **Equal division of the cytoplasm** so that each daughter cell has approximately the same amount of cytoplasm.

For example, if a parent cell has four chromosomes in its nucleus and reproduces asexually, each daughter cell that results must have four chromosomes and be approximately the same size.

Example Problems
These problems review asexual reproduction.

1. How is asexual reproduction different from sexual reproduction?

    Answer: Asexual reproduction is the formation of a new individual from one parent. Sexual reproduction is the formation of a new individual from two parents.

2. What kinds of organisms reproduce asexually?

    Answer: Unicellular organisms such as amoeba, paramecia and yeast reproduce asexually. Some multicellular organisms such as molds and green plants reproduce asexually to form new individuals.

3. What happens to the parent cell when asexual reproduction is complete?

    Answer: When the parent cell reproduces asexually, it does not exist anymore. Two daughter cells form from the parent, and they are identical to each other.

4. What are the advantages and disadvantages of asexual reproduction?

    Answer: The advantages of asexual reproduction are that only one parent is required, and all the offspring are identical to the parent. The disadvantage is that no variety exists among the offspring.

Work Problems

Use these problems on asexual reproduction for additional practice.

1. Why must multicellular organisms be capable of asexual reproduction?

2. Describe the events that occur during asexual reproduction.

3. How can a cell that has 10 chromosomes produce two daughter cells, each with 10 chromosomes?

4. A human liver cell with 46 chromosomes reproduces asexually. How many chromosomes does each daughter cell have?

5. Why must the chromosome number of an organism be the same from one generation to the next?

Worked Solutions

1. Multicellular organisms perform asexual reproduction to produce new cells for growth and to replace cells in the organism that have died.

2. During asexual reproduction, the cytoplasm divides equally; the chromosomes replicate and are distributed equally to each daughter cell.

3. During asexual reproduction, each chromosome replicates. If the parent cell has 10 chromosomes, after replication it has 20. When the cytoplasm divides, these 20 chromosomes are distributed equally so that each daughter cell has 10 chromosomes.

4. Each daughter cell has 46 chromosomes. The species number of chromosomes remains the same from one generation to the next.

5. Chromosomes contain the hereditary information of a cell. If a daughter cell was produced with missing chromosomes, this error could result in the death of the cell because the functions controlled by that chromosome would not be available. For example, suppose the chromosome with the information needed to produce digestive enzymes was missing; the cell would not have the enzymes needed to digest food and would die.

Examples of Asexual Reproduction

Binary fission is the simplest form of asexual reproduction. The organism divides into two equal-sized daughter cells, each with the same amount of cytoplasm and nuclear material (chromosomes). Unicellular organisms such as amoeba, paramecia and bacteria reproduce this way.

Binary Fission

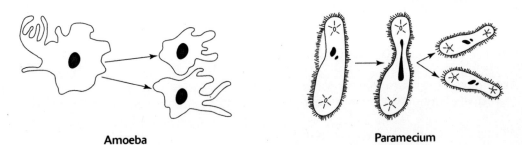

Amoeba   Paramecium

Budding results in two unequal-sized cells as a result of unequal division of the cytoplasm. The larger cell is called the parent, and the smaller one is called the bud. Yeast cells reproduce this way. In hydra, a small new hydra called a bud can form off the parent to become a new individual.

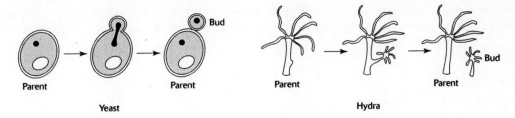

Parent   Parent   Parent   Parent

Yeast   Hydra

Reproduction from spores is a form of asexual reproduction in some molds. A spore is a small, usually circular structure that contains a nucleus, some cytoplasm and a protective outer wall. When Fungi spores fall on a suitable food source, rhizoids form that anchor the organism to the food. Thread-like filaments called hyphae develop that end in a reproductive structure called a sporangium where new spores develop. When the sporangium breaks, the spores are released and the cycle can begin again.

Spore Formation

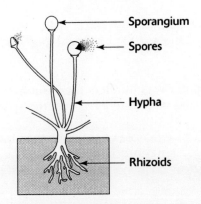

— Sporangium

— Spores

— Hypha

— Rhizoids

Vegetative propagation is a form of asexual reproduction in plants where a new plant can form from the root, stem or leaf of an already existing plant. We will study plant asexual and sexual reproduction in Chapter 10.

Regeneration is the replacement or repair of a lost or damaged part in an organism. A starfish that loses an arm can grow back the missing arm. A lobster that loses a claw can regenerate the missing claw. Planaria are small flatworms that have amazing powers of regeneration. If a planarian is cut in half, each half regenerates the missing part.

Example Problems

These problems review various examples of asexual reproduction.

1. Describe the process of binary fission.

   Answer: An organism that reproduces by binary fission has equal division of the cytoplasm and equal distribution of the nuclear material to form two identical daughter cells.

2. What parts of a plant can produce new plants?

   Answer: Plants can use their roots, stems or leaves to produce a new plant.

3. How does the ability of regeneration increase an organism's chances of survival?

   Answer: A missing part makes it difficult for an organism to find food or defend itself. The ability to replace the missing part makes the organism whole and competitive once again.

Work Problems

Use these problems on examples of asexual reproduction for additional practice.

1. How is budding in yeast cells different from binary fission?

2. How can spores survive unfavorable conditions?

3. Why is regeneration considered an example of asexual reproduction?

Worked Solutions

1. When yeast cells reproduce by budding, the division of the cytoplasm is unequal. The larger cell is the parent, and the smaller cell is the bud. In binary fission, the division of the cytoplasm is equal.

2. Spores have a protective outer wall that lets them survive unfavorable conditions. However, as soon as food and water are available, the spore can form a new organism.

3. Regeneration is considered an example of asexual reproduction because when an organism loses a part, cells in the organism reproduce to replace the missing part. Some organisms can be cut in half and still grow back the missing part, resulting in the formation of two organisms.

Mitosis

In the life cycle of a cell, several distinct stages or phases can be identified. The following diagrams show the different stages of animal cell mitosis. The length of time needed for mitosis varies with the organism and the type of cell that is dividing.

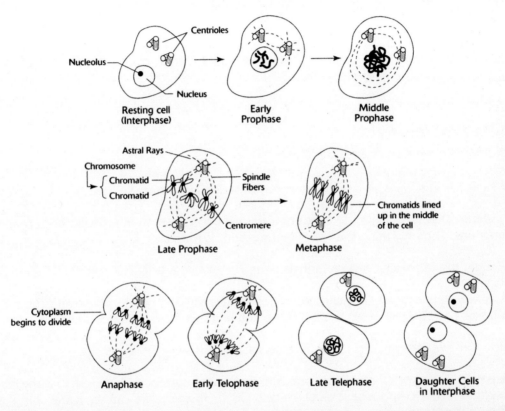

Animal Cell Mitosis

The *interphase* stage of mitosis is often called the *resting stage;* however, the cell is not truly resting. During interphase, the cell is growing and performing all its life functions. Most of the time the cell is in interphase; the other stages occur only when the cell reproduces. At some point during interphase, the chromosomes replicate, signaling the start of mitosis or cell division. Usually, when this happens, the cell enters a stage called prophase.

During the *prophase* stage of mitosis, the two halves of the replicated chromosome (now called *chromatids*) come together and are joined to each other by a structure called a *centromere*.

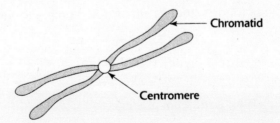

The nuclear membrane and nucleolus disappear, and the two centrioles (cylindrical structures near the nucleus) begin to move to opposite sides of the cell. The centrioles are attached to the cell membrane by star-shaped structures called *astral rays*. The chromatids attach to *spindle fibers* that form between the centrioles.

In *metaphase* the chromatids line up on the spindle fibers in the center of the cell.

During *anaphase* the chromatids separate (now called chromosomes again) and move to the opposite sides of the cell. Toward the end of anaphase and the beginning of telophase, the cell membrane begins to pinch in, and daughter cells begin to form.

During *telophase* daughter-cell formation is completed, and two cells are produced. A nuclear membrane forms around the chromosomes in each cell, the nucleolus reappears and the centrioles in each cell replicate. Mitosis is now complete.

Plant-cell mitosis differs from animal-cell mitosis in that plant cells do not have centrioles or astral rays. During telophase in plant-cell mitosis, a *cell plate* forms in the middle of the cell dividing the cell into two daughter cells.

Example Problems

These problems review the stages of mitosis.

1. Why is *resting stage* an incorrect term to describe interphase?

   Answer: During its life cycle, most of a cell's time is spent in interphase performing its life functions. The only thing the cell is resting from is mitosis.

2. Identify the event that begins mitosis.

   Answer: Replication of the chromosomes signals the end of interphase and the start of mitosis, or cell division. At this point the cell enters a stage of mitosis known as prophase.

3. How can we recognize the prophase stage of mitosis?

   Answer: During prophase the nuclear membrane and nucleolus disappear, and the centrioles begin to move to opposite sides of the cell. For the first time we can see astral rays, spindle fibers and chromatids.

4. What are the characteristics of metaphase?

   Answer: In metaphase the chromatids are lined up on the spindle fibers in the middle of the cell.

5. How can we recognize the anaphase stage of mitosis?

   Answer: In this stage of mitosis, the chromatids separate and move toward opposite sides of the cell.

Work Problems

Use these problems on mitosis for additional practice.

1. Describe the telophase stage of mitosis.

2. What is a chromatid?

3. Why is it important that the cytoplasm divide equally?

4. What is the end result of mitosis?

5. How is animal-cell mitosis different from plant-cell mitosis?

Worked Solutions

1. This is the stage of mitosis that results in daughter-cell formation. Each cell has a nucleus with a nucleolus and a pair of centrioles.

2. A chromatid is a chromosome and its duplicate held together by a centromere.

3. Equal division of the cytoplasm is important because each daughter cell must end up with enough cell organelles to carry out the life functions.

4. The end result of mitosis is two identical daughter cells, each approximately the same size and containing the same number of chromosomes.

5. In animal-cell mitosis, centrioles and astral rays are present. Plant-cell mitosis does not have these structures.

Sexual Reproduction

In sexual reproduction a new individual is formed from two parents. The male parent produces sperm cells, and the female parent produces egg cells. The sperm cell is the male sex cell or *gamete*. The egg cell (*ovum*) is the female sex cell or gamete. *Fertilization* is the process by which a sperm cell and egg cell combine. The fertilized egg cell is called a *zygote*. The new organism develops from the zygote. Fertilization is the defining process that occurs during sexual reproduction.

Chromosomes are found in the nucleus of cells as *homologous* (similar) pairs. The number of chromosomes in a somatic cell (body cell) is its *diploid number* ($2n$). The number of chromosomes in a gamete is its *haploid* number (n). The letter n stands for the number of chromosomes found in a gamete. The somatic cells of an organism have double the number of chromosomes found in a gamete.

In sexual reproduction and asexual reproduction, the chromosome number of a species must remain the same from one generation to the next. For example, humans have a diploid number of 46 chromosomes in their *somatic cells*. When a couple has a child, the number of chromosomes in the child's body cells must also be 46.

During sexual reproduction a sperm cell with n chromosomes combines with an ovum, which has n chromosomes, yielding a zygote with $2n$ chromosomes. In this way the chromosome number for a species remains the same from one generation to the next. For example, in humans a sperm cell with 23 chromosomes combines with an egg cell with 23 chromosomes to produce a zygote with 46 chromosomes, preserving the human species number of chromosomes.

| Sperm Cell | Egg Cell | Zygote | Human Sperm Cell | Human Egg Cell | Human Zygote |

Sperm cells are produced in the testes of the male. The testes are somatic cells. Ova are produced in the ovaries of the female. Ovaries are somatic cells. How do somatic cells (2n) produce gametes (n)? Meiosis is a special kind of cell division that results in the formation of gametes.

Example Problems

These problems review sexual reproduction.

1. What are gametes?

   Answer: Gametes are the male sex cells (sperm) and the female sex cells (ova).

2. Define fertilization.

   Answer: Fertilization is the union (combination) of sperm and egg cells during sexual reproduction.

3. What are somatic cells?

   Answer: Somatic cells are any cells in the body that are not gametes.

4. Which organs in the body produce gametes?

   Answer: The gonads (sex organs) produce gametes. The testes produce sperm, and the ovaries produce ova.

5. What are the advantages and disadvantages of sexual reproduction?

   Answer: The advantage of sexual reproduction is variety among the offspring, which inherit some characteristics from each parent. The disadvantage is that two parents are required, so each must find a mate.

Work Problems

Use these problems on sexual reproduction for additional practice.

1. How is the species number of chromosomes kept the same from one generation to the next in an organism that reproduces sexually?

2. How are haploid cells different from diploid cells?

3. How many chromosomes are in a stomach cell of a gorilla if its pancreas cells have 48 chromosomes?

4. How many chromosomes are in a gamete in an organism that has 30 chromosomes in its somatic cells?

5. What is the diploid number for a cell that has a haploid number of 24?

Worked Solutions

1. Organisms that reproduce sexually form gametes that have half the species number of chromosomes. When gametes combine during fertilization ($n + n$), the species number of chromosomes ($2n$) is restored.

2. Haploid cells are gametes with n chromosomes. Diploid cells are somatic cells with $2n$ chromosomes.

3. The stomach cells of the gorilla have 48 chromosomes. Stomach cells and pancreas cells are somatic cells, and all somatic cells in an organism have the same number of chromosomes.

4. A gamete in this organism has 15 chromosomes. Gametes (n) have half the number of chromosomes found in somatic cells ($2n$).

$$2n = 30$$
$$n = 15$$

5. The diploid number for this cell is 48. If the haploid number is $n = 24$, the diploid number is $2n$ ($2 \times 24 = 48$).

Meiosis

Meiosis is a process by which gametes are produced. Gamete production is called *gametogenesis;* the production of sperm cells is called *spermatogenesis;* the production of egg cells is called *oogenesis.* In biology when a process works well, it is used again and again. During meiosis a cell goes through a series of stages similar to those in mitosis. The major difference between meiosis and mitosis is that meiosis has two cell divisions, and mitosis has only one. The second cell division reduces the number of chromosomes from $2n$ to n. Meiosis is often called reduction division.

During *interphase* each chromosome in a pair of homologous chromosomes replicates to form four chromosomes.

In *prophase I* the four chromosomes come together by a process called *synapsis* to form a *tetrad* (a combination of four chromosomes).

In *metaphase I* the tetrads line up on the spindle fibers in the middle of the cell.

During *anaphase I* the chromosomes in the tetrad are pulled apart. This is called *disjunction.* Two chromosomes go into each of the two cells that form.

In *telophase I* two cells form.

Each of the two cells goes through the stages of cell division for a second time. For each cell that starts meiosis, four gametes (sex cells) are formed.

In humans each primary spermatocyte produces four sperm cells. Each primary oocyte produces one large cell called the ovum and three very small cells called polar bodies. The polar bodies do not have enough cytoplasm and die.

Animal Cell Meiosis

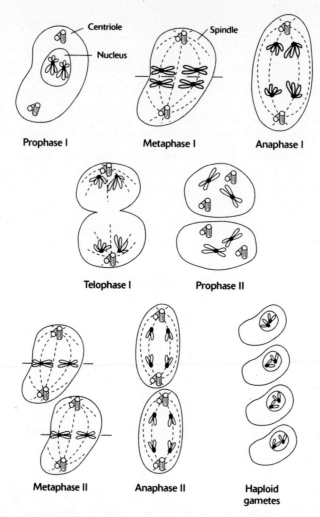

The following table summarizes the differences between mitosis and meiosis.

Mitosis and Meiosis Compared		
Characteristics	**Mitosis**	**Meiosis**
Type of reproduction	Asexual	Sexual
Synapsis	Absent	Present
Tetrad formation	Absent	Present
Disjunction	Absent	Present
Number of cell divisions	1	2
Type of cell formed	2 daughter cells	4 gametes (in males) OR 1 gamete and 3 polar bodies (in females)
Number of chromosomes in cells	$2n$	n

Example Problems

These problems review meiosis.

1. What is the end result of meiosis?

    Answer: The end result of meiosis is the formation of gametes with n chromosomes.

2. Describe the process of synapsis and disjunction.

    Answer: During synapsis each pair of homologous chromosomes and their replicas come together to form a set of four chromosomes called a tetrad. Disjunction is the breaking up of the tetrad. Two chromosomes go into one cell, and the other two go into another cell.

3. How can we tell if a cell that is just starting to divide will perform meiosis or mitosis?

    Answer: If synapsis and tetrad formation take place during prophase, the cell goes through the process of meiosis. However, if a tetrad is not formed and chromatids begin to line up on the spindle fiber, the cell enters mitosis.

Work Problems

Use these problems on meiosis for additional practice.

1. How does meiosis help maintain the chromosome number of a species?

2. How is spermatogenesis in humans different from oogenesis?

3. Why do the polar bodies produced during oogenesis in humans die?

4. Discuss the key difference between meiosis and mitosis.

Worked Solutions

1. Meiosis is reduction division; somatic cells that have $2n$ chromosomes produce gametes that have n chromosomes. After fertilization is complete, the number of chromosomes is restored ($n + n = 2n$). Without meiosis, gametes would have $2n$ chromosomes; after fertilization there would be $4n$ ($2n + 2n = 4n$). After each generation, the chromosome number for the species would double, and soon there would be no room for anything but chromosomes in the cell.

2. During spermatogenesis, four sperm cells are produced from each primary spermatocyte. In oogenesis, one ovum is produced along with three polar bodies.

3. The polar bodies produced during oogenesis die and disintegrate because they have very little cytoplasm and not enough organelles to carry out the life functions of the cell.

4. The major difference between meiosis and mitosis is that meiosis has two cell divisions, and mitosis has only one. The second cell division in meiosis reduces the number of chromosomes from $2n$ to n. Meiosis results in the production of four gametes, and mitosis produces two daughter cells.

Chapter Problems and Answers

Problems

The following is a brief paragraph based on cell reproduction. For problems 1–10, fill in the missing terms.

A student was looking through a compound microscope at a cell reproducing. He

observed that the nuclear membrane was disintegrating and that the nucleolus had

disappeared. The student concluded that the cell was in a stage of cell division known as

_____. Soon he was able to observe the presence of chromatids lining up on spindle fibers
    1

in the middle of the cell. The student also noted the absence of astral rays and centrioles.

This stage of cell division is called _____. After several minutes the chromatids began to
                                     2

separate and move to opposite sides of the cell. This is characteristic of a stage called

_____. Shortly thereafter, a _____ began to develop in the middle of the cell, beginning
    3   4

the formation of two cells. The cell was now in a stage of cell division known as _____.
                                                                                 5

The two cells that formed are called _____. After careful observation, the student noticed
                                    6

that cell division had stopped. The cell was now in the longest stage of its life cycle

known as _____. The process of cell division observed by the student was _____. The
          7   8

student was able to count six chromosomes in each of the two cells that formed. From

this he concluded that the original cell must have had _____ chromosomes. The cell the
                                                      9

student was looking at was a _____ cell.
                            10

Problems 11–15 are based on the following diagram; select the *best* answer.

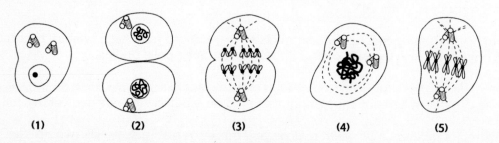

(1)   (2)   (3)   (4)   (5)

11. Which sequence shows the correct order of cell division?

  A. 1 → 4 → 5 → 3 → 2   B. 4 → 1 → 5 → 3 → 2
  C. 1 → 5 → 4 → 3 → 2   D. 2 → 1 → 4 → 5 → 3

12. Identify telophase.

    A. 1 B. 2 C. 3 D. 4 E. 5

13. Which stage is interphase?

    A. 1 B. 2 C. 3 D. 4 E. 5

14. The cell division shown is *most* probably for a:

    A. virus B. bacteria C. plant cell D. animal cell

15. The cell division shown is known as:

    A. mitosis B. meiosis C. sexual reproduction D. gametogenesis

For problems 16–20, select the process from the following list that is associated with each description. A choice can be used more than once or not at all.

    A. meiosis B. mitosis C. meiosis and mitosis D. fertilization E. synapsis

16. Select the process that restores the number of chromosomes to 2*n*. _____

17. Which process is characteristic of asexual reproduction? _____

18. Identify the type of cell division that results in a reduction in the species number of chromosomes from 2*n* to *n*. _____

19. In which process(es) does replication of the chromosomes take place? _____

20. Select the process(es) that results in spermatogenesis. _____

Answers

1. **Prophase.** During this stage or phase of cell division, the nuclear membrane disintegrates, the nucleolus disappears and chromatids begin to line up on the spindle fibers.

2. **Metaphase.** Chromatids begin to line up on the spindle fibers during prophase, but by the time the chromatids are aligned in the middle of the cell, metaphase is well under way.

3. **Anaphase.** Separation of the chromatids marks the beginning of anaphase. The end of anaphase occurs when the chromatids have reached the opposite sides of the cell.

4. **Cell plate.** The formation of a cell plate begins in the center of the cell and divides the cell in half. If the cell membrane pinches in to form two cells, a cell plate does not develop.

5. **Telophase.** The formation of two cells is the main characteristic of this stage.

6. **Daughter cells.** At the conclusion of telophase daughter cells are formed.

7. **Interphase.** This is the so-called resting stage of mitosis. The cell is performing all its life functions with the exception of reproduction.

8. **Mitosis.** Many clues in the paragraph point toward mitosis instead of meiosis. No mention is made of synapsis, tetrad formation or disjunction. These are processes that occur only in meiosis. Also, only one cell division was observed, which is characteristic of mitosis.

9. **Six.** In mitosis the number of chromosomes in a daughter cell is always equal to the number of chromosomes in the parent cell.

10. **Plant cell.** Two clues in the paragraph lead to the conclusion that this is plant-cell mitosis. First, the student observed the absence of astral rays and centrioles, structures that are found in animal cells but never in plant cells. Second, a cell plate forms. Cell plate formation occurs only in plant-cell mitosis.

11. **A.** This choice shows the progression of cell division from interphase to prophase to metaphase to anaphase to telophase.

12. **B.** Daughter-cell formation takes place during telophase.

13. **A.** This diagram shows a cell with an intact nuclear membrane, a nucleolus inside the nucleus, centrioles that are together and no spindle fibers or astral rays. All these observations are characteristics of interphase.

14. **D.** The presence of centrioles indicates an animal cell. None of the other choices contain centrioles.

15. **A.** The cell divided only one time. This is the key characteristic of mitosis or asexual reproduction. Also, we note the absence of synapsis, tetrad formation and disjunction—all characteristics that must be present to select meiosis.

16. **D. Fertilization.** During fertilization, a sperm cell (n) combines with an egg cell (n) to restore the species number of chromosomes ($2n$).

17. **B. Mitosis.** Cells reproduce asexually by mitosis.

18. **A. Meiosis.** Meiosis is also known as reduction division.

19. **C. Meiosis and mitosis.** In both of these processes, replication of the chromosomes takes place during interphase.

20. **A. Meiosis.** Meiosis results in gamete formation, spermatogenesis in the male and oogenesis in the female.

Supplemental Chapter Problems

Problems

For problems 1–20, select the *best* answer.

1. A major characteristic of asexual reproduction is that all the offspring usually:

   A. have the same hereditary material as the parent

   B. have different hereditary material from each other

   C. are different from the parent

   D. result from fertilization

2. In asexual reproduction, the number of chromosomes is kept constant from one generation to the next by:

   A. mitosis   B. meiosis   C. fertilization   D. gametogenesis

3. During mitosis, chromatids are held together by button-like structures called:

   A. chloroplasts   B. centrioles   C. chromosomes   D. centromeres

4. The star-shaped structures that form from centrioles during the process of mitosis are called:

   A. spindle fibers   B. astral rays   C. cell plates   D. vacuoles

5. Which event does *not* take place during the process of mitosis?

   A. chromosome replication   B. formation of spindle fibers

   C. tetrad formation   D. the breakdown of the nuclear membrane

6. A parent cell with four chromosomes undergoes mitosis to produce two daughter cells. How many chromosomes are in the nucleus of each daughter cell?

   A. 2   B. 4   C. 8   D. 16

7. At the completion of mitosis, the number of chromosomes in each daughter cell is:

   A. half the number of chromosomes in the parent

   B. the same as the number of chromosomes in the parent

   C. double the number of chromosomes in the parent

   D. four times the number of chromosomes in the parent

8. The cellular division of an amoeba to form two daughter cells is characteristic of a type of asexual reproduction known as:

   A. binary fission   B. budding   C. spore formation   D. vegetative propagation

9. The production by molds of many small cells, each surrounded by a protective wall, is known as:

   A. binary fission   B. budding   C. spore formation   D. vegetative propagation

10. A small hydra can develop from the side of a parent hydra by the process of:

    A. binary fission   B. budding   C. spore formation   D. mitosis

11. When a root, stem or leaf of an existing plant produces more plants of the same kind, the process is called:

    A. binary fission   B. budding   C. spore formation   D. vegetative propagation

12. A major characteristic of sexual reproduction is that all the offspring usually:

    A. have the same hereditary material

    B. have different hereditary material from their parents

    C. result from binary fission

    D. have double the hereditary material of their parents

13. In organisms that reproduce sexually, the diploid number of chromosomes is restored during the process of:

    A. mitosis   B. meiosis   C. fertilization   D. oogenesis

14. In an organism with a diploid number of chromosomes of 36, the haploid number of chromosomes equals:

    A. 9   B. 18   C. 36   D. 72

15. If a sperm cell in a male grasshopper has 12 chromosomes, how many chromosomes does an egg cell in a female grasshopper have?

    A. 6   B. 12   C. 24   D. 48

16. How many sperm cells are produced from one primary spermatocyte?

    A. 1   B. 2   C. 4   D. 8

17. In humans, the number of egg cells produced during oogenesis is usually:

    A. 1   B. 2   C. 4   D. 8

18. Homologous pairs of chromosomes are never found in:

    A. somatic cells   B. stomach cells   C. gametes   D. testes cells

19. The number of cell divisions in meiosis is:

    A. 1   B. 2   C. 4   D. 8

20. The female gonads are called the:

    A. testes   B. sperm cells   C. ovaries   D. ova

Answers

1. **A.** "Asexual Reproduction," p. 167

2. **A.** "Asexual Reproduction," p. 167

3. **D.** "Mitosis," p. 171

4. **B.** "Mitosis," p. 171

5. **C.** "Mitosis," pp. 171–172

6. **B.** "Asexual Reproduction," p. 167

7. **B.** "Asexual Reproduction," p. 167

8. **A.** "Examples of Asexual Reproduction," p. 169

9. **C.** "Examples of Asexual Reproduction," p. 169

10. **B.** "Examples of Asexual Reproduction," p. 169

11. **D.** "Examples of Asexual Reproduction," p. 169

12. **B.** "Sexual Reproduction," p. 173

13. **C.** "Sexual Reproduction," p. 173

14. **B.** "Sexual Reproduction," p. 173

15. **B.** "Sexual Reproduction," p. 173

16. **C.** "Sexual Reproduction," p. 175

17. **A.** "Sexual Reproduction, p. 175

18. **C.** "Sexual Reproduction," p. 173

19. **B.** "Meiosis," p. 175

20. **C.** Chapter 6, "Ovaries," p. 142

Chapter 9
Human Reproduction

The continuity of life depends on the life function of reproduction. For a species to survive, some of its members must reproduce; otherwise, the species will become extinct. In humans, reproduction is sexual. A sperm cell produced in the testes of a male combines with an egg cell produced in the ovaries of a female to form a fertilized egg cell called a *zygote*. Eventually the zygote develops into a baby, which grows into an adult, completing the human life cycle.

The Male Reproductive System

The human male reproductive system is specialized for the production of sperm cells and their delivery to the female's reproductive system so that an egg can be fertilized and a zygote produced. The following is a list of organs in the male reproductive system:

Male Reproductive System

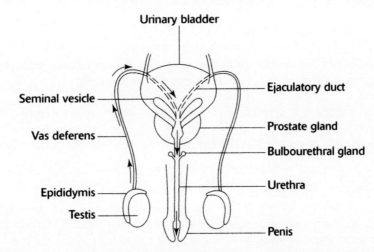

- ❏ **Testes**—The testes are the male gonads; they produce sperm cells by gametogenesis inside coiled tubes known as the *seminiferous tubules*. The testes are inside the *scrotum*, which is a sac located outside the body. The temperature in the scrotum is about 2–3 degrees (Celsius) lower than normal body temperature, which is necessary for proper sperm-cell production and development. Mature sperm cells are among the smallest cells in the body, about 60 micrometers in length. The testes also produce the male sex hormone testosterone. From the testes, sperm enter the epididymis.

- ❏ **Epididymis**—This is a coiled structure located on the top and sides of the testes that stores sperm cells until they are ready to enter the vas deferens.

❑ **Vas deferens (sperm duct)**—The vas deferens is a long tube that carries sperm cells from the epididymis to the *ejaculatory duct* and into the *urethra,* which is located in the penis. The ejaculatory duct propels sperm cells through the urethra and out of the body. However, before leaving the body the sperm cells combine with fluid from the seminal vesicles, prostate gland and bulbourethral glands to produce *semen.*

❑ **Seminal vesicles**—The seminal vesicles are two sac-like structures located above the prostate gland, which produce a fluid that goes into the ejaculatory duct. Seminal fluid activates sperm cells and gives them motility.

❑ **Prostate gland**—This gland is located below the urinary bladder. The fluid from the prostate gland neutralizes the acid pH of the vagina. This is important because acid conditions kill sperm cells.

❑ **Bulbourethral glands (Cowper's glands)**—These two glands are located below the prostate and secrete their fluid into the urethra. The fluid from Cowper's glands neutralizes the acid pH of the urethra.

❑ **Urethra**—The urethra is located in the penis and has a dual function in males. During reproduction it carries sperm cells out of the body, and during excretion it carries urine out of the body. When reproduction occurs, the urethra is pinched closed at the urinary bladder, stopping urine from entering and killing the sperm cells. Sperm cells are ejaculated from the *penis* and deposited into the *vagina* of the female reproductive system.

Example Problems

These problems review the structure and function of the male reproductive system.

1. What process produces sperm cells?

    Answer: Sperm cells are produced in the testes by the process of gametogenesis. During meiosis one primary spermatocyte produces four sperm cells.

2. Where in the reproductive system of the male are sperm cells stored?

    Answer: Sperm cells are stored in the epididymis, where they mature before entering the vas deferens.

3. What contribution do the seminal vesicles make to the reproductive process?

    Answer: The seminal vesicles produce one of the three fluids that make up semen. Prior to combining with seminal fluid, sperm cells are not motile. Seminal fluid activates sperm cells and gives them motility.

4. Why is the prostate gland important?

    Answer: The prostate gland produces the second of the three fluids that make up semen. The fluid from the prostate gland contains a substance that neutralizes the acid pH of the vagina, which would otherwise kill sperm cells.

5. State the function of bulbourethral glands.

    Answer: Bulbourethral glands produce the third fluid found in semen, which contains a substance that neutralizes the acid conditions found in the male's urethra.

Work Problems
Use these problems on the male reproductive system for additional practice.

1. Why must the testes be located in the scrotum?

2. What is the dual function of the testes?

3. Describe the dual function of the urethra.

4. What substances are normally found in semen?

5. Outline the pathway that sperm must take to leave the body of a male.

Worked Solutions

1. The scrotum is a sac located outside the body cavity. The temperature in the scrotum is a few degrees lower than normal body temperature, which is necessary for proper sperm-cell production and development.

2. The testes produce sperm cells for reproduction and the sex hormone testosterone, which regulates male secondary sex characteristics.

3. The urethra has two jobs; it carries sperm cells out of the body for reproduction, and during excretion it helps remove urine from the body.

4. Semen contains sperm cells and fluid from the seminal vesicles, prostate gland and bulbourethral glands.

5. Sperm are produced in the testes, stored in the epididymis and then move into the ejaculatory duct. The ejaculatory duct connects to the urethra, which takes the sperm out of the body.

The Female Reproductive System

The human female reproductive system is specialized for egg-cell production, fertilization and development of the zygote into a new individual. Since humans are land animals, fertilization is internal (inside the female reproductive system). The following is a list of organs in the female reproductive system:

Female Reproductive System

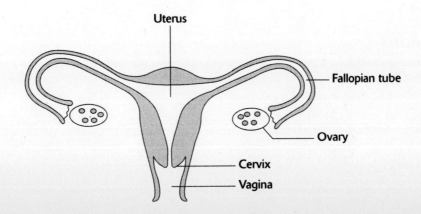

❑ **Ovaries**—A woman has two ovaries, one on each side of the uterus. The ovaries produce egg cells by the process of oogenesis. At the time of birth, a baby girl has approximately 200,000 immature egg cells in each ovary. When a young woman begins to have menstrual cycles at the time of *puberty* (sexual development), an egg is matured and released during each cycle. A woman's first menstrual cycle is known as menarche, and the last cycle is called menopause. The ovaries produce the hormones estrogen and progesterone that function in the menstrual cycle. The release of an egg cell from the ovary is called *ovulation*. The released egg enters one of the fallopian tubes.

❑ **Fallopian tubes**—These narrow tubes are lined with cilia and connect the ovaries to the uterus. Two things can happen to an egg while it is in a fallopian tube: It can be fertilized or not. Usually the egg is not fertilized, and the cilia in the tube sweep the egg toward the uterus where it disintegrates. If fertilization takes place, the zygote begins to divide and eventually implants itself into the wall of the uterus. Only one sperm cell can fertilize an egg. Immediately after a zygote is produced, a *fertilization membrane* forms that prevents other sperm cells from entering the egg.

❑ **Uterus (womb)**—The uterus is a muscular organ where an embryo forms and develops into a fetus, a process that takes about nine months from the time of fertilization. The *cervix* is the lower part of the uterus and opens into the vagina.

❑ **Vagina**—The vagina is also known as the birth canal from which a fully developed fetus is delivered. Sperm cells deposited into the vagina must swim past the cervix, into the uterus and into the fallopian tube that has the egg for fertilization to take place.

Example Problems

These problems review the structure and function of the female reproductive system.

1. What process produces an egg cell?

   Answer: Egg cells are produced in the ovaries by the process of oogenesis. During meiosis one primary oocyte produces one egg cell and three polar bodies. This process takes place before a baby girl is born. The purpose of the menstrual cycle is to mature and release an egg for fertilization.

2. Where does fertilization take place?

   Answer: In humans fertilization takes place in a fallopian tube.

3. What happens to the egg cell if it is not fertilized?

   Answer: If an egg cell is not fertilized, it disintegrates and dies.

Work Problems

Use these problems on the female reproductive system for additional practice.

1. What is the dual function of the ovaries?

2. Where does development take place?

3. Outline the pathway that sperm must take in the female reproductive system to fertilize an egg.

4. Why can't an egg cell be fertilized by more than one sperm cell?

Worked Solutions

1. The ovaries produce egg cells for reproduction and hormones for regulation. The hormone estrogen regulates female secondary sex characteristics and stimulates the uterus to start producing its capillary lining. Progesterone maintains the uterine lining during the menstrual cycle.

2. Shortly after fertilization a zygote begins to divide, and the several cells that form implant themselves into the wall of the uterus where an embryo eventually develops. In the uterus, the embryo develops into a fetus.

3. The penis deposits sperm cells into the vagina, and if an egg cell is present, the sperm swim through the cervix, into the uterus and finally into the fallopian tube that contains the egg. Fertilization takes place in the fallopian tube.

4. After a zygote is produced, a fertilization membrane forms around the zygote, preventing other sperm cells from entering the egg. The head of a sperm cell contains enzymes that can break down the membrane that surrounds the egg. However, these enzymes cannot break down the fertilization membrane because it is made of different proteins, and enzyme action is specific.

The Menstrual Cycle

The menstrual cycle is a hormonal cycle designed to achieve two goals. First, an egg cell must be matured and released. Second, a special lining of capillaries is prepared inside the uterus in anticipation of receiving a fertilized egg. This capillary lining helps supply the embryo/fetus with food, oxygen and water. The menstrual cycle averages about 28 days in humans and is characterized by four distinct stages:

Menstrual Cycle

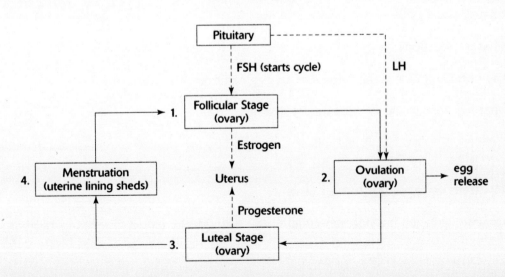

1. **Follicular stage**—During this stage of the menstrual cycle, *Follicle Stimulating Hormone (FSH)* produced by the anterior pituitary gland targets cavities in the ovaries called *follicles,* where the immature egg cells are found. One or more follicles begin to mature an egg cell. The egg cell that matures first is the one that is eventually released. FSH also stimulates the ovary to begin the production of the hormone estrogen. Estrogen targets the uterus and stimulates it to begin preparing a special lining with many capillaries to receive the fertilized egg. As the level of estrogen in the blood increases, a negative feedback mechanism occurs. The pituitary gland stops FSH production and begins to make *Luteinizing Hormone (LH).*

2. **Ovulation stage**—As the level of LH in the blood increases, it stimulates one of the follicles in the ovary to rupture and release the egg. The ovary repairs the damage by producing a mass of cells in the follicle called a *corpus luteum.*

3. **Luteal stage**—The corpus luteum produces the hormone progesterone, which targets the uterus and instructs it to maintain its lining. As the level of progesterone in the blood increases, a negative feedback mechanism occurs. The pituitary gland eventually stops LH production, which results in a sharp decrease of progesterone from the corpus luteum. The luteal stage is now over.

4. **Menstruation**—At this stage in the cycle, estrogen and progesterone levels are very low, resulting in a breakdown of the uterine lining, which passes out of the vagina. However, if the egg is fertilized in a fallopian tube and implantation occurs in the uterus, progesterone produced by the developing embryo maintains the uterine lining throughout pregnancy.

Example Problems

These problems review the menstrual cycle.

1. Which hormone starts the menstrual cycle?

   Answer: FSH produced by the anterior pituitary gland starts the menstrual cycle by stimulating follicles to mature an egg and by stimulating the ovary to begin the production of estrogen.

2. Where are follicles found?

   Answer: Follicles are cavities located in the ovaries where egg cells are matured during the menstrual cycle.

3. What is ovulation?

   Answer: Ovulation is the release of an egg cell from the ovary.

4. What happens to the follicle after ovulation?

   Answer: After ovulation, the follicle is repaired by the growth of a mass of hormone-producing cells called the corpus luteum.

5. How does estrogen get to the uterus?

   Answer: Estrogen is a hormone that travels through the blood to reach its target.

Work Problems

Use these problems on the menstrual cycle for additional practice.

1. What are the characteristics of the follicular stage?

2. Which hormones regulate the uterine lining?

3. What causes the breakdown in the uterine lining?

4. How is the uterine lining maintained during pregnancy?

5. How can we help a woman who is having difficulty ovulating?

Worked Solutions

1. The follicular stage is the first stage of the menstrual cycle. This stage is dominated by the presence of the follicle. The follicle begins to mature an egg and produce the hormone estrogen.

2. Estrogen begins the development of the uterine lining. Progesterone is responsible for maintaining the uterine lining.

3. When the level of progesterone is low, the uterine lining begins to break down.

4. Progesterone produced by the developing embryo during pregnancy maintains the uterine lining.

5. For a woman to ovulate, an egg must be present and then released. Giving a woman FSH stimulates her follicles to mature an egg; giving her LH results in the release of the egg. FSH and LH are the two hormones found in fertility pills.

Embryological Development

After fertilization, a zygote undergoes a special series of mitotic cell divisions known as *cleavage*. During cleavage the cells divide, getting smaller and smaller until they reach the size of a typical adult somatic cell. Eventually, cleavage results in the development of an embryo.

The following diagram shows the start of embryological development (fertilization and cleavage).

Fertilization and Cleavage

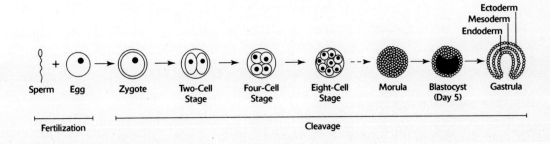

Several stages in this process merit further discussion.

- ❏ **Two-cell stage**—If the cells separate during the two-cell stage, each can develop into a fetus, resulting in the formation of *identical twins. Fraternal twins* result when two eggs are released and each is fertilized with a different sperm cell. Fraternal twins are different from each other genetically.

- ❏ **Morula**—This is the solid-ball-of-cells stage of development.

- ❏ **Blastocyst (blastula in other animals)**—This stage takes place about five days after fertilization and is basically a hollow ball of cells with fluid and an inner mass of cells at one end. The blastocyst implants itself into the wall of the uterus. The cells inside the blastocyst are *stem cells*, which can potentially form any cells or tissues in the body.

- ❏ **Gastrula**—This stage forms about 14 days after fertilization and is characterized by the presence of three cell layers. This stage is critical for the development of the embryo. The cells in each layer become different from each other (we call this *differentiation*) and give rise to various organs and systems of the body. The outer layer of cells is called the *ectoderm*, which forms the skin and nervous system. The middle layer of cells is the *mesoderm*, which produces the skeletal, circulatory and reproductive systems. The inner layer of cells is the *endoderm*, which gives rise to the organs of the digestive and respiratory systems.

Example Problems

These problems review embryological development.

1. What is cleavage?

   Answer: Cleavage is a special series of mitotic cell divisions, where each time the cells divide they get smaller and smaller. The size of the morula is about the same size as the zygote, yet the morula is a solid ball composed of many cells.

2. Where does embryological development take place?

   Answer: Embryological development begins in the fallopian tube with the first cleavage division. By the fifth day the blastocyst is implanted into the wall of the uterus, where embryological development continues until the eighth week, when the embryo is called a fetus.

3. Define differentiation.

   Answer: Differentiation occurs at some point between the blastocyst stage and the gastrula stage and results in the cells of the developing embryo becoming different from each other. Certain cells form the eye, others the brain and so on.

Work Problems

Use these problems on embryological development for additional practice.

1. How are fraternal twins formed?

2. How are the cells of the gastrula different from those of the blastocyst?

3. Why are stem cells important?

4. Which organ systems develop from each of the three layers in the gastrula?

5. When in human development does sexual reproduction take place?

Worked Solutions

1. Sometimes when a woman ovulates, two **egg** cells are released (each from its own follicle). If each egg cell is fertilized by a different sperm cell, fraternal twins result. Fraternal twins are different from each other genetically.

2. The cells of the gastrula are different from each other. Each cell is programmed to become a specific type of tissue or organ in the body. The cells of the blastocyst are undifferentiated and can form any type of tissue or organ in the body.

3. Stem cells are important because they have the potential of being removed from the blastocyst and used to create tissue and organs that can serve as replacement parts for damaged human organs.

4. The three layers of the gastrula are the ectoderm, mesoderm and endoderm. The ectoderm layer forms the skin and nervous system. The mesoderm layer forms the skeletal, circulatory and reproductive systems. The endoderm produces the organs of the digestive and respiratory systems.

5. Sexual reproduction takes place at fertilization, when a sperm cell and an **egg** cell combine to form a zygote. After fertilization, human development takes place asexually by mitosis.

Fetal Development

After eight weeks, the developing embryo is called a fetus, which in Latin means young one.

The following diagram shows a fetus developing inside the uterus.

Fetal Development

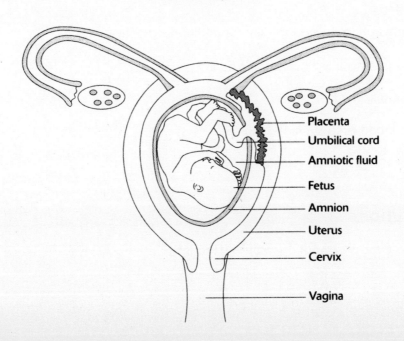

- Placenta
- Umbilical cord
- Amniotic fluid
- Fetus
- Amnion
- Uterus
- Cervix
- Vagina

The *placenta* is a combination of tissue from the mother's uterine wall and from the developing fetus. Cells in the placenta produce progesterone, which maintains the uterine lining during pregnancy. The fetus gets food, oxygen and water from the placenta. These materials are carried to the fetus by the *umbilical cord,* which connects the fetus to the placenta. The umbilical cord carries waste materials away from the fetus to the placenta. The fetus has its own blood supply and circulatory system. At no time do blood cells from the fetus or the mother pass through the placenta. The *amnion* is a sac that contains a fluid called *amniotic fluid.* The sac and its fluid protect the fetus from shock.

During birth the amnion breaks, and the amniotic fluid passes out of the mother's body. Uterine contractions force the baby out of the vagina (birth canal). The period of time from fertilization to birth is known as *gestation* and lasts approximately nine months.

Example Problems

These problems review fetal development.

1. What is a fetus?

   Answer: A fetus is the name given to an embryo after the eighth week of development.

2. Why is the amnion important?

   Answer: The amnion is a sac that contains a fluid called amniotic fluid. The sac and its fluid protect the fetus from shock.

3. What is the function of the umbilical cord?

   Answer: The umbilical cord connects the fetus to the placenta and is the fetus's *lifeline.* The fetus gets its food, oxygen and water from the placenta through the umbilical cord. Wastes from the fetus pass through the umbilical cord to the placenta.

Work Problems

Use these problems on fetal development for additional practice.

1. How is the uterine lining maintained during pregnancy?

2. What is the placenta?

3. How does blood circulate in the fetus?

4. What is meant by the term *gestation?*

Worked Solutions

1. Cells in the placenta produce progesterone, which maintains the uterine lining during pregnancy.

2. The placenta is a combination of tissue from the mother's uterine wall and from the developing fetus.

3. The fetus has its own circulatory system, heart and blood cells for the transport of materials. The baby's blood and the mother's blood never mix with each other during development.

4. Gestation is the period of time from fertilization to birth and lasts approximately nine months.

Chapter Problems and Answers

Problems

The following is a brief paragraph based on human reproduction and development. For problems 1–10, fill in the missing terms.

The continuity of life depends on the life function of reproduction. In humans sexual

reproduction begins when a sperm cell and an egg cell combine during the process of

_____. This takes place in a part of the female reproductive system known as the _____.
  1   2

The fertilized egg cell is called a _____ and begins to divide by a type of mitosis known
                                                  3

as _____. After several days, a _____ forms that implants itself in the wall of the mother's
        4   5

_____. Approximately eight weeks later, the developing embryo is called a fetus. Food,
    6

water and oxygen are transferred to the fetus by the _____. This structure is connected to
                                                                            7

the fetus by the _____. During development the fetus is surrounded by a membrane or
                      8

sac called the _____, which contains a fluid that protects the fetus from _____.
                    9   10

For problems 11–15, name the structure of the male reproductive system that is associated with each function.

11. Identify the coiled tubes in the testes where spermatogenesis takes place.
_____

12. Which tube in the male carries sperm cells to the ejaculatory duct?
_____

13. Name the organ(s) that produces a fluid capable of activating sperm cells to make them motile. _____

14. Which structure provides a suitable environment for spermatogenesis?
_____

15.   This tube carries semen out of the body. _____

For problems 16–20, select the structure of the female reproductive system from the following list that is associated with each function. A choice can be used more than once or not at all.

A. cervix   B. fallopian tubes   C. ovaries   D. uterus   E. vagina

16.   The muscular organ where the placenta forms. _____

17.   Which structure is the lower part of the uterus? _____

18.   Select the structure that is also an endocrine gland. _____

19.   Which structure contains cilia, which move egg cells? _____

20.   Identify the organ where embryological and fetal development take place. _____

For problems 21–25, select the hormone from the following list that is associated with each regulatory function. A choice can be used more than once or not at all.

A. estrogen   B. FSH   C. LH   D. progesterone   E. testosterone

21.   Identify the hormone that is responsible for maintaining the uterine lining during pregnancy. _____

22.   Which *two* hormones are produced by the pituitary gland?
      _____

23.   Select the hormone that is directly responsible for ovulation.
      _____

24.   Which hormone stimulates the follicle to begin maturing an egg?
      _____

25.   Which hormone is usually not associated with the menstrual cycle?
      _____

Answers

1.   **Fertilization.** The fusion (combining) of gametes is known as fertilization.

2.   **Fallopian tube.** In humans, fertilization takes place in the fallopian tubes. If an egg cell is not fertilized here, it dies and disintegrates.

3.   **Zygote.** Fertilization of an egg cell by a sperm cell results in the formation of a zygote.

4.   **Cleavage.** Cleavage is a special type of mitosis. With each division, the cells get smaller and smaller. However, the overall size of the cell mass remains the same as the size of the zygote.

5.   **Blastocyst.** The blastocyst contains fluid and undifferentiated cells that are capable of forming any tissues or organs of the body.

6.   **Uterus.** Embryological and fetal development take place in the uterus.

7. **Placenta.** The placenta is a combination of tissue from the mother's uterine wall and from the developing embryo or fetus.

8. **Umbilical cord.** This is the lifeline of the fetus, bringing it needed materials from the placenta.

9. **Amnion.** During birth, this membrane breaks as the uterus contracts. When this happens, a woman's "water" has broken.

10. **Shock.** If a pregnant woman falls or receives a blow to her abdomen, the amniotic sac and its fluid will protect the fetus from shock.

11. **Seminiferous tubules.** The seminiferous tubules are tightly coiled and make up the major part of each testis. Each primary spermatocyte can produce four sperm cells, which mature in the epididymis.

12. **Vas deferens.** The vas deferens carries sperm cells from the epididymis to the ejaculatory duct. The sperm cells combine with fluids produced by other organs to form semen.

13. **Seminal vesicles.** The fluid of the seminal vesicles contains an activating agent that gives sperm cells their motility.

14. **Scrotum.** The scrotum is a sac located outside the body cavity and is 2°–3° C cooler than the rest of the body. The lower temperature is needed to produce healthy sperm cells.

15. **Urethra.** The urethra is located in the penis, and during reproduction it carries semen out of the body.

16. **D. Uterus.** The placenta is a structure that forms partly from tissue in the mother's uterus and partly from tissue formed by the developing embryo.

17. **A. Cervix.** The cervix is the lower part of the uterus and extends into the upper part of the vagina.

18. **C. Ovaries.** Endocrine glands produce hormones; the ovaries produce the hormones estrogen and progesterone.

19. **B. Fallopian tubes.** After ovulation, an egg cell enters a fallopian tube where cilia sweep it along toward the uterus. Egg cells are sessile and cannot move on their own.

20. **D. Uterus.** The first few cleavage divisions take place in a fallopian tube; however, embryological and fetal development take place in the uterus.

21. **D. Progesterone.** Progesterone produced during pregnancy comes from the placenta.

22. **B. FSH** and **C. LH.** FSH and LH are the hormones produced by the pituitary gland that regulate the menstrual cycle. Estrogen and progesterone are the ovarian hormones that regulate the menstrual cycle.

23. **C. LH.** LH stimulates the follicle to release an egg on or about day 14 of the menstrual cycle.

24. **B. FSH.** FSH produced by the pituitary gland stimulates the follicles in the ovaries to begin maturing an egg cell.

25. **E. Testosterone.** Testosterone is the male sex hormone and has nothing to do with regulation of the menstrual cycle.

Supplemental Chapter Problems

Problems

For problems 1–25, select the *best* answer.

1. Select the structure that can produce motile gametes.

    A. ovary　　B. testis　　C. uterus　　D. prostate

2. Which duct do sperm travel through to leave the male reproductive system?

    A. ureter　　B. urethra　　C. uterus　　D. fallopian tube

3. Where are sperm cells never found?

    A. urethra　　B. vas deferens　　C. fallopian tube　　D. ovary

4. Select the organ that does *not* produce a fluid found in semen.

    A. prostate　　B. bulbourethral glands　　C. seminal vesicles　　D. testes

5. Which organ produces both gametes and hormones?

    A. testes　　B. prostate gland　　C. epididymis　　D. bulbourethral glands

6. The structure in the male reproductive system that provides the best temperature environment for sperm-cell production is the:

    A. penis　　B. scrotum　　C. urethra　　D. prostate

7. All the following can take place in the testes *except:*

    A. meiosis　　B. mitosis　　C. gametogenesis　　D. fluid production for semen

8. A woman's first menstrual cycle is known as:

    A. menarche　　B. menopause　　C. ovulation　　D. oogenesis

9. Which structure in the female reproductive system is the first to receive sperm cells from the male?

    A. vagina　　B. cervix　　C. fallopian tube　　D. uterus

10. Where in the female reproductive system does fertilization take place?

    A. vagina　　B. cervix　　C. fallopian tube　　D. uterus

11. Which structure in the female reproductive system is the first to receive an egg cell following ovulation?

    A. vagina　　B. cervix　　C. fallopian tube　　D. uterus

12. Regulation of the menstrual cycle depends on the interaction of hormones produced by the pituitary and the:

    A. vagina　　B. ovary　　C. uterus　　D. cervix

13. Which stage of the menstrual cycle does *not* take place if the egg cell is fertilized?

    A. ovulation　　B. luteal stage　　C. follicular stage　　D. menstruation

14. Select the stage of the menstrual cycle characterized by egg-cell maturation.

A. ovulation   B. luteal stage   C. follicular stage   D. menstruation

15. Select the sequence that represents the correct order of stages in the menstrual cycle.

A. menstruation → luteal stage → ovulation → follicular stage

B. ovulation → follicular stage → luteal stage → menstruation

C. follicular stage → ovulation → luteal stage → menstruation

D. menstruation → ovulation → follicular stage → luteal stage

16. Identify the structure in the female reproductive system that is specialized for embryological and fetal development.

A. vagina   B. cervix   C. fallopian tube   D. uterus

17. How does a developing human embryo obtain food, water and oxygen?

A. directly from the external environment

B. by absorption from the placenta into the umbilical cord

C. directly from the mother's blood, which is circulating through the embryo

D. by diffusion from the cervix

18. Which stage of cleavage is characterized by the presence of stem cells?

A. two-cell stage   B. gastrula   C. blastocyst   D. morula

19. Fraternal twins develop from:

A. two eggs, each fertilized by a separate sperm cell

B. two eggs fertilized by the same sperm cell

C. one egg fertilized by one sperm cell

D. one egg fertilized by two sperm cells

20. Differentiation of cells takes place during a stage of cleavage known as the:

A. gastrula   B. blastocyst   C. morula   D. two-cell stage

21. Select the embryonic layer produced during cleavage that develops into the skeletal and muscular systems of the body.

A. epidermis   B. ectoderm   C. endoderm   D. mesoderm

22. The period of time from zygote formation to birth is known as:

A. gastrulation   B. gestation   C. regulation   D. circulation

23. Which structure connects the fetus to the placenta?

A. aorta   B. cervix   C. umbilical cord   D. fallopian tube

24. This structure is formed from a combination of tissue from the mother and the developing fetus.

A. umbilical cord   B. ovary   C. placenta   D. vagina

25. Identify the structure that is a protective membrane surrounding the fetus.

A. scrotum   B. amnion   C. vagina   D. uterus

Answers

1. **B.** "The Male Reproductive System," p. 183

2. **B.** "The Male Reproductive System," p. 183

3. **D.** "The Male Reproductive System," pp. 183–184

4. **D.** "The Male Reproductive System," p. 184

5. **A.** "The Male Reproductive System," p. 183

6. **B.** "The Male Reproductive System," p. 183

7. **D.** "The Male Reproductive System," pp. 183–184

8. **A.** "The Female Reproductive System," p. 186

9. **A.** "The Female Reproductive System," p. 186

10. **C.** "The Female Reproductive System," p. 186

11. **C.** "The Female Reproductive System," p. 186

12. **B.** "The Menstrual Cycle," p. 186

13. **D.** "The Menstrual Cycle," p. 188

14. **C.** "The Menstrual Cycle," p. 188

15. **C.** "The Menstrual Cycle," p. 188

16. **D.** "Embryological Development," pp. 185–186, 191

17. **B.** "Embryological Development," pp. 191–192

18. **C.** "Embryological Development," p. 190

19. **A.** "Embryological Development," p. 190

20. **A.** "Embryological Development," p. 190

21. **D.** "Embryological Development," p. 190

22. **B.** "Fetal Development," p. 192

23. **C.** "Fetal Development," p. 192

24. **C.** "Fetal Development," p. 192

25. **B.** "Fetal Development," p. 192

Chapter 10
Plant Reproduction

Plants can reproduce asexually or sexually. Plants that reproduce asexually utilize a form of reproduction known as *vegetative propagation*. All plants that reproduce sexually have a type of reproduction known as *alternation of generations*. For plants in the phylum Anthophyta, the *gametophyte generation* is characterized by the production of male and female haploid (n) gametes. The alternate generation is the *sporophyte generation*, which is characterized by the fertilization of gametes and formation of a zygote (2n) that eventually becomes a sporophyte or plant. The sections below explain plant asexual and sexual reproduction.

Vegetative Propagation

Vegetative propagation is a form of asexual reproduction in plants, where a new plant develops from the root, stem or leaf of an already existing plant. No seeds are involved in this process. Vegetative propagation can occur naturally or can be artificially applied by humans to produce new plants.

Propagation from Roots

The sweet potato and kudzu are just two examples of plants propagated from roots. The sweet potato is an enlarged underground root. When placed in water or moist soil, *shoots* (young plants) sprout from this root. These shoots can be removed and replanted to form new sweet potato plants.

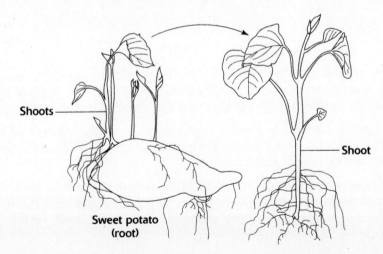

Shoots

Shoot

Sweet potato
(root)

Kudzu is a climbing vine in the pea family and is found in many southeastern states. Kudzu roots are thick and can exceed 1 meter in length. Each root can have up to 30 vines growing out of it. Kudzu vines can grow about a foot per day during the growing season, and many vines exceed 10 meters in length.

Propagation from Stems

Several kinds of propagation from stems are: runners (stolons), rhizomes, tubers and bulbs.

❏ **Runners (stolons):** A runner is an aboveground stem that grows horizontally along the surface when the stem touches the ground. Buds from the stem form roots and leaves, and a new plant develops. Strawberry and spider plants can reproduce this way.

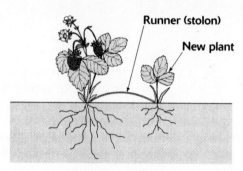

Strawberry Runner

❏ **Rhizomes:** A rhizome is an underground stem that grows horizontally along the surface; shoots from buds at *nodes* (enlarged areas) of the stem develop into a new plant. Examples are irises and ferns.

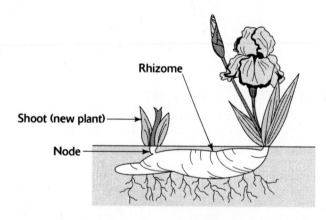

Iris Rhizome

❏ **Tubers:** Tubers are enlarged and thickened underground stems. The white potato is an example of a tuber. If pieces of potato with buds ("eyes") are cut out and planted in soil, new potato plants can form. Each potato plant can develop many underground tubers.

❏ **Bulbs:** Bulbs have a short, underground stem and thick, fleshy leaves that are colorless. A bulb can produce several smaller bulbs, each of which can grow into a new plant. Tulips, onions and daffodils can reproduce this way.

❏ **Corms:** A corm is an enlarged, short, underground stem that stores food. Corms resemble bulbs but do not have thick, fleshy leaves. New corms develop from the stem of an old corm to produce new plants. Gladiolus and crocus reproduce from corms.

Propagation from Leaves

Some plants have the ability to form a new plant from a leaf. Two examples are *Bryophyllum* and the African violet. If a *Bryophyllum* leaf falls on moist soil, new plants can develop from the edges of the leaf. An African violet leaf placed in moist soil or sand can grow into a new plant.

Example Problems

These problems review vegetative propagation.

1. What is vegetative propagation?

   Answer: Vegetative propagation is a form of asexual reproduction by plants. New plants are produced from the root, stem or leaf of an already existing plant.

2. How can one sweet potato produce many new plants?

   Answer: A sweet potato placed in water develops many young plants called shoots from the enlarged root. Each of these can produce a new sweet potato plant.

3. How can the eyes of a potato produce new potato plants?

   Answer: The eyes of a white potato are actually buds. When one of these buds is removed along with a section of the potato and put into the ground, a new potato plant forms. The starch in the potato serves as food for the developing plant until leaves form and the process of photosynthesis begins, enabling the young plant to produce its own food.

4. What is the difference between a corm and a bulb?

   Answer: Bulbs have a short underground stem with thick, fleshy leaves; corms do not have thick, fleshy leaves.

Work Problems

Use these problems on vegetative propagation for additional practice.

1. Why is vegetative propagation considered asexual reproduction?

2. How is a runner different from a rhizome?

3. How can one onion bulb produce many new plants?

4. How does the *Bryophyllum* plant reproduce asexually?

Worked Solutions

1. Vegetative propagation is considered asexual reproduction because new plants are produced from part of an existing plant. Gamete production and seed formation do not take place in vegetative propagation.

2. A runner is an aboveground stem that grows horizontally along the surface when the stem touches the ground; buds from the stem produce a new plant. A rhizome is an underground stem that grows horizontally along the surface; shoots from buds in the stem develop into new plants.

3. A bulb can produce several smaller bulbs at the base of its short stem. Each of these small bulbs can develop into a new plant.

4. When a *Bryophyllum* leaf falls on moist soil, new "baby" plants can form from the edges of the leaf.

Artificial Propagation

Artificial propagation occurs when humans take part of an already existing plant to produce new plants. Many plants in the United States that are bought each year by consumers have been produced by artificial propagation. In the United States the green industry is a multibillion-dollar industry that relies in large part on the advantages of vegetative propagation.

Cuttings

A root, stem or leaf is cut off an existing plant and put into water, moist sand or soil; after several weeks a new plant develops. Shoots from a sweet potato plant grown in water can be removed and planted in soil to get new sweet potato plants. A branch cut from a geranium plant and placed in water eventually grows roots. At this point the cutting can be placed in soil. A *Bryophyllum* leaf can be detached from an existing plant and put on top of soil to grow many new plants.

Layering

Plants such as roses, blackberries and raspberries produce long branches that can be bent into the ground and covered with a layer of soil. Roots and new stems form on a small part of the bent branch that is left sticking out of the soil. The bent branch is now cut to separate the new plant from the "parent."

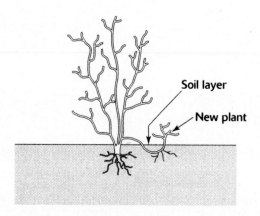

Grafting

Grafting is one of the most important methods of artificial vegetative propagation. Grapes, apples, oranges and many varieties of fruit trees are produced using this technique. In this process a stem is cut off a plant and attached to the stem of a plant that is rooted in the ground. The rooted stem is called the stock, and the stem that is attached to the stock is called the scion. The scion is attached to the stock so that their cambium layers are in contact with each other. The fruit produced by a plant developed from grafting is the fruit of the scion. For example, a scion from a seedless orange tree grafted onto a seeded orange stock produces only seedless oranges.

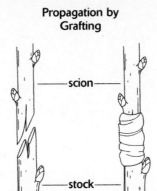

Propagation by
Grafting

scion

stock

Advantages of Vegetative Propagation

The following list shows some advantages of vegetative propagation.

❏ **New plants are identical to the parent.** Vegetative propagation is a form of asexual reproduction that produces offspring that are identical to the parent. For example, a stem cutting
from a geranium that has pink flowers produces a new plant that has exactly the same kind
of pink flowers. An apple tree produced by grafting bears exactly the same type of apples
as the tree that the scion came from.

❏ **New plants are produced quickly.** A plant produced by grafting matures and bears fruit
sooner than one produced from a seed because it starts from a stock that is already a young
plant. In layering and cutting, the plant gets a head start by developing from the stem of an
already existing plant. In all cases, plants grown by vegetative propagation develop and
mature faster than plants started from seed.

❏ **Seedless varieties of fruits can be propagated.** Taking a scion from a seedless plant and
grafting it onto a seeded variety is the only way to propagate seedless varieties of fruits.

Example Problems

These problems review artificial propagation and the advantages of vegetative propagation.

1. What is a cutting?

    Answer: A cutting is a root, stem or leaf that is cut off an existing plant and put into
    water, moist sand or soil to produce a new plant.

2. How are rosebushes produced by layering?

    Answer: Roses have long branches that can be bent into the ground and covered with a
    layer of soil so that a small part of the branch is left sticking out. Roots and stems form
    from the bent branch. When the bent branch is cut, two separate rose plants are obtained.

3. How can grafting be used to propagate new plants?

    Answer: In grafting, a stem (scion) is cut off a plant and attached to the stem of a plant
    that is rooted in the ground (stock). The fruit produced by this combination is the fruit of
    the scion.

Work Problems

Use these problems on artificial propagation and the advantages of vegetative propagation for additional practice.

1. How is natural vegetative propagation different from artificial vegetative propagation?

2. How are seedless grapes propagated?

3. Why would a farmer prefer to plant apple trees produced by grafting rather than trees produced from seed?

4. What is a disadvantage of vegetative propagation?

Worked Solutions

1. In natural vegetative propagation, a plant reproduces asexually on its own from a part of itself. In artificial vegetative propagation, humans take roots, stems or leaves from existing plants to produce new plants.

2. Seedless grapes are propagated vegetatively by grafting. A scion from a seedless grape vine is grafted onto a stock from a seeded grape vine to produce only seedless grapes.

3. Farmers always use apple trees or other fruit trees produced by grafting because the fruit of the scion is genetically the same as the plant the scion was taken from. If the tree the scion is taken from produces great-tasting apples, the grafted apple tree produces exactly the same quality of apple. Also, grafted trees mature and produce fruit sooner than those started from seed.

4. A disadvantage of vegetative propagation is the lack of variation among the offspring. The offspring can never be better than the parent.

Plant Sexual Reproduction

Anthophyta (angiosperms) are flowering plants. The flower is the reproductive part of an anthophyte that produces gametophytes (gametes). After fertilization, seeds develop in the ovary of the flower, which enlarges to become a fruit. In the sections that follow, we will study the life cycle of anthophytes beginning with the flower.

The life cycle of an anthophyte plant:

Flower formation → gametophyte formation → pollination → fertilization → seed formation → seed dispersal → germination → sporophyte → growth and development → flower formation

Flower Structure and Function

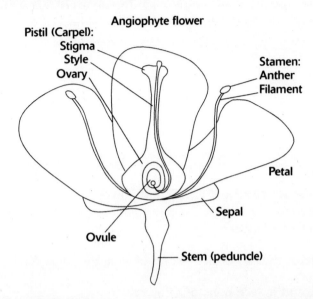

The reproductive parts of a flower are the stamen and the pistil (carpel). The stamen is the male part of the flower and consists of the filament and anther. The pistil is the female part of the flower and consists of the stigma, style and ovary. The remaining parts of the flower are the petals and sepals. Many flowers have both stamens and pistils, while some contain stamens *or* pistils but not both.

Stamen

The stamen is the male part of the flower, consisting of:

❑ **Filament:** The filament is a thin stalk that supports the anther.

❑ **Anther:** Each anther of a flower produces thousands of pollen grains, which are the male gametophytes. Many pollen grains contain three monoploid nuclei surrounded by a protective outer wall (one tube nucleus and two sperm nuclei). Some pollen grains have a tube nucleus and a generative nucleus that later produce two sperm nuclei.

Pistil

The pistil or carpel is the female part of the flower, consisting of:

❑ **Stigma:** The stigma is sticky and holds pollen grains that fall on it.

❑ **Style:** This is a narrow stalk-like structure that connects the stigma to the ovary.

❑ **Ovary:** The ovary produces the female gametophytes of the flower. An ovule is an embryo sac that contains seven cells. One of these is the monoploid (*n*) egg cell, and another is a cell with two monoploid nuclei that helps form the seed. In many flowers, the ovary contains several ovules.

Petals

Petals are modified leaves that are white or colored. The petals of a flower attract insects, birds and other animals to the flower. As these organisms visit the flower, they help with pollination.

Sepals

The sepals are modified leaves that are usually green in color and protect the bud. The sepals are found at the base of the flower.

Example Problems

These problems review the structure and function of flowers.

1. What is the function of a flower?

   Answer: The flower is the reproductive part of an anthophyte plant.

2. What are the male and female parts of a flower?

   Answer: The male part of the flower is the stamen. Remember the *men* in stamen to help you recall that this is the male part of the flower. The female part of the flower is the pistil.

3. Where are ovules produced?

   Answer: The ovules of a flower are produced in the ovary.

4. Where are pollen grains produced?

   Answer: Pollen grains are produced in the anther of the stamen.

Work Problems

Use these problems on the structure and function of flowers for additional practice.

1. What kind of plant produces flowers?

2. Why are the petals of a flower important?

3. Why must the stigma be sticky?

4. Describe the structure and function of a pollen grain.

5. Describe the structure and function of the ovule.

Worked Solutions

1. Flowers are produced by plants in the phylum Anthophyta.

2. The petals of a flower are often colored to attract insects, birds and other animals that help in the process of pollination.

3. The stigma is sticky so that when a pollen grain falls on it, the pollen grain doesn't fall off.

4. Pollen grains contain the male gametophytes of the flower that fertilize the female gametophytes. Many pollen grains contain three monoploid nuclei, a tube nucleus and two sperm nuclei. Some pollen grains have a tube nucleus and a generative nucleus that divides to produce two sperm nuclei when a pollen tube is formed. A protective outer wall surrounds all pollen grains.

5. An ovule is an embryo sac that contains seven cells. One of these is the monoploid (n) egg cell, and another is a cell with two monoploid nuclei that help form the seed. An ovary can contain one or more ovules.

Pollination and Fertilization

Pollen grains are produced in the anther of the stamen. For fertilization to take place, the pollen grain must be transferred to a stigma.

Pollination

Pollination is the transfer of pollen from an anther to a stigma. Three kinds of pollination exist: self-pollination, cross-pollination and artificial pollination.

❑ **Self-pollination** is the transfer of pollen from an anther to a stigma within the same flower. This often occurs when the anthers of the flower are higher than the stigma, and the pollen grains fall down onto the stigma.

❑ **Cross-pollination** is the transfer of pollen from the anther of one flower to the stigma of another flower. Wind, insects, birds and other animals are the most common agents of cross-pollination.

❑ **Artificial pollination** occurs when humans transfer pollen from an anther to a stigma.

Fertilization

After a pollen grain lands on a stigma, its protective wall breaks down, and a pollen tube begins to form through the style. The tube nucleus goes down the pollen tube followed by the two sperm nuclei. The role of the tube nucleus and how the pollen tube is formed is not clearly understood. The two sperm nuclei enter the ovule through an opening called the *micropyle,* and double fertilization takes place. One sperm nucleus (n) fertilizes the egg cell (n) to produce a diploid zygote ($2n$). This zygote becomes the embryo of the seed. The other sperm nucleus (n) combines with the cell that has the two monoploid polar nuclei to form the *endosperm nucleus* ($3n$). This $3n$ cell becomes the endosperm of the seed. The endosperm contains the food (starch) that nourishes the embryo during seed germination and subsequent development until it can produce its own nutrition (and become autotrophic).

After fertilization the petals fall off, and the ovary of the flower increases in size to become a fruit. A *fruit* is defined as the ripened ovary of a flower that contains seeds. Apples, oranges and peaches are fruits. Tomatoes, cucumbers and peppers are also fruits because they are ripened ovaries that have seeds.

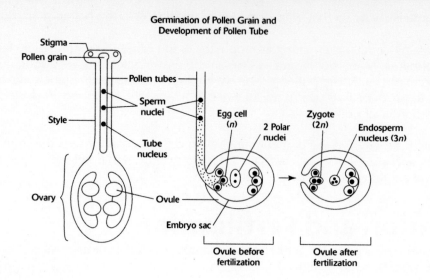

Germination of Pollen Grain and
Development of Pollen Tube

Example Problems

These problems review pollination and fertilization.

1. What is pollination?

    Answer: Pollination is the transfer of pollen from an anther to a stigma.

2. What are the different kinds of pollination?

    Answer: The three kinds of pollination are self-pollination, cross-pollination and artificial pollination. Self-pollination is the transfer of pollen from an anther to a stigma within the same flower. Cross-pollination is the transfer of pollen from the anther of one flower to the stigma of another flower. Artificial pollination occurs when humans transfer the pollen.

3. What happens to a flower after fertilization?

    Answer: After fertilization takes place, the petals of the flower fall off and the ovary begins to enlarge, eventually becoming a fruit.

Work Problems

Use these problems on pollination and fertilization for additional practice.

1. Describe how fertilization takes place.

2. What is meant by double fertilization?

3. Why is the $3n$ cell important?

4. What happens to the zygote after fertilization?

5. What parts of a plant are not fruits?

Worked Solutions

1. After a pollen grain lands on a stigma, its protective wall breaks down and a pollen tube forms. The pollen tube grows through the style to the ovary. The two sperm nuclei enter the ovule through an opening called the micropyle to fertilize the egg and the cell that contains the two monoploid nuclei.

2. Two fertilizations take place in the ovule. One sperm nucleus fertilizes the egg cell to produce a diploid zygote ($2n$). The other sperm nucleus combines with a cell that has two monoploid nuclei to form a cell ($3n$).

3. This $3n$ cell becomes the endosperm of the seed. The endosperm contains starch and provides food for an embryo during germination.

4. After fertilization, the zygote becomes the embryo of the seed.

5. The parts of a plant that are not fruits are the roots, leaves and stems. In some plants these structures can produce new plants by vegetative propagation. Also, all parts of a flower except the ripened ovary are not fruits.

Seed Structure and Function

The Anthophyta (angiosperms) are the largest phylum or division in the plant kingdom. The two classes of Anthophyta are monocotyledons and dicotyledons.

❏ **Monocotyledons (monocots):** Monocots have one seed part (cotyledon) that cannot be split. Examples are corn, grass and wheat.

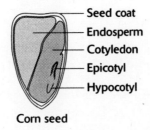

Seed coat
Endosperm
Cotyledon
Epicotyl
Hypocotyl

Corn seed

❏ **Dicotyledons (dicots):** Dicots have two seed parts (cotyledons) that can be split in two. Examples are peas, beans and maple tree seeds.

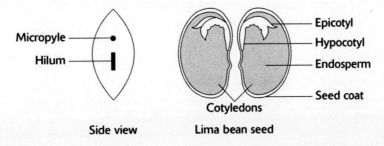

Micropyle
Hilum

Epicotyl
Hypocotyl
Endosperm
Seed coat

Cotyledons

Side view   Lima bean seed

The seed is important because it contains the embryo that becomes the new plant or sporophyte. Many seeds are a major source of food for humans and animals. Corn, wheat and rice are seeds that are used for food by people all over the world.

The following list describes the various parts of a seed.

- **Seed coat:** The seed coat is a thin outer covering that protects the embryo and endosperm of the seed.

- **Micropyle:** The micropyle is a tiny opening in the seed through which the sperm nuclei enters to fertilize the egg of the ovule.

- **Hilum:** The hilum is a scar that shows where the stalk attached the seed to the inside of the ovary.

- **Cotyledon(s):** These are the seed leaves that contain the endosperm. Monocots have one cotyledon, and dicots have two.

- **Endosperm:** This structure forms from the $3n$ cell of the ovule. The endosperm provides food for the germinating seed.

- **Embryo:** The embryo develops from a diploid zygote ($2n$) and contains the epicotyl and the hypocotyl.

    Epicotyl: The epicotyl consists of two leaves that become the upper part of the stem and the leaves of the plant.

    Hypocotyl: Upon germination, the hypocotyl develops into the roots and the lower part of the stem. Many biologists refer to the lower part of the hypocotyl as the radicle. The radicle of the hypocotyl forms the roots.

Example Problems

These problems review seed structure and function.

1. Why are seeds important?

    Answer: Seeds are important in plant reproduction because they contain the embryo that becomes a new plant. Seeds are important sources of food for humans and animals.

2. What does the epicotyl of a seed become?

    Answer: The epicotyl becomes the upper part of the stem and the leaves of the plant.

3. What does the hypocotyl of a seed become?

    Answer: The hypocotyl develops into the roots and the lower part of the stem.

Work Problems

Use these problems on seed structure and function for additional practice.

1. How is a monocot seed different from a dicot seed?

2. What is the function of the endosperm?

3. Why is the seed coat important?

4. What is the significance of the micropyle on a seed?

Worked Solutions

1. A dicot seed has two cotyledons, or seed parts, that can be separated from each other. A monocot seed has only one cotyledon.

2. The endosperm of the cotyledon supplies the embryo with food during germination.

3. The seed coat protects the seed and its embryo prior to germination.

4. The micropyle is a tiny opening in the seed through which the sperm nuclei enter to fertilize the ovule.

Seed Dispersal and Germination

In Anthophyta, seeds are found inside the ripened ovary of the flower called the fruit. Fruits help protect and disperse the seeds of the plant. Seed dispersal is necessary to reduce competition between the parent and young plant for scarce resources such as light, soil and water. Animals eat the fruits; the seeds pass through the digestive system unharmed and are eliminated at a new location. Other agents of seed dispersal are wind, water and insects. *Germination* is the development of a seed into a new plant, and it signals the end of dormancy. Prior to germination a seed is dormant. *Dormancy* is a period of inactivity for a seed, where life functions have been slowed or temporarily suspended. Dormancy allows a seed to survive until it encounters conditions that favor germination. Conditions that favor seed germination are oxygen, moisture and warmth.

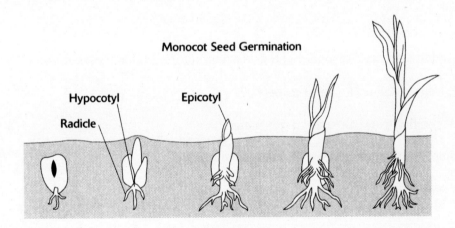

Monocot Seed Germination

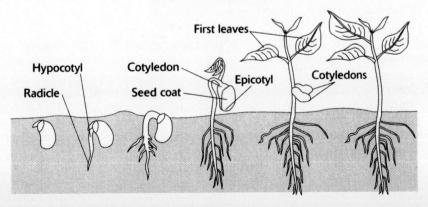

Dicot Seed Germination

The first part of the seed to germinate is the hypocotyl, which forms the roots and the lower stem of the new plant. When the epicotyl breaks through the surface of the soil and is exposed to light, the process of photosynthesis begins, and the plant manufactures its own food. By this time the endosperm in the cotyledon is used up; the cotyledon(s) shrivel and fall off the stem.

Example Problems

These problems review seed dispersal and germination.

1. Why must seeds be dispersed?

   Answer: Seed dispersal is necessary to reduce competition between the parent and new plant for scarce resources such as light, soil and water.

2. What are the agents of seed dispersal?

   Answer: The agents of seed dispersal are wind, water, insects and other animals. The agents for seed dispersal are the same as those for pollination.

3. Define germination.

   Answer: Germination is the breaking of dormancy and the development of a seed into a new plant.

Work Problems

Use these problems on seed dispersal and germination for additional practice.

1. Why is fruit important for seed dispersal?

2. What is the first part of a seed to germinate?

3. What happens to the endosperm after germination?

4. Define dormancy.

Worked Solutions

1. Fruit helps disperse the seeds of a plant by attracting animals that eat the fruit and then deposit the seeds at a new location.

2. The first part of a seed to germinate is the hypocotyl, which forms the roots needed to anchor the new plant in the soil.

3. The endosperm provides the embryo with food during germination. After all the endosperm has been used up, the cotyledons that contain the endosperm fall off the stem.

4. Dormancy is a period of inactivity for a seed, where life functions have been slowed down or temporarily suspended.

Chapter Problems and Answers

Problems

The following is a brief paragraph based on plant reproduction. For problems 1–10, fill in the missing terms.

The flower is the reproductive structure of plants known as _____. Many flowers have
<u> </u>
1

both male and female parts. The male part of a flower is called the _____, and the female

2

part of the flower is called the _____. Pollen grains produced by the anther of a flower

3

contain the _____ gametes. The female gametes of a flower can be found in a structure

4

called the _____. For fertilization to take place, a pollen tube forms from the pollen grain

5

that grows through the _____ of the flower. The sperm nuclei from the pollen grain enter

6

the ovule through a small opening called the _____. In the ovule, _____ fertilization takes

7   8

place. One sperm nucleus fertilizes the egg cell to produce a 2n _____. The other sperm

9

nucleus fertilizes a 2n cell (the two polar nuclei) to produce a 3n cell. This cell produces

the _____, which becomes the food source for the embryo.

10

Problems 11–20 are based on the following diagram. Answer the question or identify the term
or letter of the structure described.

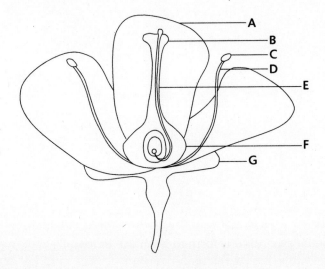

11.   Which phylum of plants produces this structure? _____

12.   Select the structure that attracts insects and birds. _____

13.   Which structure is capable of producing a female gametophyte? _____

14. Where can ovules be found? _____

15. Which structure produces the male gametophyte? _____

16. Identify the part that can become a fruit. _____

17. Which part is sticky and able to hold pollen? _____

18. Where does fertilization take place? _____

19. Where does a seed develop? _____

20. Identify the two structures that make up the male reproductive part. _____

For problems 21–25, select the plant part from the following list that is associated with each description. A choice can be used more than once or not at all.

A. bulb   B. corm   C. rhizome   D. runner   E. tuber

21. An underground stem that grows horizontally along the surface. _____

22. An aboveground stem that grows horizontally along the surface. _____

23. A short underground stem with thick, fleshy, colorless leaves. _____

24. Onions and tulips often use this plant part to reproduce asexually. _____

25. An enlarged underground stem with many buds that are often called eyes. _____

Problems 26–30 are based on the following diagram. Select the letter that is associated with each statement.

26. Select the part of the plant that is formed from the epicotyl. _____

27. Which part of the plant contains the endosperm? _____

28. Select the part of the plant that is formed from the hypocotyl. _____

29. Identify the two parts of a plant that develop from the embryo. _____

30. Name the class that this plant belongs to. _____

Answers

1.   **Anthophyta.** Anthophytes are flowering plants, coniferophytes are plants whose seeds are in cones and pterophytes are ferns that reproduce by spores.

2.   **Stamen.** The *men* in stamen can help you recall that this is the male part of the flower. The stamen consists of the anther and the filament.

3.   **Pistil.** The pistil or carpel consists of the stigma, style and ovary.

4.   **Male.** Each pollen grain has two sperm nuclei, which participate in double fertilization of the ovule.

5.   **Ovary.** The ovules of the ovary contain the female gametophytes.

6.   **Style.** The style connects the stigma to the ovary, which contains the ovule(s). The pollen tube grows through the style, allowing the sperm nuclei to reach the ovule.

7.   **Micropyle.** If you look at a dicot seed, you can often see the micropyle as a tiny opening either above or below the hilum (depending on how you hold the seed).

8.   **Double.** Double fertilization is a characteristic of sexual reproduction in flowering plants.

9.   **Zygote.** The egg cell has n chromosomes, and the sperm nucleus has n chromosomes. When the two cells unite, a $2n$ zygote is produced.

10.   **Endosperm.** The endosperm cell has $3n$ chromosomes. Each of the two polar nuclei has n chromosomes, and the second sperm nucleus has n chromosomes. Thus, $n + n + n = 3n$, which is the number of chromosomes in the endosperm cell.

11.   **Anthophyta.** The drawing shown is of a flower. Anthophyta (angiosperms) are flowering plants.

12.   **A.** The petals of a flower can have brilliant colors that attract insects, birds and other animals to the flower. These organisms aid in pollinating the flower.

13.   **F.** The inside wall of the ovary produces the ovule, which contains the gametophyte.

14.   **F.** The ovary of a flower can contain one or more ovules.

15.   **C.** The anther of the stamen produces the male gametophyte.

16.   **F.** After fertilization takes place, the ovary enlarges and becomes a fruit.

17.   **B.** The stigma is the top part of the pistil that has a sticky surface adapted to catch and hold pollen.

18.   **F.** The ovary of the flower contains the ovules, which are the site for double fertilization. One sperm nucleus fertilizes the egg cell to produce a $2n$ zygote. The other sperm nucleus fertilizes a $2n$ cell to produce a $3n$ cell, which becomes the endosperm.

19.   **F.** In flowering plants seeds usually develop inside the ovary. However, there are some exceptions to this rule.

20. **C** and **D.** The male part of the flower consists of the anther **C** and the filament **D.**

21. **C. Rhizome.** Shoots from buds on the underground stem develop into a new plant.

22. **D. Runner.** When a runner or stolon touches the ground, buds from the stem form roots and leaves, and a new plant develops.

23. **A. Bulb.** A bulb can produce several smaller bulbs, each of which can grow into a new plant.

24. **A. Bulb.** A bulb can produce several smaller bulbs, each of which can grow into a new plant.

25. **E. Tuber.** The white potato is a tuber that has buds, each of which can develop into a new plant.

26. **A.** The epicotyl produces the upper stem and the leaves.

27. **B.** These structures are the cotyledons that contain the endosperm.

28. **C.** The hypocotyl produces the roots and the lower part of the stem.

29. **A** and **C.** The epicotyl, **A,** is the part of the embryo that produces the upper stem and the leaves. The hypocotyl, **C,** is the part of the embryo that produces the lower stem and the roots.

30. **Dicotyledons.** This plant is a dicot because two cotyledons are attached to the stem. Also, Chapter 3 describes the netted veins that dicots have in their leaves. (Monocots have parallel veins in their leaves.)

Supplemental Chapter Problems

Problems

For problems 1–30, select the *best* answer.

1. The formation of a new plant from the root, stem or leaf of an already existing plant is known as:

    A. regeneration   B. sexual reproduction   C. meiosis   D. vegetative propagation

2. A thickened, enlarged, underground stem with buds (often called eyes) is a:

    A. bulb   B. rhizome   C. tuber   D. cutting

3. The white potato is an example of a:

    A. bulb   B. rhizome   C. tuber   D. cutting

4. This is an aboveground stem that grows horizontally along the surface.

    A. runner   B. rhizome   C. tuber   D. cutting

5. Which of the following can develop into a new plant from a root?

   A. tulip   B. sweet potato   C. white potato   D. onion

6. Which of the following can develop into a new plant from a leaf?

   A. strawberry   B. *Bryophyllum*   C. geranium   D. onion

7. A detached stem is placed in water and allowed to develop roots. This process involves a:

   A. bulb   B. rhizome   C. tuber   D. cutting

8. All the following are examples of artificial propagation *except:*

   A. propagation by seed   B. grafting   C. layering   D. cutting

9. A vine producing sweet, seedless grapes is grafted onto a stock that produces sour, seeded grapes; the grafted vine (scion) produces grapes that are:

   A. sweet and seedless   B. sour and seeded
   C. sweet and seeded   D. sour and seedless

10. All the following are advantages of vegetative propagation *except:*

    A. The offspring are identical to the parent.
    B. The reproduction of new plants is rapid.
    C. Seedless varieties of fruits can be propagated.
    D. Genetic variation is produced in the offspring.

11. The female part of the flower is called the:

    A. stamen   B. ovary   C. pistil   D. ovule

12. A pollen grain contains the:

    A. male gametophyte   B. female gametophyte   C. ovule   D. ovum

13. The ovule develops into a part of the flower called the:

    A. ovary   B. stigma   C. style   D. seed

14. Which of the following structures are in the female part of the flower?

    A. fallopian tube, ovary, epicotyl   B. stigma, style, ovary
    C. pollen tube, filament, anther   D. epicotyl, hypocotyl, endosperm

15. In which of the following two structures can meiosis take place?

    A. stigma and style   B. filament and anther
    C. anther and ovule   D. style and ovary

16. The transfer of pollen from one flower to another is called:

    A. self-pollination   B. cross-pollination   C. ovulation   D. germination

17. The sperm nuclei of a pollen grain reach the ovule by moving through the:

    A. fallopian tube   B. pollen tube   C. filament   D. stem

18. How many chromosomes does the zygote in an ovule have?

    A. *n*   B. 2*n*   C. 3*n*   D. 4*n*

19. The endosperm forms from a cell in the zygote. How many chromosomes does this cell have?

    A. *n*   B. 2*n*   C. 3*n*   D. 4*n*

20. The sperm nuclei of a pollen grain enters the ovule through the:

    A. micropyle   B. hilum   C. cotyledon   D. endosperm

21. The embryo of a seed contains the:

    A. cotyledon and endosperm   B. seed coat and endosperm
    C. micropyle and hilum   D. epicotyl and hypocotyl

22. The 3*n* cell of the ovule becomes the:

    A. zygote   B. embryo   C. endosperm   D. micropyle

23. Which structure of a seed represents the point of attachment between the seed and the wall of the ovary?

    A. micropyle   B. hilum   C. epicotyl   D. hypocotyl

24. Food for the developing embryo is found in the:

    A. epicotyl   B. hypocotyl   C. endosperm   D. hilum

25. The development of a seed into a new plant is known as:

    A. germination   B. fertilization   C. pollination   D. vegetative propagation

26. The first leaves of a germinating seed develop from the:

    A. epicotyl   B. hypocotyl   C. cotyledon   D. endosperm

27. The period of time when a seed is characterized by reduced performance of life functions is known as:

    A. germination   B. dormancy   C. fertilization   D. growth and development

28. From which part of a flower does a fruit develop?

    A. anther   B. stamen   C. petal   D. ovary

29. All the following are agents of seed dispersal *except:*

    A. wind   B. water   C. vegetative propagation   D. animals

30. Select the sequence that correctly represents the life cycle of a flowering plant.

    A. germination → fertilization → seed formation → pollination
    B. fertilization → pollination → seed formation → germination
    C. pollination → fertilization → seed formation → germination
    D. pollination → seed formation → fertilization → germination

Answers

1. **D.** "Vegetative Propagation," p. 199

2. **C.** "Propagation from Stems," p. 200

3. **C.** "Propagation from Stems," p. 200

4. **A.** "Propagation from Stems," p. 200

5. **B.** "Propagation from Roots," p. 199

6. **B.** "Propagation from Leaves," p. 200

7. **D.** "Cuttings," p. 202

8. **A.** "Artificial Propagation," p. 202

9. **A.** "Grafting," p. 202

10. **D.** "Advantages of Vegetative Propagation," p. 203

11. **C.** "Flower Structure and Function," p. 205

12. **A.** "Flower Structure and Function," p. 205

13. **D.** "Flower Structure and Function," p. 207

14. **B.** "Flower Structure and Function," p. 205

15. **C.** "Flower Structure and Function," pp. 204–205

16. **B.** "Pollination and Fertilization," p. 207

17. **B.** "Pollination and Fertilization," p. 207

18. **B.** "Pollination and Fertilization," p. 207

19. **C.** "Pollination and Fertilization," p. 207

20. **A.** "Pollination and Fertilization," p. 207

21. **D.** "Seed Structure and Function," p. 209

22. **C.** "Seed Structure and Function," p. 209

23. **B.** "Seed Structure and Function," p. 210

24. **C.** "Seed Structure and Function," p. 210

25. **A.** "Seed Dispersal and Germination," p. 211

26. **A.** "Seed Dispersal and Germination," pp. 211-212

27. **B.** "Seed Dispersal and Germination," p. 211

28. **D.** "Seed Dispersal and Germination," p. 207

29. **C.** "Seed Dispersal and Germination," p. 211

30. **C.** "Plant Sexual Reproduction," p. 207–211

Chapter 11
Mendelian Genetics

In 1856 Gregor Mendel began researching the inheritance of traits in peas, and in 1866 he published his now famous article titled "Experiments in Plant Hybridization." In this article he proposed that peas contained *factors* that determined the characteristics of a trait. Today we call these factors *genes* or *alleles*. Mendel is often referred to as the father of genetics.

Genetics is the branch of biology that studies heredity and the transmission of traits from one generation to the next. Before Mendel, people used to believe that the characteristics of parents were blended in the offspring. This explained why offspring resembled their parents. But if inherited characteristics were determined by blending, eventually all organisms in a species would become *the blend*. Blending fails to explain the diversity of characteristics that exist within a species. Mendel's experiments with pea plants showed that hereditary factors (genes/alleles) are inherited from one generation to the next, with one allele coming from each parent. Mendel's laws of heredity explain what blending fails to.

Mendel's experiments over approximately eight years led to his discovery of three principles or laws of heredity: the law of dominance, the law of segregation and the law of independent assortment. The sections that follow explain each of these laws in detail.

Genetic Vocabulary

Genetics has its own terminology. The following list defines some of the key terms and concepts that help to better understand this branch of biology. Later in this chapter, these terms will be applied to Mendel's laws.

❑ **Trait:** A trait is some distinguishing feature or property of an organism.

❑ **Characteristic:** This term refers to the different versions of a trait.

We often see the terms *trait* and *characteristic* used interchangeably.

The following chart gives some examples of common human traits and their characteristics.

Traits and Their Characteristics	
Trait	*Characteristics*
Height	Tall or short
Eye color	Brown eyes or blue eyes
Earlobe attachment	Detached earlobes (free) or attached earlobes
Eyelashes	Long eyelashes or short eyelashes

❑ **Gene:** The gene is the hereditary unit or factor that is passed down from one generation to the next. Genes determine the characteristics of traits in an organism. Today we know that genes are composed of DNA molecules, but this was not known in Mendel's time.

❑ **Allele:** Different forms of genes for a trait. Alleles occupy the same location on a pair of homologous chromosomes.

We often use a letter to indicate the allele for the characteristic of a trait. For example, in the trait for height, the uppercase letter T represents the allele for tall, and the lowercase letter t represents the allele for short. The uppercase letter is used for the dominant characteristic of the trait. The lowercase letter is used for the recessive characteristic of the trait. Look at the following definitions of *dominant* and *recessive*. Usually two alleles determine a trait.

❑ **Homologous chromosomes:** Homologous chromosomes are pairs of similar chromosomes found in diploid cells.

❑ **Homozygous (pure):** Two of the same kinds of alleles for a trait, one on each chromosome in a pair of homologous chromosomes. Example: TT (two alleles for the tall characteristic of height).

❑ **Heterozygous (hybrid):** Two different kinds of alleles for a trait, one on each chromosome in a pair of homologous chromosomes. Example: Tt (an allele for tall and an allele for short).

❑ **Dominant:** In a hybrid, we have two different kinds of alleles for a trait; the characteristic of the trait that we see is called dominant. Example: For the trait of height, if we have an allele for tall and an allele for short (Tt) and the organism is tall, tall is dominant and short is recessive. In a hybrid situation the letter representing the dominant allele is always written first, and the letter for the recessive allele is written second.

❑ **Recessive:** In a hybrid, we have two different kinds of alleles for a trait; the characteristic of the trait that we *don't* see is called recessive. Example: For the trait of height, if we have an allele for tall and an allele for short (Tt) and the organism is *not* short, short is recessive and tall is dominant. Note that the dominant allele for a trait is written first, and the recessive allele is written second.

The only way to get the recessive characteristic of a trait is to have two of the recessive alleles. Example: tt or homozygous short.

❑ **Genotype:** The type of allele combination for a trait (homozygous or heterozygous).

❑ **Phenotype:** This is the characteristic of the trait that is determined by observation. What you see is what the organism has. If we look at a plant and we see that the plant is tall, we say its phenotype is tall. If a person has blue eyes, the person's phenotype for eye color is blue. Phenotype does not tell us the genotype of an organism. We cannot determine whether an organism is homozygous or heterozygous for the dominant characteristic of a trait just by looking at the organism.

Example Problems

These problems review genetic vocabulary.

1. What are homologous chromosomes?

    Answer: Homologous chromosomes are pairs of similar chromosomes found in diploid cells.

2. Define the term dominant.

    Answer: In a hybrid or heterozygous individual, where we have two different kinds of alleles for a trait, the characteristic of the trait that we see is called dominant.

3. What is a gene?

   Answer: The gene is the hereditary unit or factor that is passed down from one generation to the next. Genes are composed of DNA molecules and determine the characteristics of traits in an organism.

4. What is an allele?

   Answer: An allele is different forms of genes for a trait. Alleles occupy corresponding positions on homologous chromosomes. Two alleles determine a trait.

Work Problems

Use these problems on genetic vocabulary for additional practice.

1. How is a trait different from a characteristic?

2. Explain how the homozygous genotype differs from the heterozygous genotype.

3. What genotype produces the recessive characteristic of a trait?

Worked Solutions

1. A trait is some distinguishing feature or property of an organism. A characteristic refers to the different versions or expressions of the trait. For example, height is a trait in pea plants; tall or short stems are characteristics.

2. The homozygous or pure genotype has two of the same kinds of alleles for a trait (for example, TT, two alleles for the tall characteristic of height). The heterozygous or hybrid genotype has two different kinds of alleles for a trait (for example, Tt, an allele for tall and an allele for short).

3. The only way to get the recessive characteristic of a trait is to have two of the recessive alleles (for example, tt, or homozygous short).

The Punnett Square and Probability

The Punnett square is a diagram of a square box that is subdivided into four smaller boxes and used by biology students to solve genetic problems (see the figure in the section on Mendel's Law of Dominance). Reginald Punnett was an English geneticist who studied pea plants and the comb shape of chickens. The Punnett square is used to show different allelic combinations in gametes and to predict the probability of offspring ratios in the next generation. The Punnett square never gives actual results; it only predicts probability. If you flip a coin 100 times, the probability of heads to tails is 50:50. However, actual results usually vary from predicted results. The more times the coin is flipped, the closer we come to the predicted result of 50:50. Thus, the larger the sample size, the more accurate the results. Try it and see for yourself.

Mendel's Law of Dominance

Mendel artificially cross-pollinated pea plants with a variety of different traits (height, seed color, seed texture and others). When Mendel crossed homozygous tall plants with homozygous short plants, all the peas from these plants produced tall pea plants in the next generation. When

Mendel crossed homozygous yellow-seeded plants with homozygous green-seeded plants, all the peas from these plants produced yellow-seeded pea plants in the next generation. When Mendel crossed homozygous round plants with homozygous wrinkled plants, all the peas from these plants produced plants with wrinkled pea seeds in the next generation. This led Mendel to conclude that when two different characteristics of a trait are crossed, the *dominant* one tends to be seen in the next generation, and the *recessive* one tends not to show up. This concept is called *Mendel's Law of Dominance* and is used to define dominance in modern terms. With the help of the Punnett square, we will solve some genetic problems that demonstrate the Law of Dominance.

Let us use height in pea plants to illustrate the preceding discussion. T represents the allele for the tall characteristic of the trait, and t represents the allele for the short characteristic.

The following problem illustrates Mendel's Law of Dominance: A homozygous tall pea plant is crossed with a homozygous short pea plant: (TT × tt). Predict the results of this cross by phenotype and genotype. Look at the following Punnett square for the solution.

TT x tt

Results:

Phenotype: 100% tall

Genotype: 100% heterozygous tall (Tt)

Begin by placing one genotype (TT) on one side of the Punnett square and the other genotype (tt) on top of the square. Now distribute the alleles into the boxes and analyze the results. All boxes contain the Tt allele combination. The results can be reported as follows: phenotype—100% tall, genotype—100% heterozygous tall (Tt).

Note: It is not necessary to mention that the heterozygous individual is "tall." A heterozygous individual has the dominant characteristic. However, the dominant characteristic is stated here and in some future examples so that the student will be aware of it.

The heterozygous genotype demonstrates Mendel's Law of Dominance: When an organism has two different alleles for a trait, the characteristic we see is the dominant one.

Mendel's Law of Segregation

Solve the following problem: A heterozygous (hybrid) tall pea plant is crossed with a heterozygous (hybrid) tall pea plant: (Tt × Tt). Predict the results of this cross by phenotype and genotype. Look at the following Punnett square for the solution.

Tt x Tt

Results:

Phenotypic ratio of offspring:
3 tall : 1 short
or by percent
75% tall : 25% short

Genotypic ratio of offspring:
1TT : 2Tt : 1tt
or by percent
25% homozygous tall
50% heterozygous tall
25% homozygous short

The preceding genetic cross is often called a *hybrid cross* (or *monohybrid cross*) and demonstrates Mendel's Law of Segregation: When two hybrids are crossed, each allele segregates (separates) during gamete formation so that new allele combinations (genotypes) can be formed at fertilization. A hybrid cross produces a phenotype ratio of 3:1, dominant to recessive. A hybrid cross always results in a genotype ratio of 1:2:1, one homozygous dominant to two heterozygous to one homozygous recessive.

Solve the following problem: A heterozygous tall pea plant is crossed with a homozygous short pea plant: (Tt × tt). Predict the results of this cross by phenotype and genotype. Look at the following Punnett square for the solution.

Tt x tt

Results:

Phenotypic ratio of offspring:
2 tall : 2 short
or by percent
50% tall : 50% short

Genotypic ratio of offspring:
2Tt : 2tt
or by percent
50% heterozygous tall
50% homozygous short

Note: It is not necessary to show the arrows distributing the alleles in a Punnett square.

Example Problems

These problems review genetic crosses and Mendel's Laws of Dominance and Segregation.

1. State Mendel's Law of Dominance.

    Answer: When an organism has two different alleles for a trait, the characteristic we see is the dominant one.

2. State Mendel's Law of Segregation.

    Answer: When two hybrids are crossed, each allele segregates (separates) during gamete formation so that new allele combinations (genotypes) can be formed at fertilization.

3. How is the Punnett square used to help solve genetic problems?

    Answer: The Punnett square is used to show different allelic combinations of gametes and to predict the probability of offspring ratios.

Example problems 4–6 are based on the following information: Brown fur color is dominant over white fur color in mice.

4. What is the genotype for a homozygous brown mouse?

    Answer: The genotype for a homozygous brown mouse is BB.

    Begin solving this problem by assigning the uppercase letter B to represent the allele for brown fur color and the lowercase letter b to represent the allele for white fur color. A homozygous brown mouse has two alleles for brown fur color.

5. What is the genotype for a heterozygous brown mouse?

 Answer: The genotype for a heterozygous brown mouse is Bb.

 The heterozygous brown mouse has two different alleles for brown fur color, one allele for brown fur and one allele for white fur.

6. What are the results by phenotype of crossing two hybrid brown mice?

 Answer: 75% brown fur and 25% white fur.

 The genotype for a hybrid (heterozygous) brown mouse is Bb. The genetic cross that gives the answer is Bb × Bb. Use a Punnett square to find the solution.

Work Problems

Use these problems on genetic crosses and Mendel's Laws of Dominance and Segregation for additional practice.

1. What are the results of a hybrid cross by genotype?

2. Why do the actual results of a genetic cross differ from the predicted results of the Punnett square?

Work problems 3 and 4 are based on the following information: Assume that in humans brown eye color is dominant over blue eye color.

3. If one parent is heterozygous for brown eyes and the other parent is homozygous for brown eyes, what are their chances of having a blue-eyed child?

4. Can two brown-eyed parents have a blue-eyed child? Show your answer with a Punnett square.

Worked Solutions

1. The result of a hybrid cross by genotype is always 1:2:1, one homozygous dominant to two heterozygous to one homozygous recessive. As a percentage, this ratio is expressed as 25% homozygous dominant to 50% heterozygous to 25% homozygous recessive.

2. The Punnett square can be used as a mathematical tool to predict the possible results that can be obtained from a genetic cross. To get actual results, the genetic cross has to be performed. Real-world results are always different from predicted results. However, the larger the sample used, the closer the actual result is to the predicted result. In his experiments, Mendel used large numbers of pea plants.

3. They have a 0% chance of having a blue-eyed child. The genetic cross that gives the answer is Bb × BB. Use a Punnett square to find the solution.

4. Yes, two brown-eyed parents can have a blue-eyed child. If both parents are heterozygous for brown eyes (Bb × Bb), their chances of having a blue-eyed child is 25%. Look at the following Punnett square for the solution.

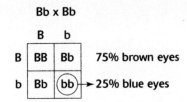

Bb x Bb

75% brown eyes

25% blue eyes

Mendel's Law of Independent Assortment

All the genetic problems that we have investigated up to this point involved studying one trait at a time. However, organisms possess many traits. What happens if we perform genetic crosses involving two traits at one time? One of Mendel's experiments involved studying the traits of seed color and seed texture in pea plants. We will use the following key to assign letters to all the alleles involved:

- ❏ Trait 1 (seed color): Y allele for yellow seeds and y allele for green seeds
- ❏ Trait 2 (seed texture): R allele for round (smooth) seeds and r allele for wrinkled seeds

The genotype YYRR is homozygous for yellow and round seeds and produces gametes that are YR. The genotype yyrr is homozygous for green and wrinkled seeds and produces gametes that are yr. When YR and yr gametes combine, a dihybrid genotype (YyRr) is produced.

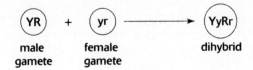

What happens when two dihybrids (YyRr × YyRr) are crossed with each other?

During gametogenesis, each dihybrid can produce four kinds of gametes: YR, Yr, yR and yr. Thus, we need a Punnett square with 4 boxes on each side (16 boxes total) to determine the different phenotypes. The dihybrid cross produces a phenotype ratio of 9:3:3:1. Nine individuals have both dominant characteristics for the two traits, three have one dominant and one recessive characteristic, three have the other dominant and the other recessive characteristic and one has both recessive characteristics.

From this experiment and others like it, Mendel developed his *Law of Independent Assortment*. It states that in a cross between two dihybrid organisms, alleles for different traits separate and are inherited independently of each other.

Example Problems

These problems review Mendel's Law of Independent Assortment.

1. State Mendel's Law of Independent Assortment.

   Answer: In a dihybrid cross, alleles for the different traits separate and are inherited independently of each other.

2. What is a dihybrid?

   Answer: A dihybrid has the hybrid (heterozygous) genotype for two different traits. For example, YyRr is a dihybrid. This individual is hybrid for seed color and seed texture.

3. What kind of seeds are produced by a pea plant with the Yyrr genotype?

    Answer: Yellow and wrinkled. Yy is hybrid yellow, and rr is homozygous wrinkled.

Work Problems

Use these problems on Mendel's Law of Independent Assortment for additional practice.

1. How many kinds of gametes does a dihybrid individual produce?

2. How many boxes must a Punnett square have to solve a dihybrid cross?

3. What is the phenotype ratio that results from the dihybrid cross?

Worked Solutions

1. A dihybrid individual can produce four kinds of gametes.

2. To solve a dihybrid cross, a Punnett square must have 16 boxes. The 4 boxes on each side represent the four possible gametes that can be produced by each dihybrid parent.

3. The dihybrid cross produces a phenotype ratio of 9:3:3:1. If you memorize this ratio, you will not have to spend the time needed to solve dihybrid cross problems by the Punnett square method.

The Test Cross

To determine the phenotype of an organism, we just need to look at the organism, but how can we determine the genotype of an organism? Let's use rabbits as an example to help answer this question. In rabbits, black fur (B) is dominant over white fur (b). If a rabbit has white fur, its genotype must be bb. (The only way to get the recessive characteristic of a trait is to have two of the recessive alleles for that characteristic.) However, if a rabbit has black fur, its genotype could be BB or Bb. Which is it? We can find out the answer by crossing the organism having the unknown genotype with a homozygous recessive, in this case a white rabbit (bb). This type of cross is called a *test cross*. If even one of the offspring is a white rabbit, the unknown organism is heterozygous black (Bb). If all the baby rabbits are black, the unknown organism is probably homozygous black (BB). If we repeat the cross and all the baby rabbits are black again, the probability of the unknown organism being homozygous black increases.

Incomplete Dominance

Incomplete dominance occurs when two different alleles exist for a trait and neither one is dominant or recessive. Examples of incomplete dominance can be seen in Japanese four-o'clocks, snapdragons and Andalusian fowl. We will use snapdragons to illustrate incomplete dominance. The letter R represents the allele for a red snapdragon, and the letter W represents the allele for a white snapdragon. The genotype for a plant with red flowers is RR, the genotype for a plant with white flowers is WW and a plant with pink flowers has the genotype RW. Look at the following sample crosses:

RR x WW

	W	W
R	RW	RW
R	RW	RW

100% pink

RW x RW

	R	W
R	RR	RW
W	RW	WW

Genotype ratio	Phenotype ratio
25% homozygous red	1 red
50% heterozygous	2 pink
25% homozygous white	1 white

Codominance

In *codominance*, both alleles for a trait are dominant, and organisms produced from these crosses have both characteristics of the trait. Roan horses and roan cattle are two examples of codominance. When a horse with a red coat is crossed with a horse with a white coat, a roan is produced. The roan horse has red hairs and white hairs in its coat. Genetic problems involving codominance are similar to those of incomplete dominance. Let's use the roan horse as an example: C^R represents the allele for red hair, and C^W represents the allele for white hair. The letter C is used to indicate that we are dealing with codominance. The genotype for a horse with red hair is $C^R C^R$, the genotype for a horse with white hair is $C^W C^W$ and the genotype for the roan is $C^R C^W$. Look at the following sample crosses.

Note: Different texts use different "keys" to indicate alleles involved in examples of incomplete and codominance problems. Use any key that helps you find the answer.

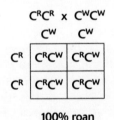

100% roan

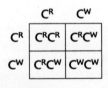

Genotype ratio	Phenotype ratio
25% homozygous red	1 red
50% heterozygous	2 roan
25% homozygous white	1 white

Example Problems

These problems review the test cross, incomplete dominance and codominance.

1. How can we determine the genotype of an organism?

    Answer: To determine the genotype of an organism, we do a test cross. We cross the organism having the unknown genotype with a homozygous recessive. If any of the offspring have the recessive characteristic of the trait, the unknown organism is heterozygous. If all the offspring have the dominant characteristic, the unknown organism is most likely homozygous.

2. When red Japanese four o'clock flowers are crossed with white Japanese four o'clock flowers, plants with pink flowers are produced. What phenotype ratio does crossing two pink-flowered plants produce?

Answer: 25% red, 50% pink and 25% white. This is a 1:2:1 ratio. If you remember the ratio produced when two individuals with the new characteristic are crossed, you do not need to work out the solution with a Punnett square. If you wish to check your answer with a Punnett square, you can use R for the red allele and W for the white allele. The genetic cross that gives the solution is RW × RW.

3.   What are the possible results of crossing a Japanese pink four o'clock plant with a white plant?

    Answer: 50% pink and 50% white. The genetic cross that gives the answer is RW × WW. Use a Punnett square to find the solution.

Work Problems

Use these problems on the test cross, incomplete dominance and codominance for additional practice.

1.   Why is the homozygous recessive always used in a test cross?

2.   How is incomplete dominance different from codominance?

3.   In cattle, an animal with a red coat crossed with one that has a white coat produces offspring that are roan (animals with both red and white hair). What phenotype ratio is produced in the offspring when two roan animals are crossed?

Worked Solutions

1.   The homozygous recessive is always used in a test cross because this organism has a known genotype, two recessive alleles for a trait. An organism with the dominant characteristic cannot be used because we don't know if it is homozygous or heterozygous. We would not know the genotype of the test organism or the genotype of the organism it is being crossed with. The results obtained from crossing two unknown genotypes cannot lead to any conclusions as to their allele combinations.

2.   In incomplete dominance, we have two different alleles for a trait and neither is dominant. In codominance, both alleles are dominant.

3.   The genotype ratio produced is 25% red, 50% roan and 25% white. To solve this problem, use C^R for the red allele and C^W for the white allele. The genetic cross that gives the answer is $C^R C^W \times C^R C^W$. Use a Punnett square to find the solution.

Chapter Problems and Answers

Problems

The following is a brief paragraph based on the concepts of Mendelian genetics. For problems 1–10, fill in the missing terms. Use the Punnett square to help find the solutions to some of the problems.

A student was given a rare guinea pig with black fur and was told that black fur was dominant over white fur. The student wanted to know whether her pet was homozygous or heterozygous for black fur. To determine the answer, she decided to mate her pet with a second guinea pig that had _____ fur. The genotype of this guinea pig is _____. The

1   2
type of cross the student made is known as a _____. As a result of this cross, nine baby

             3
guinea pigs were produced; six had black fur, and three had white fur. From the results of this cross, she concluded that her pet guinea pig was _____ for black fur. The ratio that

      4
she obtained was approximately _____. According to the law of probability, the student

     5
should have obtained a ratio of _____. The student took a third guinea pig with black fur

     6
and mated it with her pet guinea pig. This time twelve guinea pigs were born; nine had black fur, and three had white fur. She concluded that this third guinea pig must have been _____for black fur. If the third guinea pig had a genotype that was _____, all the

   7   8
babies would have had _____ fur. To determine the phenotype of any of the guinea pigs,

     9
the student had to _____ the guinea pigs.

   10

For problems 11–15, fill in the term that *best* completes each sentence.

11. The hereditary factors that Mendel studied are today known as _____.

12. In a hybrid, the characteristic of the trait that we see is called _____.

13. When two heterozygous brown rabbits are crossed, a 1:2:1 ratio results. This illustrates Mendel's Law of _____.

14. In pea plants, tall is dominant over short, and yellow seeds are dominant over green seeds. When two dihybrid plants are crossed, a 9:3:3:1 ratio results. This illustrates Mendel's Law of _____.

15. The square used to solve genetic problems was developed by _____.

For problems 16–20, select the genetic term from the following list that is associated with each description. A choice can be used more than once or not at all.

    A. allele   B. heterozygous   C. homologous   D. homozygous   E. recessive

16. This term describes a pair of similar chromosomes found in the somatic cells of an organism. _____

17. Which of the preceding terms is often used to describe the hybrid genotype? _____

18. Identify the term used to describe the YY allele combination. _____

19. Identify the term that indicates two different kinds of alleles for a trait (Yy). _____

20. This term identifies one of the two kinds of genes for a trait that occupy the same location on a pair of similar chromosomes. _____

For problems 21–25, select the genetic cross from the following list that best fits each description. A choice can be used more than once or not at all.

A. TT × TT   B. TtYy × TtYy   C. Tt × Tt   D. tt × tt   E. Tt × tt

21. This cross demonstrates Mendel's Law of Segregation. _____

22. Which is the test cross? _____

23. Which is the dihybrid cross? _____

24. Organisms produced as a result of this cross are 100% homozygous dominant. _____

25. Organisms produced as a result of this cross are 100% homozygous recessive. _____

Answers

1. **White.** A guinea pig with white fur has two alleles for this characteristic.

2. **Homozygous recessive.** The genotype for the guinea pig with white fur is bb.

3. **Test cross.** This cross is always used to determine an unknown genotype. In this problem the cross is BB × bb if the unknown genotype is homozygous black fur, or Bb × bb if the unknown genotype is heterozygous black fur.

4. **Heterozygous** or **hybrid.** The genotype of the student's pig was Bb, or heterozygous. This is the only genotype that can produce a guinea pig with the recessive characteristic (white). The genetic cross that gives the answer is Bb × bb. This cross gives a 50:50 chance of having a guinea pig with white fur. Use a Punnett square to find the solution.

5. **2:1.** Twice as many guinea pigs with black fur (6) were born, compared to guinea pigs with white fur (3). This gives us a ratio of 2:1.

6. **1:1.** The chance of having guinea pigs with black fur compared to guinea pigs with white fur is 50:50. Another way to write this ratio is 1:1.

7. **Heterozygous** or **hybrid.** This time the student obtained a phenotype ratio of 3:1, black to white fur. The genetic cross that gives this result is Bb × Bb. The student had already determined that her pet guinea pig was heterozygous. Use a Punnett square to find the solution.

8. **Homozygous (pure).** The genetic cross that gives this result is Bb × BB.

9. **Black.** The genetic cross that gives this result is Bb × BB.

10. **Look at** or **observe.** The phenotype of an organism is determined by observation.

11. **Alleles.** The term *alleles* is used to indicate different forms of genes for a trait.

12. **Dominant.** Another term for hybrid is heterozygous. This genotype has two different alleles for a trait (for example, Tt). The characteristic of the allele that we see is called dominant; the one that we do not see is called recessive.

13. **Segregation.** When two hybrids are crossed, each allele segregates during gamete formation so that new allele combinations can be formed at fertilization. A hybrid cross always results in a genotype ratio of 1:2:1.

14. **Independent assortment.** In a dihybrid cross, alleles for different traits separate and are inherited independently of each other, resulting in a phenotype ratio of 9:3:3:1.

15. **Punnett.** The Punnett square is used to show different allelic combinations of gametes and to predict the probability of offspring ratios in the next generation.

16. **C. Homologous.** Somatic cells are diploid. Chromosomes are found in pairs inside the nucleus of these cells. During gametogenesis, one chromosome from each pair is in a gamete.

17. **B. Heterozygous.** The terms *hybrid* and *heterozygous* can be used interchangeably. These terms describe two different kinds of alleles for a trait, one on each chromosome in a pair of homologous chromosomes (for example, Tt, an allele for tall and an allele for short).

18. **D. Homozygous.** The genotype YY indicates two of the same kinds of alleles for a trait, one on each chromosome in a pair of homologous chromosomes.

19. **B. Heterozygous.** The genotype Yy indicates two different kinds of alleles for a trait.

20. **A. Allele.** This term is used to indicate different forms of genes for a trait.

21. **C. Tt × Tt.** Mendel's Law of Segregation is demonstrated by a hybrid cross.

22. **E. Tt × tt.** In a test cross, an unknown genotype is crossed with a homozygous recessive.

23. **B. TtYy × TtYy.** Both of these genotypes are dihybrid (two hybrids).

24. **A. TT × TT.** Each allele in this cross is a dominant allele. Therefore, the offspring from this cross can only have dominant alleles.

25. **D. tt × tt.** Each allele in this cross is a recessive allele. Therefore, the offspring from this cross can only have recessive alleles.

Supplemental Chapter Problems

Problems

For problems 1–20, select the *best* answer.

1. The laws of heredity developed by Gregor Mendel were based on:

   A. experiments performed by cross-pollinating pea plants

   B. microscopic observation of snapdragon pollen cells

   C. dissection of Japanese four o'clock flowers

   D. solving genetic problems with Punnett squares

2. Which of the following is *not* one of Mendel's laws?

   A. Law of Codominance   B. Law of Dominance

   C. Law of Segregation   D. Law of Independent Assortment

3. Genetics is a branch of biology that deals with:

   A. classification   B. transmission of traits from one generation to the next

   C. regulation   D. diversity of living things

4. Before Mendel, people believed that:

   A. all the characteristics of the male parent were passed on to the offspring

   B. only the dominant characteristics of the male parent were passed on to the offspring

   C. all traits were inherited from the female parent

   D. the characteristics of parents were blended in the offspring

5. To determine the phenotype of a cat:

   A. look at the cat

   B. look at the cat's parents

   C. cross the cat with a homozygous dominant cat

   D. cross the cat with any cat of the opposite sex

6. Which of the following terms describes the allele combination that an organism has for a trait?

   A. dominant   B. homologous   C. genotype   D. phenotype

7. In an organism with the Bb genotype, the characteristic of the trait that does not show up is called:

   A. dominant   B. recessive   C. hybrid   D. pure

8. If a guinea pig has the genotype BB for coat color, the guinea pig is:

   A. homologous   B. homozygous   C. heterozygous   D. hybrid

9. The ratios predicted by the Punnett square can be achieved:

   A. when the number of offspring produced is large

   B. when the number of offspring produced is small

   C. if a square with 64 boxes is used

   D. if the results are analyzed mathematically

10. In a cross between brown mice and white mice, only white offspring are produced. This is an example of Mendel's Law of:

    A. Independent Assortment   B. Segregation

    C. Incomplete Dominance   D. Dominance

11. How can two organisms have the same phenotype but different genotypes?

    A. One is homozygous dominant, and the other is homozygous recessive.

    B. Both are homozygous for the dominant trait.

    C. One is heterozygous, and the other is homozygous dominant.

    D. Both are heterozygous for the dominant trait.

12. Two heterozygous brown-eyed parents have a blue-eyed child. This can best be explained by:

    A. independent assortment   B. differentiation and incomplete dominance

    C. incomplete dominance and codominance   D. segregation and recombination

13. How many alleles usually determine a trait?

    A. 1   B. 2   C. 3   D. 4

14. When two dihybrid plants are crossed, the ratio that results is:

    A. 9:3:3:1   B. 1:2:1   C. 2:1   D. 1:1

15. A girl has blond hair and blue eyes, and her sister has brown hair and brown eyes. This different combination of traits can best be explained by:

    A. codominance   B. independent assortment

    C. dominance   D. blending inheritance

16. To determine the genotype of a brown rabbit, cross the rabbit with:

    A. a homozygous brown rabbit   B. a heterozygous brown rabbit

    C. a homozygous white rabbit   D. a hybrid brown rabbit

17. A cross that is performed to determine an unknown genotype is called a:

    A. test cross   B. monohybrid cross   C. dihybrid cross   D. codominant cross

18. When two Japanese four o'clock plants are crossed, 24 pink and 26 white four o'clock plants are produced. What were the flower colors of the two parents?

    A. pink and pink   B. pink and red   C. pink and white   D. red and white

19. A cross between purple petunias and white petunias produces 100 offspring, all blue. This is an example of:

    A. dominance   B. segregation

    C. incomplete dominance   D. independent assortment

20. When two blue petunias are crossed, offspring are produced in a ratio of:

    A. 75% white, 25% purple   B. 75% purple, 25% white

    C. 100% blue   D. 25% purple, 50% blue, 25% white

Answers

1. **A.** "Gregor Mendel," p. 221

2. **A.** "Gregor Mendel," pp. 223–227

3. **B.** "Gregor Mendel," p. 221

4. **D.** "Gregor Mendel," p. 221

5. **A.** "Genetic Vocabulary," p. 222

6. **C.** "Genetic Vocabulary," p. 222

7. **B.** "Genetic Vocabulary," p. 222

8. **B.** "Genetic Vocabulary," p. 222

9. **A.** "The Punnett Square and Probability," p. 223

10. **D.** "Mendel's Law of Dominance," pp. 223–224

11. **C.** "Mendel's Law of Dominance," pp. 224–225

12. **D.** "Mendel's Law of Segregation," pp. 224–225

13. **B.** "Genetic Vocabulary," pp. 222–223

14. **A.** "Mendel's Law of Independent Assortment," p. 227

15. **B.** "Mendel's Law of Independent Assortment," p. 227

16. **C.** "The Test Cross," p. 228

17. **A.** "The Test Cross," p. 228

18. **C.** "Incomplete Dominance," p. 228

19. **C.** "Incomplete Dominance," pp. 228–229

20. **D.** "Incomplete Dominance," pp. 228–229

Chapter 12
Patterns of Inheritance

The 20th century was the age of genetics. The geneticists of the 20th century built on the work of Mendel and discovered a great deal about the patterns of inheritance in organisms. In the early 1900s, geneticists learned how sex is determined, how blood types are inherited, and how certain traits are linked to the sex chromosome. They also discovered the importance of the interaction between heredity and environment.

Sex Determination

Humans have 46 chromosomes (23 pairs) in the somatic cells of the body. In a *karyotype,* the chromosomes are removed from the nucleus of a cell, arranged in decreasing size order and photographed. The karyotype reveals that humans have 22 pairs of homologous chromosomes called *autosomes* and one pair of chromosomes called *sex chromosomes.* In a female, the two sex chromosomes are XX, and in a male they are XY. Logic indicates that, like flipping a coin, the chance of a child being male or female is 50:50. This can be proven using knowledge of chromosomes and cell reproduction. During gametogenesis, half the sperm cells produced carry an X chromosome, and the other half carry a Y chromosome. An egg cell can only have an X chromosome. If the sperm cell that fertilizes the egg is an X-carrying sperm cell, the new individual is female. If the sperm cell that fertilizes the egg is a Y-carrying sperm cell, the new individual is male.

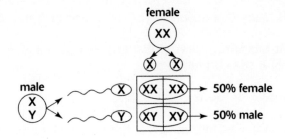

Example Problems

These problems review sex determination.

1. How many pairs of chromosomes are in a human somatic cell?

   Answer: Human somatic cells have 23 pairs of chromosomes (22 pairs of autosomes and 1 pair of sex chromosomes).

2. What process produces gametes with the haploid number of chromosomes?

   Answer: Gametogenesis produces sperm cells and egg cells that are haploid.

3. What is the chance of producing a male or female child at fertilization?

Answer: The chance of producing a male or female child is one in two or 50:50.

4. How are sex chromosomes different from autosomes?

Answer: Sex chromosomes determine whether an individual is male (XY) or female (XX). Autosomes have no role in sex determination; however, they carry alleles for many other traits.

Work Problems

Use these problems on sex determination for additional practice.

1. What is a karyotype?

2. How many human autosomes are in a somatic cell?

3. How many sex chromosomes are in a gamete?

4. Which parent determines the sex of the child?

5. When is the sex of a new individual determined?

Worked Solutions

1. A karyotype is an arrangement of chromosomes that were removed from a nucleus and arranged in decreasing size order.

2. Human somatic cells have 44 autosomes (22 pairs) and two sex chromosomes (1 pair).

3. All gametes have one sex chromosome. A sperm cell can have an X or Y sex chromosome. Egg cells contain only X sex chromosomes.

4. The male parent determines the sex of the child.

5. The sex of a new individual is determined at the instant of fertilization when the sperm and egg cells combine.

Blood Types and Multiple Alleles

Humans have four blood types determined by multiple alleles. Blood types and multiple alleles are explained in the following sections.

Blood Types

In the early 1900s, Karl Landsteiner classified human blood into four groups: A, B, AB, and O. People with type A blood have a protein on their red blood cells called antigen A and an antibody in their plasma called anti-B. People with type B blood have a protein on their red blood cells called antigen B and an antibody in their plasma called anti-A. Type AB individuals have both antigen A and B but no antibody A or B. Type O individuals have red blood cells that do not have antigen A or B, but their plasma contains antibodies A and B. Landsteiner showed that certain transfusions between people with different blood types result in agglutination (clumping). Clumping of the blood can result in the death of the individual due to blocked blood vessels. Blood agglutinates, or clumps together, when the antigen on the red blood cell corresponds to the antibody in the plasma. For example, if a person with type A blood is given a transfusion of type B blood, the anti-B antibodies in the plasma (of the type A individual) interact with the anti-gen B (of the type B individual), resulting in agglutination. In an emergency, a person with type O blood can donate blood to anyone. This individual is called the *universal donor*. In an emergency, a person with blood type AB can receive blood from anyone. This individual is called the *universal recipient*.

The following table shows the different blood types, antigens on the red blood cell, antibodies in the plasma, and transfusion possibilities.

Blood Types: Antigens, Antibodies, and Transfusion Possibilities				
Blood Type	*Antigen on Red Blood Cell*	*Antibody in Plasma*	*Can Donate Blood to:*	*Can Receive Blood from:*
A	A	Anti-B	A and AB	A and O
B	B	Anti-A	B and AB	B and O
AB	A and B	None present	AB	A, B, AB, and O
O	None present	Anti-A and anti-B	A, B, AB, and O	O

Multiple Alleles

Human blood types are inherited according to Mendel's Laws of Dominance and Segregation. Blood type is a trait with four possible characteristics: A, B, AB, and O. Three alleles produce all four blood types, but only two alleles at a time determine the trait. *Multiple alleles* refers to the control of a trait by more than two alleles. I^A is a dominant allele that produces antigen A, I^B is a dominant allele that produces antigen B, and i is a recessive allele that does not produce antigens.

The following table shows the four different blood types and their possible genotypes.

Human Blood Types and Their Genotypes		
Blood Type (Phenotype)	*Antigen on Red Blood Cell*	*Genotype*
A	A	$I^A I^A$ or $I^A i$
B	B	$I^B I^B$ or $I^B i$
AB	A and B	$I^A I^B$
O	None present	ii

Questions about blood types and multiple alleles can be answered with the aid of a Punnett square. Consider the following problem: A man with blood type A whose genotype is $I^A i$ marries a woman with blood type O. What is the probability that they can have a child with blood type A? Look at the following Punnett square for the solution.

$$I^A i \times ii$$

	i	i	
I^A	$I^A i$	$I^A i$	50% blood type A
i	ii	ii	50% blood type O

Example Problems

These problems review blood types and multiple alleles.

1. What are the four different blood types?

    Answer: The four different blood types are A, B, AB, and O. The type of antigen present on the surface of the red blood cell determines the blood type of the individual.

2. What types of antigens are on the red blood cells of people with type A, B, AB, and O blood?

    Answer: People with type A blood have antigen A on the red blood cell. Individuals with type B blood have antigen B. Those with type AB blood have antigens A and B. Finally, people with type O blood do not have any antigens on their red blood cells.

3. What types of antibodies are in the plasma of people with type A, B, AB and O blood?

    Answer: People with type A blood have antibody B in their plasma. Individuals with type B blood have antibody A. Those with type AB blood do not have antibodies A or B in their plasma. People with type O blood have antibodies A and B.

4. What is a blood transfusion?

    Answer: A blood transfusion is the transfer of blood from one individual to another. The donor's blood is matched to the recipient's blood whenever possible.

5. How many alleles control the different blood types?

    Answer: Three alleles control human blood types. Six different genotypes are possible for the four blood types. Look at the preceding table for the different genotype combinations.

Work Problems

Use these problems on blood types and multiple alleles for additional practice.

1. Why can a person with type O blood donate to anyone in an emergency?

2. Why can't a person with type A blood donate to a person with type B blood?

3. A man with blood type B ($I^B i$) is married to a woman with blood type A ($I^A i$). What is the probability that they will have a child with blood type O?

4. A man with blood type A, whose genotype is $I^A i$, is married to a woman with the same blood type and genotype. What is the probability that they will have a child with blood type A?

5. A woman with blood type B has a child with blood type A. She accuses a man with blood type O of being the father. Is this man the father?

Worked Solutions

1. A person with type O blood can donate to anyone in an emergency because type O blood does not contain antigens on the red bloods cell that can cause agglutination or clumping of the blood.

2. A person with blood type A cannot donate blood to a person with blood type B. The type A individual has anti-B antibodies that cause agglutination with antigen B in a person with blood type B.

3. This couple's chance of having a child with blood type O is 25%. The genetic cross that gives the answer is $I^B i \times I^A i$. The only way to get the recessive phenotype is to have two recessive alleles. Look at the following Punnett squares for the solution. The Punnett square on the right shows a shortcut. The letters I and i are omitted; the results are the same.

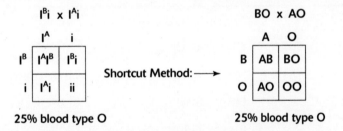

4. This couple's chance of having a child with blood type A is 75%. The genetic cross that gives the answer is $I^A i \times I^A i$. Use a Punnett square to find the solution.

5. The man with blood type O is not the father. In this problem, the woman's genotype is not given. She could be homozygous for blood type B ($I^B I^B$) or she could be heterozygous ($I^B i$). Therefore, we need to try both possibilities. Because the man has blood type O, his genotype is ii. The two genetic crosses that give the answer are: $I^B I^B \times ii$ and $I^B i \times ii$. Neither of these two genetic crosses produces a child with blood type A. Look at the following Punnett squares for the solution.

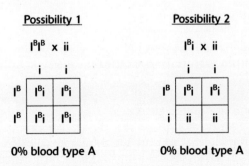

Sex Linkage

A *sex-linked trait* is controlled by an allele located on the X chromosome and sometimes on the Y chromosome. In humans, hemophilia and color blindness are two examples of sex-linked traits. Hemophilia is often called *the bleeder's disease* and is caused by a mutant allele that is defective (recessive) and fails to produce a protein necessary for blood clotting. Color blind people cannot see red and green. A female (XX) that has a defective allele on an X chromosome for hemophilia or color blindness does not have the disease because she has a functioning backup on the other X chromosome. A *carrier* is a female with an allele for a sex-linked disease who does not have the disease. A male (XY) who has a defective allele on an X chromosome for either hemophilia or color blindness does have the disease because he has only one X chromosome and lacks a functioning backup. Three possible alleles are involved in determining sex-linked traits:

❑ X^N represents a normal X chromosome.

❑ X^n represents a mutant or defective X chromosome that results in hemophilia or color blindness.

❑ Y represents a male sex chromosome.

The following table shows the five possible genotypes and their phenotypes.

Possible Genotypes and Phenotypes for Sex-Linked Traits	
Genotype	*Phenotype*
$X^N X^N$	Normal female
$X^N X^n$	Carrier female
$X^n X^n$	Female with a sex-linked trait (hemophilia or color blindness)
$X^N Y$	Normal male
$X^n Y$	Male with a sex-linked trait (hemophilia or color blindness)

Note: The allele for normal clotting can also be indicated as (X^H) and the recessive allele as (X^h). When discussing color vision, the allele for normal vision can be indicated as (X^C) and the allele that causes color blindness as (X^c).

Example Problems

These problems review sex-linked traits.

1. What are sex-linked traits?

   Answer: Sex-linked traits are traits that have an allele located on the X chromosome and sometimes on the Y chromosome. Two examples of sex-linked traits are hemophilia and color blindness.

2. Who can be a carrier of a sex-linked disease?

   Answer: A carrier is a female that has an allele for a sex-linked disease but does not have the disease.

3. How can we tell whether a person has hemophilia?

   Answer: A person with hemophilia who gets a cut or bruise does not stop bleeding because the individual is missing a clotting factor in the blood.

4. What is the genotype of a woman who is color blind?

   Answer: The genotype of a color blind woman is $X^n X^n$, or homozygous recessive.

5. From which parent does a son inherit a sex-linked allele?

   Answer: A son inherits a sex-linked allele from his mother. If a woman is heterozygous, her son has a 50% chance of having a sex-linked disease. A daughter can inherit a sex-linked allele from her mother or father.

Work Problems

Use these problems on sex-linked traits for additional practice.

1. Why are almost all people with sex-linked traits males?

Problems 2 and 3 are based on the following information: A man with hemophilia is married to a woman with normal blood clotting whose genotype is $X^N X^N$.

2. What percentage of this couple's female children will be carriers?

3. What is the probability that this couple will have a child with hemophilia?

Problems 4 and 5 are based on the following information: A man with normal blood clotting is married to a woman who is a carrier for hemophilia.

4. What is the probability that this couple will have a female child who is a carrier?

5. What is the probability that this couple will have a child with hemophilia?

Worked Solutions

1. Most individuals that have sex-linked traits are men because men lack a second X chromosome that normally has a backup allele. Approximately 1 in 7,000 men have the defective allele for hemophilia and fail to produce clotting factor VIII. A woman must have two defective alleles to have a sex-linked trait.

2. All the female children are carriers. The genetic cross that gives the answer is $X^n Y \times X^N X^N$. The Punnett square shows that both possible females are carriers ($X^N X^n$). If the question had asked what the probability is that this couple could have a child who is a carrier, the answer would be 50% (two out of the four possible offspring that could result). Look at the following Punnett square for the solution.

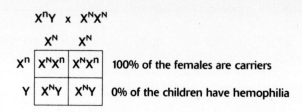

3. None of the children will have hemophilia. Look at the preceding Punnett square for the solution.

4. The chance of this couple having a female child who is a carrier is 50%. A man with normal blood clotting has the genotype: X^NY. A woman who is a carrier for hemophilia has the genotype X^NX^n. The genetic cross that gives the answer is $X^NY \times X^NX^n$. One out of every two possible females (50%) is a carrier. Look at the following Punnett square for the solution.

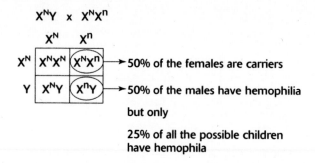

5. The chance of this couple having a child with hemophilia is 25%. One out of every four possible children (25%) has hemophilia. Look at the preceding Punnett square for the solution.

Pedigree Charts

A pedigree chart is used in genetics to track the inheritance of a trait in a family through several generations. Pedigree charts use symbols to indicate females, males, the trait that is being tracked, and the relationships between the different family members. Many pedigree charts have a "key" with legends that indicate what each symbol on the chart represents. Look at the following pedigree chart. A circle is used to represent a female, and a square represents a male. The horizontal line connecting a female and male indicates marriage or mating. A vertical line from a couple to a horizontal line with circles and squares shows the offspring of the next generation. Each generation is indicated with a Roman numeral beginning with I for the parent generation, II for the second generation, and so on. In this pedigree, individual 1 is a female, 2 is a male, and they represent the parent generation. Individuals 1 and 2 have three children (4, 5, and 6) who are in the second generation. Individual 3 is a female that has married into this family.

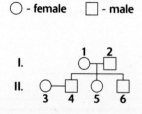

Example Problems

These problems review pedigree charts.

1. What is the function of a pedigree chart?

    Answer: In genetics a pedigree chart is used to keep track of a trait in a family from one generation to the next.

2. How are different generations indicated on a pedigree chart?

    Answer: On a pedigree chart, different generations are indicated by the use of Roman numerals. The parent generation is generation I, and their offspring are in generation II.

3. How is a married couple indicated on a pedigree chart?

    Answer: To indicate that a couple is married or has mated, a horizontal line is drawn between a circle (female) and a square (male).

Work Problems

Use these problems on pedigree charts for additional practice.

Use the following pedigree chart to answer problems 1–5.

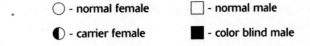

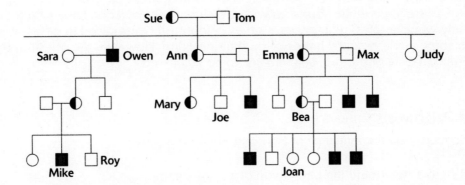

1. How many children do Sue and Tom have?

2. To which generation of offspring do Mike, Roy, and Joan belong?

3. What percentage of Sue and Tom's children are color blind?

4. Why isn't Emma color blind?

5. Why is Mike color blind?

Worked Solutions

1. Sue and Tom have four children: Owen, Ann, Emma, and Judy. Sara and Max are individuals that have married into this family.

2. Mike, Roy, and Joan are in generation IV. Sue and Tom are in generation I. Owen, Ann, Emma, and Judy are in generation II. Mary, Joe, and Bea are in generation III.

3. Twenty-five percent (one out of four) of Sue and Tom's children are color blind. We determine this by looking at the pedigree chart. Owen is the only color-blind child of Sue and Tom. Ann and Emma are carriers.

4. Emma is a carrier and is not color blind because she has only one defective allele for color vision. A woman must have two defective alleles for color vision to be color blind.

5. Mike is color blind because he inherited a defective allele for this disease from his mother. Males need only one defective sex-linked allele to have a sex-linked disease.

The Influence of Environment on Heredity

The expression of a trait in an organism often depends on the interaction of heredity (genotype) and environment. The Himalayan rabbit has white fur except for its extremities (ears, nose, feet and tail) where the fur color tends to be black. The Himalayan rabbit has an allele that produces a heat-sensitive enzyme. This enzyme produces melanin (a dark pigment found in fur, skin and hair). In warm areas, the enzyme is less active; less melanin is produced, and the fur in the rabbit is white. In the cooler extremities, the enzyme is more active; more melanin is produced, and the fur in these areas is darker. If a patch of fur is shaved off the rabbit and a cold pack is applied to the area, the fur that grows back is dark in color. A similar effect can be seen in Siamese cats, which also have darker fur in their extremities. The arctic fox produces enzymes that cause its fur to be reddish brown in the summer. However, in the winter, these enzymes don't function, and the coat color of the fox is white. Fruit flies with the curly wing mutation have straight wings at 16°C and curly wings at 25°C. In humans, studies of identical twins raised in different homes have shown some environmental influence on their heredity. However, the question of which is more important, nature or nurture, still remains to be answered.

Example Problems

These problems review the influence of environment on heredity.

1. What factors determine the phenotype of an organism?

   Answer: The phenotype of an organism is determined by its genotype and any environmental factors that can influence the expression of certain alleles.

2. Why do the Himalayan rabbit and the Siamese cat have darker fur on their ears, nose, feet, and tail?

   Answer: Some organisms have darker extremities because they have an allele that produces a heat-sensitive enzyme, which promotes melanin production. In warm areas, the enzyme is less active; less melanin is produced, and the fur of the animal is lighter. In the cooler extremities, the enzyme is more active; more melanin is produced, and the fur in these areas is darker.

Work Problems
Use these problems on the influence of environment on heredity for additional practice.

1. Why are identical twins used in studies that investigate the influence of environment on heredity?

2. When *Serratia marcescens* bacteria are grown at a temperature of 30°C, the colonies that form are cream colored. The same bacteria grown at 25°C produce colonies that are red. How can we explain this difference in color?

3. Why is it advantageous for the arctic fox to be white in color during the winter?

Worked Solutions

1. Identical twins are used in studies that investigate the influence of environment on heredity because they have the same heredity (genes). Thus, the only variable in the study is the specific environmental influence that is being investigated.

2. The difference in the color of the bacteria can be explained by the effect of temperature on the expression of the allele that controls color.

3. The white fur color in winter allows the fox to blend in with the snow in its environment. This helps the fox be a more efficient predator and, at the same time, enables the fox to avoid animals that might consider it prey.

Chapter Problems and Answers

Problems
The following is a brief paragraph based on blood types and multiple alleles. For problems 1–10, fill in the missing terms.

Before 1900, many people died when receiving blood transfusions. Karl Landsteiner

discovered that humans have four blood groups known as blood types. The four blood

types are _____. Humans have a protein on the red blood cell called an _____. This
          1   2

protein determines the blood type of the individual. However, people with blood type

_____ do not have any proteins on their red blood cells. After careful research,
  3

Landsteiner learned that most people have a protein in their plasma called an _____.
                                                                              4

However, people with blood type _____ do not have these proteins in their blood.
                                  5

Landsteiner discovered that when giving a blood transfusion, the blood type of the

recipient must _____ that of the donor; otherwise, the blood _____, and the recipient
                6   7

dies. In an emergency, a person with blood type _____ can receive blood from anyone,
8

and a person with blood type _____ can donate blood to anyone. The number of alleles
9

that produce all four blood types is _____.
10

For problems 11–15, answer the question or fill in the missing term.

11.   How many chromosomes are inside a human egg cell? _____

12.   A woman is about to have a child. What is the probability that her child will be female?_____

13.   What is the blood type of a person with the $I^B i$ genotype? _____

14.   An allele that is located on an X chromosome is called _____.

15.   The effect of temperature on the development of wings in certain fruit flies demonstrates
      the influence of _____ on heredity.

For problems 16–20, select the term from the following list that is associated with each description. A choice can be used more than once or not at all.

      A. karyotype   B. pedigree   C. Punnett square   D. genotype   E. phenotype

16.   Which term describes a chart used to track a trait from one generation to the next? _____

17.   An arrangement that shows the chromosomes removed from the nucleus of a cell. _____

18.   This term describes the combination of alleles that determine a trait. _____

19.   Select the chart that is used to show the inheritance of blood types in a family. _____

20.   A tool used to help answer genetic questions. _____

Answers

1.   **A, B, AB and O.** These four blood types are determined by multiple alleles.

2.   **Antigen.** People with blood type A have antigen A on their red blood cells. Individuals
     with blood type B have antigen B. People with blood type AB have antigens A and B on
     their red blood cells.

3.   **O.** Individuals with blood type O do not have antigen A or B on their red blood cells.

4.   **Antibody.** The antibodies that can be found in plasma are called anti-A and anti-B. People
     with blood type A have anti-B antibodies; people with blood type B have anti-A antibodies.

5.   **AB.** People with blood type AB do not have anti-A or anti-B antibodies in their plasma.

6.   **Match.** When performing a transfusion, the blood type of the donor should be matched
     to the recipient whenever possible to avoid the antigen-antibody interaction that causes
     blood to agglutinate.

7.   **Agglutinates.** Blood agglutinates, or clumps together, when the antigen on the red blood
     cell corresponds to the antibody in the plasma.

8.   **AB.** A person with type AB blood does not have antibodies in their plasma that can agglutinate with antigens on red blood cells. Therefore, in an emergency, a person with blood type AB can receive blood from anyone.

9.   **O.** A person with blood type O does not have antigens on the red blood cells that could agglutinate with anti-A or anti-B antibodies found in another individual's blood. Therefore, in an emergency, type O blood can be donated to anyone.

10.   **Three.** The three alleles that control blood type are I^A (which produces antigen A), I^B (which produces antigen B) and i (which is recessive and does not produce antigens).

11.   **Twenty-three (23).** The human egg cell is a gamete, and gametes are haploid and have n chromosomes.

12.   **Fifty percent.** If an X-carrying sperm cell fertilizes the egg, the child is female. If a Y-carrying sperm cell fertilizes the egg, the child is male.

13.   **Blood type B.** I^B is a dominant allele, and i is recessive.

14.   **Sex-linked.** The X chromosome is called a sex chromosome because it plays a role in sex determination. An allele located on the X chromosome is called sex linked.

15.   **Environment.** The factors that determine the expression of a trait are heredity and environment.

16.   **B. Pedigree.** A pedigree chart is used in genetics to track the inheritance of a trait in a family through several generations.

17.   **A. Karyotype.** In a karyotype, the chromosomes are removed from the nucleus of a cell, arranged in decreasing size order and photographed.

18.   **D. Genotype.** A genotype can be heterozygous or homozygous for a trait.

19.   **B. Pedigree.** This question is similar to question 20. Here, blood type is the trait that is being tracked from one generation to the next.

20.   **C. Punnett square.** The Punnett square shows the genotype for a trait in the gametes of each parent and predicts the probability of certain genotypes and phenotypes occurring in the offspring.

Supplemental Chapter Problems

Problems

For problems 1–20, select the *best* answer.

1.   How many sex chromosomes are in a human skin cell?

  A. 0   B. 2   C. 23   D. 46

2.   Chromosomes that are *not* involved in sex determination are called:

  A. autosomes   B. centrosome   C. chromatids   D. sex chromosomes

3. A couple has three children, two girls and one boy. What is the probability that their next child will be a boy?

   A. 25%   B. 50%   C. 75%   D. 100%

4. The removal of paired, homologous chromosomes from the nucleus of a cell and their arrangement in decreasing size order is known as:

   A. a pedigree   B. a genome   C. blood screening   D. a karyotype

5. Which combination of gametes at fertilization results in the production of a male?

   A. Y sperm cell and Y egg cell   B. X sperm cell and Y egg cell

   C. Y sperm cell and X egg cell   D. X sperm cell and X egg cell

6. What is the blood type of a person that does *not* have A or B antigens in the blood?

   A. A   B. B   C. AB   D. O

7. A person with blood type O can only receive a transfusion of blood from a person with blood type:

   A. A   B. B   C. AB   D. O

8. A couple has four children, and each child has a different blood type. What is the genotype of the parents?

   A. I^AI^A and I^Ai B. I^AI^B and I^Bi C. I^Ai and I^Bi D. I^AI^B and ii

9. A man with type A blood is married to a woman with type B blood. How is it possible that they have a child with type O blood?

   A. Both parents are homozygous for their blood type.

   B. Both parents are heterozygous for their blood type.

   C. The father is homozygous for type A blood, and the mother is heterozygous for type B blood.

   D. The father is heterozygous for type A blood, and the mother is homozygous for type B blood.

10. A person with hemophilia:

   A. has abnormal hemoglobin that gives the red blood cells a crescent shape

   B. has a low red blood cell count

   C. produces a protein in the blood that clogs the blood vessels

   D. is missing a protein that helps the blood clot

11. A color-blind man can transmit the sex-linked allele for his disease to:

   A. his daughters only   B. his sons only

   C. both his sons and daughters   D. 50% of his male children

12. Color blindness is more common in men than in women because:

   A. men are carriers

   B. women have a natural immunity to the disease

   C. the allele for color blindness is linked to the Y chromosome

   D. the allele for color blindness is linked to the X chromosome

13. A man with hemophilia is married to a woman who does *not* have an allele for this disease. What are their chances of having a daughter with the disease?

    A. 0%   B. 25%   C. 50%   D. 100%

14. A carrier can best be defined as:

    A. a female who has two alleles for a genetic disease

    B. a male who has an allele for a disease and has the disease

    C. a female who has an allele for a disease but does not have the disease

    D. a male who has two sex-linked chromosomes for a genetic disease

Problems 15–18 are based on the following diagram, which shows the pattern of inheritance for hemophilia. Select the *best* answer.

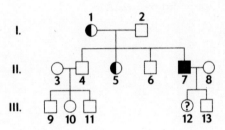

15. The diagram is a:

    A. pedigree   B. genome   C. nucleoprotein   D. karyotype

16. How many alleles for the sex-linked trait does individual 1 have?

    A. 1   B. 2   C. 3   D. 4

17. What percentage of the children from individuals 1 and 2 have hemophilia?

    A. 0%   B. 25%   C. 50%   D. 100%

18. What is the probability that individual 12 has hemophilia?

    A. 0%   B. 25%   C. 50%   D. 100%

19. The expression of a trait in an organism is influenced by:

    A. genotype only   B. phenotype only

    C. genotype and phenotype   D. the interaction between genotype and environment

20. Identical twins separated at birth and raised in different families had different grades on their SAT exams. This demonstrates:

    A. Mendel's Law of Segregation   B. the influence of the environment on gene expression

    C. the role of mutations   D. codominance

Answers

1. **B.** "Sex Determination," p. 237

2. **A.** "Sex Determination," p. 237

3. **B.** "Sex Determination," p. 237

4. **D.** "Sex Determination," p. 237

5. **C.** "Sex Determination," p. 237

6. **D.** "Blood Types and Multiple Alleles," p. 238

7. **D.** "Blood Types and Multiple Alleles," p. 238

8. **C.** "Blood Types and Multiple Alleles," p. 238

9. **B.** "Blood Types and Multiple Alleles," p. 238

10. **D.** "Sex Linkage," p. 242

11. **A.** "Sex Linkage," p. 242

12. **D.** "Sex Linkage," p. 242

13. **A.** "Sex Linkage," p. 242

14. **C.** "Sex Linkage," p. 242

15. **A.** "Pedigree Charts," p. 244

16. **A.** "Pedigree Charts," p. 244

17. **B.** "Pedigree Charts," p. 244

18. **A.** "Pedigree Charts," p. 244

19. **D.** "The Influence of Environment on Heredity," p. 246

20. **B.** "The Influence of Environment on Heredity," p. 246

Chapter 13
Modern Genetics

In the middle of the 20th century, James Watson and Francis Crick discovered the structure of the DNA molecule and opened the field of molecular genetics. By the close of the century, the human genome (the sequence of nitrogen base pairs in DNA) was largely mapped out, and genetic researchers started looking at the link between genes, disease, and cancer. In this chapter we will investigate some landmark discoveries.

Molecular Genetics

In 1869, Johann Friedrich Miescher, a Swiss physician, was able to extract an acidic molecule from the nucleus of cells that he called *nuclein*. This *nucleic acid* was eventually named DNA. In 1944, Oswald Avery reported that DNA is the molecule in the cell that contains the hereditary information. In 1953, Watson and Crick published a paper in the journal *Nature,* which revealed the structure of the DNA molecule for the first time. In the sections that follow, we will study the structure of the DNA molecule and its replication.

DNA Structure

Deoxyribonucleic acid or DNA is a double helix (spiral staircase) shaped structure that is composed of thousands to millions of repeating, paired nucleotides. Each *nucleotide* consists of three components:

- ❑ **Nitrogen base** (one of the following): adenine (A), guanine (G), Cytosine (C), or thymine (T).
- ❑ **Deoxyribose** is a five-carbon sugar.
- ❑ **Phosphate group** contains phosphorous with four atoms of oxygen (PO_4).

The deoxyribose molecules form the outside of the helix and are connected to each other by phosphate groups. Each nitrogen base on a nucleotide pairs with a complementary (partner) base on another nucleotide to form the "steps" of the DNA molecule. The nitrogen bases are held together by weak hydrogen bonds. Adenine (A) pairs with thymine (T) to form A-T. Cytosine (C) pairs with guanine (G) to form C-G.

Structure of a DNA Molecule

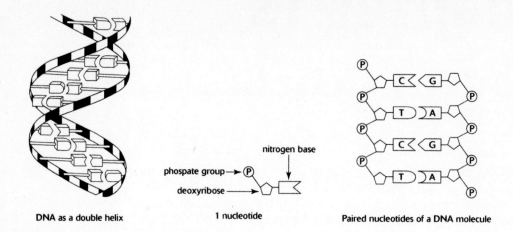

DNA as a double helix   1 nucleotide   Paired nucleotides of a DNA molecule

DNA Replication

Each chromosome in a nucleus has one DNA molecule. When a cell undergoes mitosis, the chromosomes in the nucleus replicate (duplicate), and the DNA molecule in each chromosome replicates. Enzymes separate (unzip) the two strands of the DNA molecule. Each original strand of DNA serves as a *template* (pattern) for the formation of a new strand. Nucleotides in the cell pair up with their complements on each original strand. At the completion of DNA replication, two molecules are formed. Each molecule is composed of one old (original) strand and one new strand. This kind of replication is called *semiconservative replication* because only one of the old strands is "conserved" in the original DNA molecule.

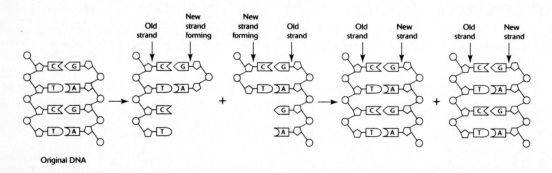

Original DNA

Example Problems

These problems review DNA structure and replication.

1. What is the shape of the DNA molecule?

   Answer: The DNA molecule is a double helix. Watson and Crick discovered this in 1953.

2. What is the structure of a DNA molecule?

   Answer: The DNA molecule is composed of thousands to millions of repeating nucleotides.

3. What are the parts of a DNA nucleotide?

    Answer: A nucleotide contains deoxyribose, a phosphate group and one of four possible nitrogen bases.

4. What are the names of the four possible nitrogen bases in a nucleotide?

    Answer: The four possible nitrogen bases in a nucleotide are adenine (A), guanine (G), cytosine (C), and thymine (T).

5. When does the DNA molecule replicate?

    Answer: The DNA molecule replicates when the chromosome that it is in replicates. During mitosis, replication takes place at the very end of interphase.

Work Problems

Use these problems on DNA structure and replication to give yourself additional practice.

1. How are the paired nucleotides of the DNA molecule held together?

2. How are the deoxyribose molecules in DNA held together?

3. What is the nitrogen-base sequence of a DNA strand if the sequence of bases in the complementary strand is A-C-T-G-T-C?

4. How does a DNA molecule replicate?

5. What is semiconservative replication?

Worked Solutions

1. The paired nucleotides of the DNA molecule are held together by weak hydrogen bonds. The weak hydrogen bonds allow the molecule to easily separate or unzip during replication and to recombine when replication is complete.

2. The deoxyribose molecules in DNA are held together by phosphate groups.

3. The complementary strand to A-C-T-G-T-C is T-G-A-C-A-G. In a DNA molecule, adenine pairs with thymine, and cytosine pairs with guanine.

4. A DNA molecule replicates when its two strands unzip, and nucleotides in the cell pair up with their complements on each original strand. At the completion of replication, two molecules of DNA are formed. Each molecule is composed of one old (original) strand and one new strand.

5. In semiconservative replication, each molecule that is formed has one old (original) strand and one new strand.

Protein Synthesis

Chapter 4 discussed proteins, which are organic molecules that are used for growth, repair and reproduction of new cells. Some of the most important molecules in living things such as hemoglobin, antibodies, hormones and enzymes are proteins. The DNA molecule is a nucleoprotein. The kinds of proteins that a person has determine their traits.

Protein synthesis takes place at ribosomes, which are in the cytoplasm outside the nucleus. Proteins are made of amino acids that are combined together by the process of dehydration synthesis. The genes in the DNA molecule have the information needed to put together a protein, but DNA is in the nucleus. How do the instructions for synthesizing a protein get to the ribosome? How do amino acids get to the ribosome? Two steps are involved in protein synthesis: transcription and translation.

Transcription

Transcription is a process that produces ribonucleic acid (RNA) nucleoproteins. Three kinds of RNA molecules exist: messenger RNA (mRNA), transfer RNA (tRNA), and ribosomal RNA (rRNA). RNA differs from DNA in several ways: DNA is a double helix that contains deoxyribose, and one of its nitrogen bases is thymine. RNA is a single helix that contains ribose, and instead of the nitrogen base thymine, it has uracil. DNA is found in the nucleus, and RNA is found in the cytoplasm.

The following table summarizes the differences between DNA and RNA.

Differences Between DNA and RNA		
	DNA	*RNA*
Number of strands	Two (double helix)	One (single helix)
Sugar molecule present	Deoxyribose	Ribose
Nitrogen bases	Cytosine, guanine, adenine, and *thymine (T)*	Cytosine, guanine, adenine, and *uracil (U)*
Location	Nucleus	Cytoplasm

Transcription begins when a segment of the DNA molecule containing the gene with the information needed to make a specific protein unzips and acts as a template for the formation of mRNA. A single-stranded mRNA molecule forms off a DNA strand by complementary base pairing in a manner similar to DNA replication. When formed, mRNA leaves the nucleus and enters the cytoplasm. The mRNA molecule has the genetic code needed to build a protein. The sequence of nitrogen bases that are transcribed from DNA to an mRNA molecule determine the genetic code. Each set of three nitrogen bases on an mRNA molecule is called a *codon*. Given four different nitrogen bases, 64 different codons are possible (4^3). Sixty-one of the codons correspond to specific amino acids. One codon acts as a *start* signal indicating where the protein begins, and 3 codons act as *stop* signals indicating where a protein ends. The 64 codons are the genetic code needed for protein synthesis in all organisms.

Translation

Translation is the process by which the ribosome combines amino acids to produce proteins. After mRNA enters the cytoplasm, it attaches to a ribosome. Now tRNA molecules *transfer* or bring amino acids to the mRNA at the ribosome. At this point the ribosome moves down the mRNA molecule, and the amino acids are attached to each other by peptide bonds. When the ribosome reaches a stop codon, the end of the newly forming protein has been reached. After the protein has formed, the tRNA molecules separate from the ribosome (as does the newly formed protein), and the process can begin again.

One side of a tRNA molecule attaches to the amino acid that is brought to the ribosome. The other side of the tRNA molecule is the *anticodon* and is complementary to the codon on the mRNA that it attaches to. For example, if a codon on an mRNA molecule is G-C-U, the anticodon on the tRNA would be C-G-A. (RNA has the nitrogen base uracil instead of thymine.)

Protein Synthesis
Parts Involved in the Process of Protein Synthesis

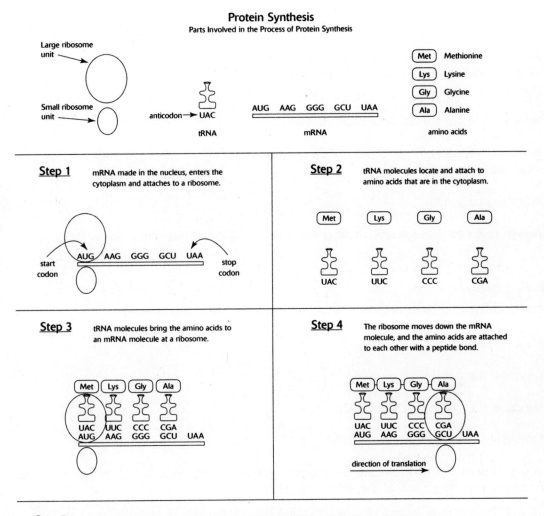

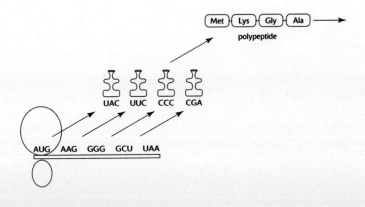

Example Problems

These problems review protein synthesis.

1. Where does protein synthesis take place?

    Answer: Protein synthesis takes place at the ribosome. Ribosomes can be found attached to the endoplasmic reticulum or floating free in the cytoplasm.

2. Why are proteins important for a cell?

    Answer: Proteins are used for growth, repair and reproduction of new cells. DNA and RNA are important nucleoproteins. The kinds of proteins that we have determine our traits.

3. Where is messenger RNA (mRNA) made?

    Answer: mRNA is made in the nucleus and then travels into the cytoplasm where it attaches to a ribosome.

4. What is a codon?

    Answer: A codon is genetic code, a set of three nitrogen bases in DNA or mRNA.

Work Problems

Use these problems on protein synthesis to give yourself additional practice.

1. How does DNA differ from RNA?

2. Describe the process of transcription.

3. Describe the process of translation.

4. How does mRNA differ from tRNA?

5. A segment of a DNA molecule has the nitrogen base sequence T-A-C. What would be the nitrogen base sequence of the tRNA anticodon?

Worked Solutions

1. DNA is a double helix, contains deoxyribose, and one of its nitrogen bases is thymine. RNA is a single helix, contains ribose, and has uracil instead of the nitrogen base thymine. DNA is found in the nucleus, and RNA is found in the cytoplasm.

2. Transcription is a process that produces RNA nucleoproteins. Transcription begins when a segment of DNA unzips, and one strand acts as a template for the formation of single-stranded mRNA. Complementary nitrogen bases pair up with bases on DNA to form mRNA. When the mRNA molecule is complete, it separates from the DNA and leaves the nucleus. At this point, each DNA strand recombines with its complement.

3. Translation is the process by which the ribosome combines amino acids to produce proteins. The function of transfer RNA (tRNA) molecules is to bring amino acids to mRNA at the ribosome. The ribosome moves down the mRNA molecule, and the amino acids are attached to each other by peptide bonds that form by dehydration synthesis. When the protein has been formed, the tRNA molecules separate from the ribosome (as does the newly formed protein).

4. mRNA and tRNA differ in both structure and function. mRNA is a long, single-stranded nucleoprotein with many codons that have the information needed to direct the order of amino acids in a protein. tRNA molecules have two sides. One side of a tRNA molecule attaches to an amino acid; the other side has an anticodon that is complementary to a codon on mRNA. tRNA molecules bring amino acids to the correct location on an mRNA molecule.

5. The sequence of nitrogen bases on the tRNA anticodon is U-A-C. This answer is obtained by first finding the complementary nitrogen base sequence of the mRNA molecule, which is A-U-G. The complement of T is A, the complement of A is U (RNA has the nitrogen base uracil instead of thymine), and the complement of C is G. The tRNA anticodon is the complement of mRNA. mRNA is A-U-G. The complement of A is U, the complement of U is A, and the complement of G is C. This gives us the answer U-A-C.

Mutations

A *mutation* is a change or error in a gene or chromosome. Any change or error to a gene or allele is a *gene mutation*. If the change or error involves the chromosomes of the organism, it is called a *chromosomal mutation*. Mutations that occur in gametes can be passed on to the next generation, if the gamete with the mutation takes part in the fertilization process. Mutations that take place in somatic cells cannot be inherited and usually have little, if any, effect on the organism. Most mutations are not beneficial to the organism. However, if a mutation is beneficial, it can be a source of variation in a population that might eventually lead to the formation of a new species.

Gene Mutations

Gene mutations sometimes occur during the process of translation. One type of gene mutation is called a *point mutation*. In this type of mutation, one nitrogen base is replaced by another nitrogen base. For example, a codon on an mRNA molecule with the nitrogen base sequence G-A-A has the middle nitrogen base replaced with U to give G-U-A. The codon G-A-A corresponds to the amino acid glutamic acid; however, G-U-A corresponds to the amino acid valine. The use of valine in a polypeptide chain results in a change in the protein. In humans, the hemoglobin molecule normally contains glutamic acid; the substitution of valine changes the structure of hemoglobin and results in a disease called sickle-cell anemia.

Another type of mutation is called a *frameshift mutation*. In this type of mutation, a nitrogen base is deleted from or added to a codon on an mRNA molecule. In a *deletion,* a nitrogen base is lost—all the remaining nitrogen bases shift to the left (see the following figure), and every codon after the deletion is changed, resulting in new sets of three nitrogen bases for each codon. In an *addition,* a nitrogen base is inserted—the nitrogen bases shift to the right, and every codon after the insertion is changed, resulting in new sets of three nitrogen bases. Frameshift mutations change many codons. This translates to many different amino acids forming the protein. As a result, the changed protein has a very different structure from the originally intended protein.

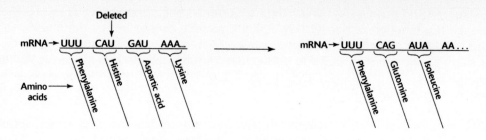

Uracil (U) is deleted, and the nitrogen bases shift to the left.
New codons are formed, and different amino acids
are brought to mRNA.

Chromosome Mutations

Chromosome mutations occur when the chromosome number or structure changes. *Nondisjunction* is the failure of a pair of chromosomes to separate during meiosis, and the result is a change in chromosome number. One gamete has an extra chromosome ($n + 1$); the other gamete is missing a chromosome ($n - 1$). After fertilization, a zygote has $2n + 1$ or $2n - 1$ chromosomes. Zygotes that have $2n - 1$ chromosomes usually do not survive. Zygotes that have $2n + 1$ chromosomes often survive but with problems. For example, Down's syndrome in humans results when pair number 21 has an extra chromosome. Children born with Down's syndrome have a chromosome count of 47 and are mentally retarded.

The following are several kinds of structural mutation:

❏ **Deletion:** A segment of a chromosome is lost or missing. This results in lost genes.

❏ **Addition:** A segment of a chromosome breaks off and is added to its homologous chromosome.

❏ **Translocation:** A segment of chromosome breaks off and is added to a nonhomologous chromosome.

❏ **Inversion:** A segment of a chromosome breaks off and is put back into the same position but in reverse.

Causes of Mutation

Some mutations occur randomly and spontaneously during the process of mitosis or meiosis. Other mutations are caused by *mutagenic agents,* which are environmental or man-made factors that can increase the frequency of mutation. Environmental factors such as ultraviolet rays of the sun can cause changes in DNA that can lead to skin cancer. Cosmic rays from space that penetrate the earth's ozone layer are also a cause of mutation. Ionizing radiation such as X-rays, gamma rays, and nuclear radiation produce *free radicals* (atoms with unpaired electrons) in the cell that can cause breaks in the DNA molecule. *Carcinogens* are chemical mutagens that cause gene mutations, which can result in cancer. A few examples of carcinogens are asbestos, benzene, some food dyes, and PCBs.

Example Problems

These problems review mutations.

1. What are mutations?

    Answer: A mutation is a change or error in a gene or chromosome.

2. What are the different types of mutations?

   Answer: The different types of mutations are gene mutations and chromosomal mutations. Gene mutations can be point or frameshift mutations. Chromosomal mutations that result from nondisjunction change the chromosome number of the organism. Structural mutations can occur because of deletions, additions, translocations, or inversions.

3. List some of the causes of mutations.

   Answer: Mutations can be caused by environmental factors such as ultraviolet rays and cosmic rays. Another cause of mutations results from mutagenic agents such as X-rays, gamma rays, nuclear radiation, and carcinogens.

4. How can a mutation be passed down from one generation to the next?

   Answer: Mutations can be passed down from one generation to the next if a gamete containing a mutated allele or chromosome takes part in the fertilization process.

5. What is a carcinogen? Give some examples.

   Answer: Carcinogens are chemicals that cause gene mutations, which can result in cancer. Some examples of carcinogens are asbestos, benzene, some food dyes, and PCBs.

Work Problems

Use these problems on mutations to give yourself additional practice.

1. Explain how a point mutation takes place.

2. Explain how a frameshift mutation takes place.

3. How does nondisjunction result in chromosomal mutations?

4. How is the type of chromosomal mutation known as an addition different from a translocation?

5. What are mutagenic agents? Give some examples.

Worked Solutions

1. When a point mutation occurs, one nitrogen base in a codon is replaced by another nitrogen base. The number of bases in the codon remains the same (three), and only one codon is affected. The changed codon results in a different amino acid in the polypeptide that is being synthesized.

2. In a frameshift mutation, one nitrogen base is added or deleted from a codon on an mRNA molecule. In a deletion, a nitrogen base is removed; all the remaining nitrogen bases shift to the left, and every codon after the deletion has a new set of three nitrogen bases. In an addition, a nitrogen base is inserted. This forces the nitrogen bases to shift to the right, and each codon then has a new set of three nitrogen bases.

3. When nondisjunction occurs during meiosis, chromosomes fail to separate, and a change in chromosome number results. One gamete has an extra chromosome, and the other gamete is missing one chromosome. After fertilization a zygote is produced that has $2n + 1$ or $2n - 1$ chromosomes.

4. In an addition, a segment of chromosome breaks off and is added to its homologous chromosome. However, in translocation, a segment of chromosome breaks off and is added to a different (nonhomologous) chromosome.

5. Mutagenic agents are environmental or man-made factors that can cause mutations. Examples of some mutagenic agents are ultraviolet rays, cosmic rays, X-rays, gamma rays, nuclear radiation, and carcinogens.

Genetic Diseases

Unfortunately, mutations sometimes result in the production of defective alleles that can be inherited. These defective alleles are responsible for a number of different genetic diseases in humans. Fortunately, most human genetic diseases are caused by recessive alleles. Therefore, an individual must be homozygous and have two defective alleles to have one of these diseases. Cystic fibrosis, phenylketonuria (PKU), Tay-Sachs disease, and sickle-cell anemia are examples of genetic diseases caused by recessive alleles. Huntington's disease is an example of a genetic disease caused by a dominant allele.

Cystic fibrosis is the most common genetic disease in white people. In the United States, 1 in 28 Caucasians (10 million people) is a carrier of this defective allele. Approximately 1,000 individuals are diagnosed with the disease each year, and the median age of survival is about 33.5 years of age. People with cystic fibrosis produce a thick mucus that clogs the air passages of the lungs and ducts in the pancreas. The end result of cystic fibrosis is death.

PKU is caused by the missing enzyme phenylalanine hydroxylase (PAH). PAH changes the amino acid phenylalanine into the amino acid tyrosine, which is needed by the body. Individuals that lack PAH cannot break down phenylalanine. High levels of phenylalanine lead to nervous system problems, mental retardation, and decreased body growth. In many countries, babies are given a blood test a few days after birth that can screen for PKU. Babies found to have PKU are put on a diet low in phenylalanine to control this disease.

Tay-Sachs disease was first described by Warren Tay and Bernard Sachs. This disease is found in Jewish people of Eastern European origin. Tay-Sachs disease is caused by a missing enzyme called hexosaminidase A (Hex-A). Hex-A is needed to break down a lipid called GM2 ganglioside. Without Hex-A, GM2 ganglioside accumulates in the nerve cells of the brain, and eventually the cells die. Children with Tay-Sachs disease usually die by age five. A blood test can be used to detect which individuals are carriers of the defective allele for this disease.

Sickle-cell anemia is the most common genetic disease in people of African descent. One in 12 African Americans are heterozygous for the defective allele. One in 500 African Americans (or 72,000 people) are homozygous for this incurable blood disease. Earlier, we stated that the hemoglobin molecule normally contains glutamic acid; the substitution of valine results in red blood cells with sickle shapes. These cells tend to block circulation in small blood vessels. The complications and symptoms of sickle-cell anemia are many and include pain, strokes, infections, kidney damage, bone damage, shortness of breath, and low red blood cell counts (anemia). Sickle-cell anemia can be treated with vitamins, penicillin, fluids, painkillers, and transfusions when needed. A blood test can be used to detect which individuals have the disease and which individuals are carriers of the defective allele. People with the *sickle-cell trait* (carriers) are codominant and produce both normal and sickled red blood cells.

Huntington's disease is unusual in that it is caused by a defective dominant allele. This allele produces a protein that damages nerve cells in the brain. As the disease progresses, the nervous system deteriorates, and the person eventually dies. The symptoms for this disease first develop between the ages of 30 and 50. By this time a person is likely to have reproduced. This explains why the allele is passed on to future generations. No cure exists for Huntington's disease.

Example Problems

These problems review genetic diseases.

1. What causes genetic diseases?

    Answer: Mutations result in the production of defective alleles that can be inherited. These defective alleles are responsible for many genetic diseases in humans.

2. What is the probability of a child having cystic fibrosis if both parents are heterozygous for the allele that causes the disease?

    Answer: The chance of this child having cystic fibrosis is 25%. This is a hybrid cross, and the ratio produced by a hybrid cross is always 3:1. The chance of having a child with cystic fibrosis is one in four (25%). A person must have two recessive alleles to have this genetic disease. If N represents the normal allele, and n represents the defective recessive allele, the cross that gives the result is Nn × Nn. Use a Punnett square to solve this problem.

3. How can a child born with PKU disease avoid the symptoms of this disease?

    Answer: A child that tests positive for PKU disease is put on a special diet that is low in phenylalanine to control this disease. High levels of phenylalanine lead to nervous system problems, mental retardation and decreased body growth. Preventing high levels of phenylalanine avoids these complications.

4. Why doesn't an individual that is heterozygous for Tay-Sachs disease have the disease?

    Answer: Tay-Sachs disease results when an individual has two recessive alleles that are defective for the production of the enzyme called Hex-A. A heterozygous individual has one normal allele for the production of this enzyme. As long as one allele is working to produce Hex-A, the individual does not have the disease.

5. How are sickle-cell red blood cells different from normal red blood cells?

    Answer: Normal red blood cells are round, and sickle-cell red blood cells are crescent shaped. The sickle cells tend to block the flow of blood in small blood vessels.

Work Problems

Use these problems on genetic diseases to give yourself additional practice.

1. How is Huntington's disease different from most other genetic diseases?

2. A man has Huntington's disease, and his wife does not have an allele for the disease. What are their chances of having a child that will develop Huntington's disease?

3. What is the probability that a child will have sickle-cell anemia if both parents have the sickle-cell trait?

4. Can a couple have a child with Tay-Sachs disease if one parent is heterozygous for the disease and the other parent is homozygous normal?

5. How can a couple determine whether they are at risk of having a child with a genetic disease?

Worked Solutions

1. Huntington's disease is caused by one defective dominant allele. Most other human genetic diseases are caused by the presence of two recessive alleles. In Huntington's disease, the presence of a protein damages nerve cells in the brain. In some recessive genetic diseases, the absence of a protein (usually an enzyme) causes the disease. Examples of this can be seen with Tay-Sachs disease and PKU.

2. Their chances of having a child that will develop Huntington's disease is 50%. To have Huntington's disease a person needs to have only one defective dominant allele for the disease. If H represents the allele for Huntington's disease, and h represents the normal allele, the genotype for the man is Hh and for the woman is hh. The cross that gives the result is Hh × hh. Use a Punnett square to find the solution.

3. The probability that this child will have sickle-cell anemia is 25%. This is the hybrid cross again. The ratio produced by a hybrid cross is always 3:1. The chance is one in four (25%) of this child having sickle-cell anemia. A person must have two recessive alleles to have this genetic disease. The key used to represent the alleles in this problem is different from others that we have seen. Hb^A represents the normal allele, and Hb^S represents the defective recessive allele. The letters Hb represent the hemoglobin molecule. Actually, any letter combination can be used that makes sense to you when solving genetic problems. The cross that gives the result is $Hb^A Hb^S \times Hb^A Hb^S$. Use a Punnett square to solve this problem.

4. No. This child of these parents has a 0% chance of having Tay-Sachs disease. The child must have two defective recessive alleles to have the disease. This cannot occur given the genotypes of the parents. If N represents the normal allele, and n represents the defective recessive allele, the cross that gives the result is Nn × NN. Use a Punnett square to find the solution.

5. A simple blood test can be done to determine whether either of the prospective parents carries an allele for a genetic disease. Genetic counseling of the parents can help them determine the risk factors they face when having children.

Biotechnology

Biotechnology is the use of engineering and technology applied to biology. Biotechnology has been used to determine the human genome, move genes from one species to another and clone organisms. One day this technology might even make it possible to produce replacement body parts.

Genomics

Genomics is the study of genes, their structure and function. In 1990, the U.S. Department of Energy and the National Institutes of Health began the Human Genome Project. The goals of this project were to determine the sequences of the 3 billion nitrogen base pairs in DNA, to identify the 30,000 genes in human DNA and to store this information in a database that would be available to researchers. In September 1999, the Celera corporation entered the race to map the human genome. By April 2000, Celera scientists completed the sequencing of the nitrogen bases and began the mapping phase of genes, which was completed by the end of 2000. The Human Genome Project completed its work in April 2003, about 50 years after Watson and Crick's discovery of the structure of DNA. The Human Genome Project is important because it will lead to increased understanding of the role of genetics in the maintenance of health and the combat of disease. We hope to enhance our knowledge of gene function and to discover new methods of diagnosing disease. As a result of the Human Genome Project, new drugs and therapeutic measures for the treatment of disease will be developed. The Human Genome Project has produced a new area of genetics called genomics.

Genetic Engineering

Genetic engineering is a process by which genes from the chromosomes of one species are inserted into the chromosomes of another species. This is accomplished with the help of two enzymes. *Restriction endonuclease* is used to cut out a segment of DNA or a gene from the chromosome of one species, and the enzyme *DNA ligase* is used to splice the inserted segment into the chromosome of another species. This technique is called *gene splicing* and results in the production of a *recombinant DNA molecule*. Gene splicing is often done to pieces of circular DNA found in bacteria called *plasmids*. The plasmid is then used as a *vector* (carrier) to bring the new DNA or gene into other cells, where it can reproduce. Viruses can also be used as vectors to carry spliced genes into other cells. The reproduction of genes (inserted) in host cells is a form of gene cloning. When the host cell reproduces, the inserted recombinant DNA also reproduces. Using bacteria as the host for recombinant DNA has produced human insulin, growth hormone, and interferon. Bacteria reproduce rapidly, and the inserted recombinant DNA can be used to economically produce large quantities of a desired protein.

Genetic Engineering

Plasmid

Plasmid with a segment removed by restriction endonuclease

Foreign DNA

Plasmid with foreign DNA spliced by DNA ligase

Plasmid with foreign DNA inserted into a bacterial cell

Bacterial DNA

Cloning

A clone can be DNA, genes, cells, tissues or an organism that is genetically identical to the ancestor it is derived from. With genetic engineering, genes can be cloned by inserting them into host cells. Also, the nucleus can be removed from a somatic cell (muscle cell, stomach cell, skin cell) and inserted into an unfertilized egg cell that has had its nucleus removed. The organism that is produced is identical to the one that donated the nucleus. To remove a nucleus from a cell, microdissection tools must be used. When cloning, it is not just one or two genes that are moved into the DNA of another organism; an entire set of chromosomes and genes are moved. Also, it is not just one protein that is produced but an entire organism. In 1997, the Scottish researcher Dr. Ian Wilmut produced a sheep that he named Dolly using this technique.

Stem Cells

Stem cells are undifferentiated cells that are removed from a blastocyst. Stems cells have the potential to produce other stem cells, or they can differentiate and specialize to produce tissues or organs that can potentially be used as replacement human body parts.

Parthenogenesis

Parthenogenesis is the development of an unfertilized egg into an embryo that eventually produces a new individual. Parthenogenesis occurs naturally in some insects, fish, lizards, and plants. Parthenogenesis can be artificially induced in an egg cell by the use of chemicals or electricity. Gregory Pincus first demonstrated this in 1936 when he induced parthenogenesis in rabbit eggs. In February 2002, scientists from the Mayo Clinic, Sloan Kettering Cancer Center, and Wake Forest University were able to stimulate a monkey egg into a blastocyst from which they were able to extract stem cells that they grew into heart, brain, and smooth muscle cells. At this point, it just seems to be a question of time as to who will be first to produce human embryos by parthenogenesis.

All the technologies discussed previously raise certain ethical and moral questions for which no easy answers exist. For example, should the cloning of organisms be permitted? What if someone decides to clone a human? Should new foods be produced using recombinant DNA technology? Many countries refuse to important corn grown from seeds that were produced using recombinant DNA technology. Are we killing a potential human being when a blastocyst is destroyed to harvest its stem cells for research? How can we deny a replacement organ for an individual that might need a heart or liver? These are difficult questions with no easy answers.

Example Problems

These problems review biotechnology.

1. What is biotechnology?

   Answer: Biotechnology involves the use of engineering and technology to advance biological research. Some applications of biotechnology are in the areas of gene sequencing, genetic engineering, cloning, and stem cell research.

2. What were the goals of the Human Genome Project?

   Answer: The three key goals of the Human Genome Project were to determine the sequences of the 3 billion nitrogen base pairs in DNA, identify the 30,000 genes in human DNA, and create a database to store this information that would be available to researchers.

3. Why is the Human Genome Project important?

   Answer: The Human Genome Project is important because it will enhance our knowledge of gene function and lead to the discovery of new methods of diagnosing and treating disease.

4. What is genetic engineering?

   Answer: Genetic engineering is a technique used to transfer genes from the chromosomes of one species into the chromosomes of another species.

5. What is a vector?

   Answer: A vector is a carrier used to bring DNA into other cells. Viruses and bacterial plasmids are examples of commonly used vectors.

Work Problems

Use these problems on biotechnology to give yourself additional practice.

1. How can genes be removed from the chromosome of one species and inserted into the chromosome of another species?

2. What are plasmids?

3. How can an organism be cloned?

4. Why is stem cell research controversial?

5. What are three ways to produce an embryo?

Worked Solutions

1. The restriction enzyme endonuclease can be used to cut a gene out of the DNA from one chromosome, and the enzyme DNA ligase can be used to splice the inserted gene into another chromosome. This technique results in the production of a recombinant DNA molecule.

2. Plasmids are pieces of circular DNA in bacteria that are often used in gene splicing. A spliced plasmid can be inserted into other cells.

3. One method of cloning involves removal of the nucleus from a somatic cell and its insertion into an unfertilized egg cell that has had its nucleus removed. An embryo develops that becomes a new individual, identical to the one that donated the nucleus.

4. Stem cells are obtained from the blastocyst stage of embryological development. To harvest stem cells, the embryo is destroyed. Many people feel human life begins at conception and the destruction of an embryo prevents the formation of a new individual. Others feel that an embryo is not a human and that the benefits to be derived from stem cell research are of primary importance.

5. The traditional method of embryo production involves the union of egg and sperm cells at fertilization to produce a zygote. This zygote becomes an embryo as a result of cleavage. Cloning and parthenogenesis are two biotechnological methods of embryo production.

Chapter Problems and Answers

Problems

The following is a brief paragraph based on human molecular genetics. For problems 1–10, fill in the missing terms.

The DNA molecule is important because it contains the _____ information of the cell.
                                                                              1

The DNA molecule is found in chromosomes that are located inside the _____. In 1953,
                                                                                    2

Watson and Crick discovered that the DNA molecule was a _____. The DNA molecule
                                                              3

is composed of paired _____ that are held together by weak _____ bonds. During the
                        4   5

process of cell division or _____, the chromosomes replicate, and the DNA molecules
                              6

in the chromosomes also _____. The DNA molecules unzip, and each strand of a DNA
                          7

molecule serves as a _____ for the formation of a new strand of DNA. At the completion
                      8

of replication, each DNA molecule that formed consists of one original and one new

_____. This type of replication is known as _____.
  9   10

For problems 11–16, answer the question or fill in the missing answer.

11. Amino acids combine to form a protein by the process of _____.

12. Which sugar molecule is present in RNA? _____.

13. A set of three nitrogen bases on an mRNA molecule is called a _____.

14. Where is mRNA produced? _____.

15. The addition of a nitrogen base to a codon or the deletion of a nitrogen base from a codon
     on an mRNA molecule is called a _____.

16. What tools are needed to remove a nucleus from a cell? _____

Problems 17–20 are based on the following diagram. Answer the question, or fill in the missing answer.

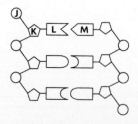

17. The diagram shows a section of a molecule known as _____.

18. Structures J, K, and L in the molecule make up a _____.

19. Structures L and M are examples of _____.

20. What is the name of structure K? _____.

For problems 21–25, select the process from the following list that is associated with each description. A choice can be used more than once or not at all.

A. translation   B. transcription   C. mutation   D. genetic engineering   E. genomics

21. A change or error in an allele. _____

22. This process produces mRNA, tRNA, and rRNA. _____

23. The process by which the ribosome combines amino acids to produce proteins. _____

24. The study of genes, their structure, and their function. _____

25. Genes from the chromosomes of one species are inserted and spliced into the chromosomes of another species. _____

For problems 26–30, select the disease from the following list that is associated with each description. A choice can be used more than once or not at all.

A. hemophilia   B. Huntington's disease   C. phenylketonuria
D. sickle-cell anemia   E. Tay-Sachs disease

26. Identify the disease caused by the missing enzyme called hexosaminidase, which is needed to break down a lipid called GM2 ganglioside. _____

27. This is a sex-linked disease. _____

28. This disease is characterized by the presence of crescent-shaped red blood cells. _____

29. This disease is caused by a defective dominant allele. _____

30. People with this disease experience difficulty with blood clotting. _____

Answers

1. **Hereditary.** Genes control heredity, and sections of a DNA molecule make up a gene.

2. **Nucleus.** In eukaryotic cells, chromosomes are found in the nucleus, where they are protected by the nuclear membrane.

3. **Double helix.** The double helix resembles a spiral staircase.

4. **Nucleotides.** Each nucleotide contains deoxyribose, a phosphate group, and one of four possible nitrogen bases.

5. **Hydrogen.** The hydrogen bonds form between the nitrogen bases of each nucleotide.

6. **Mitosis.** Each chromosome in the nucleus replicates at the very end of interphase. Replication signals the start of mitosis, or cell division.

7. **Replicate** or **duplicate.** The DNA molecule must replicate for the hereditary information contained in the genes to be passed on to the next generation.

8. **Template.** A template is a pattern.

9. **Strand.** A DNA strand is half of a double helix.

10. **Semiconservative replication.** In semiconservative replication one original DNA strand is "conserved" in the DNA molecule that replicated.

11. **Dehydration synthesis.** Dehydration synthesis is the formation of a large molecule from two or more smaller molecules by the removal of water.

12. **Ribose.** DNA contains the sugar deoxyribose.

13. **Codon.** Some examples of codons are U-U-U, A-C-G, G-U-C, and A-A-A. There are 64 possible combinations (4^3) of the three nitrogen bases that form a codon.

14. **mRNA is produced in the nucleus.** After being produced in the nucleus, mRNA travels into the cytoplasm to a ribosome where it functions in protein synthesis.

15. **Gene mutation** or **frameshift mutation.** A frameshift mutation results in the formation of new codons. Most mutations are not beneficial to the organism.

16. **Microdissection tools.** These tools are extremely small and are used with the aid of a microscope to remove or add parts of cells. Cloning requires the use of microdissection tools.

17. **DNA.** This molecule must be DNA because paired nucleotides are visible.

18. **Nucleotide.** Each nucleotide contains deoxyribose (K), a phosphate group (J), and a nitrogen base (L).

19. **Nitrogen bases.** The sequence of nitrogen bases determines the genetic code. The four nitrogen bases in DNA are adenine, thymine, cytosine, and guanine.

20. **Deoxyribose.** Deoxyribose is the sugar molecule in DNA, and ribose is the sugar molecule in RNA.

21. **C. Mutation.** A change or error in an allele is called a gene mutation.

22. **B. Transcription.** Transcription is the process that produces RNA nucleoproteins.

23. **A. Translation.** In the process of translation, tRNA brings amino acids to mRNA at the ribosome, which aids in the formation of a protein.

24. **E. Genomics.** Genomics is a new area in genetics that may one day lead to the discovery of new methods of diagnosing and treating disease.

25. **D. Genetic engineering.** Genetic engineering is an example of biotechnology.

26. **E. Tay-Sachs disease.** Without hexosaminidase (Hex-A), GM2 ganglioside accumulates in the nerve cells of the brain causing those cells to die.

27. **A. Hemophilia.** The allele responsible for hemophilia is linked to the X chromosome.

28. **D. Sickle-cell anemia.** The recessive genotype for sickle-cell anemia produces an error in the hemoglobin molecule that results in red blood cells that have a sickle shape.

29. **B. Huntington's disease.** Only one defective allele is needed to have this disease because the allele is dominant.

30. **A. Hemophilia.** Hemophilia is often referred to as the bleeders disease.

Supplemental Chapter Problems

Problems

For problems 1–30, select the *best* answer.

1. The shape of the DNA molecule can be described as:

   A. a single helix   B. a double helix   C. rod shaped   D. circular

2. All the following are nitrogen bases *except:*

   A. glycine   B. adenine   C. thymine   D. cytosine

3. One nucleotide consists of:

   A. ribose, deoxyribose and a phosphate group

   B. ribose, a nitrogen base and a hydrogen group

   C. deoxyribose, a nitrogen base and a hydrogen group

   D. deoxyribose, a nitrogen base and a phosphate group

4. The nitrogen bases of a DNA molecule are held together by:

   A. hydrogen bonds   B. oxygen bonds   C. nitrogen bonds   D. carbon bonds

5. A section of one strand of DNA has the following nitrogen base sequence: C-G-A-G-C-T. What is the nitrogen base sequence of the complementary strand?

   A. G-C-T-C-G-A   B. G-C-U-C-G-A   C. C-G-A-G-C-T   D. G-C-A-C-G-T

6. Which of the following is the correct method of DNA replication?

   A. DNA acts as a template for the formation of a new molecule directly from the original.

   B. The DNA molecule unzips, and new nucleotides form on each separated strand. The two new strands combine, and the two old strands recombine. This gives two DNA molecules, one with two old strands and one with two new strands.

   C. The DNA molecule unzips, new nucleotides form on each separated strand and old and new strands remain combined. This gives two DNA molecules, each with one old and one new strand.

   D. The DNA molecule unzips, and new nucleotides form on each separated strand. The two old strands recombine. The new strands remain separate. This gives one DNA molecule with two old strands and two molecules each with one strand.

7. Which organelle in a cell is the site for protein synthesis?

    A. mitochondrion   B. ribosome   C. nucleus   D. centriole

8. Proteins are composed of:

    A. carboxyl groups   B. polysaccharides   C. amino acids   D. fatty acids

9. Which nitrogen base is *not* found in DNA?

    A. uracil   B. thymine   C. adenine   D. cytosine

10. Which process produces RNA nucleoproteins?

    A. transpiration   B. translocation   C. translation   D. transcription

11. Which of the following molecules is *not* a nucleoprotein?

    A. mRNA   B. rRNA   C. DNA   D. ATP

12. The genetic code of the DNA molecule is determined by its:

    A. number of deoxyribose molecules   B. sequence of nitrogen bases

    C. sequence of amino acid molecules   D. sequence of sugar and phosphate groups

13. The tRNA molecule functions in the transfer of:

    A. ATP   B. electrons   C. amino acids   D. fatty acids

14. If the codon on an mRNA molecule is C-C-A, the anticodon on the tRNA molecule is:

    A. G-G-U   B. G-G-T   C. G-G-C   D. A-A-U

15. For a mutation to be inherited, it must occur in a:

    A. sex cell   B. liver cell   C. skin cell   D. kidney cell

16. A codon on an mRNA molecule has the following nitrogen base sequence: A-A-C. Which of the following choices shows a point mutation?

    A. A-A-C   B. A-A-G   C. A-A   D. A-A-C-G

17. A segment of mRNA has the following nitrogen base sequence: C-U-G-C-C-G-C-A-G. A mutation occurs, and the sequence of the nitrogen changes as follows: C-U-G-C-C-C-A-G. What do we call this type of mutation?

    A. translocation   B. inversion

    C. nondisjunction   D. frameshift mutation

18. What do we call a mutation where the sequence of nitrogen bases changes?

    A. a gene mutation   B. a chromosome mutation

    C. polyploidy   D. translocation

19. All the following are examples of a chromosome mutation *except:*

    A. translocation   B. inversion

    C. frameshift mutation   D. nondisjunction

20. A mutation where a segment of a chromosome is lost or missing is called a(n):

    A. addition   B. inversion   C. deletion   D. translocation

21. Which of the following is *not* a cause of mutations?

    A. X-rays   B. transcription   C. asbestos   D. ultraviolet rays

22. Which of the following genetic diseases is *not* caused by the presence of a defective recessive allele?

    A. sickle-cell anemia   B. Tay-Sachs disease

    C. cystic fibrosis   D. Huntington's disease

23. Select the disease that is caused by a missing enzyme.

    A. phenylketonuria or PKU   B. Huntington's disease

    C. Down's syndrome   D. sickle-cell anemia

24. Identify the disease that produces a thick mucus that clogs the air passages of the lungs and ducts in the pancreas.

    A. cystic fibrosis   B. phenylketonuria or PKU

    C. Tay-Sachs disease   D. sickle-cell anemia

25. What is the name of the process by which genes from the chromosomes of one species are inserted into the chromosomes of another species?

    A. parthenogenesis   B. genomics   C. genetic engineering   D. cloning

26. Select the process that uses chemicals or electricity to start cleavage in an unfertilized egg.

    A. parthenogenesis   B. genomics   C. genetic engineering   D. cloning

27. As a result of the Human Genome Project, scientists were able to determine the:

    A. cause of all human genetic diseases

    B. sequences of the 3 billion nitrogen base pairs in DNA

    C. cause of cancer

    D. reason for structural chromosome mutations

28. The insertion of a gene into a chromosome involves a technique known as:

    A. deletion   B. transformation   C. translocation   D. gene splicing

29. The removal of a nucleus from a somatic cell and its insertion into an unfertilized egg cell that has had its nucleus removed can result in the production of an organism that is identical to the organism that contributed the donor nucleus. The process described here is called:

    A. cloning   B. transformation   C. translocation   D. parthenogenesis

30. Undifferentiated cells that are removed from inside the blastocyst stage of embryological development are known as:

    A. translocated cells   B. deleted cells   C. stem cells   D. mutated cells

Answers

1. **B.** "DNA Structure," p. 253

2. **A.** "DNA Structure," p. 253

3. **D.** "DNA Structure," p. 253

4. **A.** "DNA Structure," p. 253

5. **A.** "DNA Structure," p. 253

6. **C.** "DNA Replication," p. 254

7. **B.** "Protein Synthesis," p. 256

8. **C.** "Protein Synthesis," p. 256

9. **A.** "Transcription," p. 256

10. **D.** "Transcription," p. 256

11. **D.** "Transcription," p. 256

12. **B.** "Transcription," p. 256

13. **C.** "Translation," p. 256

14. **A.** "Translation," p. 256

15. **A.** "Mutations," p. 259

16. **B.** "Gene Mutations," p. 259

17. **D.** "Gene Mutations," p. 259

18. **A.** "Gene Mutations," p. 259

19. **C.** "Chromosome Mutations," p. 260

20. **C.** "Chromosome Mutations," p. 260

21. **B.** "Causes of Mutation," p. 260

22. **D.** "Genetic Diseases," p. 262

23. **A.** "Genetic Diseases," p. 262

24. **A.** "Genetic Diseases," p. 262

25. **C.** "Genetic Engineering," p. 265

26. **A.** "Parthenogenesis," p. 266

27. **B.** "Genomics," p. 265

28. **D.** "Genetic Engineering," p. 265

29. **A.** "Cloning," p. 265

30. **C.** "Stem Cells," p. 266

Chapter 14
Evolution

Charles Darwin was born in 1809, in England. In 1825, Darwin started medical school but never finished. In 1828, he enrolled in divinity school at Cambridge and graduated in 1831. Instead of concentrating on his studies, he was out hiking with naturalists, collecting insects, and reading books on nature. These activities gave Darwin the tools that he eventually used to develop his theory of evolution. In 1831, Darwin signed on as the naturalist on the HMS *Beagle* and traveled around the world for the next five years. During his travels, Darwin gathered the data that in later years were used as the basis for his books and for his ideas on evolution. This chapter covers evolution, presents evidence for evolution and looks at several concepts that attempt to explain how evolution takes place.

Evolution Defined

Evolution can be defined as the change that has occurred in a species of organism with the passage of time. *The passage of time* is used here to mean from the distant past to the present. The great diversity of living things on earth today is a direct result of biological evolution. The traits in all species and in individual organisms developed over time by evolution. Evolution unifies all the different areas of study in the science of biology.

Fossil Evidence for Evolution

Fossils are the remains of organisms that lived in the distant past. Since life first developed on earth, countless numbers of organisms have lived and died. Many of these organisms have left behind some remains of their existence. By examining structures in a fossil species and comparing them to structures in similar species of today, the changes that have occurred with the passage of time can be seen. These changes represent evolution. Fossils are strong evidence of evolution because they are tangible and show that change has taken place. Some examples of fossils and their method of formation are hard parts, imprints, molds, petrifaction, refrigeration, amber, and tar pits.

Hard Parts

Hard parts are bones, teeth, or shells of organisms that lived in the past. The hard parts of an organism under dry climatic conditions often survive the natural process of decay. By comparing these remains with organisms of today, the change that that has taken place can be seen. Evolution is change. For example, fossil skeletons of horses show that this organism has changed (evolved) from a small horse, *Eohippus*, about the size of a fox, to the modern-day large horse, *Equus*.

Imprints

Imprints are impressions made by plants and animals in soft soil or mud that with the passage of time harden to become rock. Dinosaur footprints and the outlines of leaves in rock are good examples of imprints.

Molds

Imagine a fish swimming in the ocean that suddenly dies and sinks to the bottom. As time goes by, the body of the fish becomes covered with sand and mud. As the fish decays, its outline remains, forming a *mold* of the organism. Meanwhile, the sand and mud harden to form rock. Millions of years later, this rock can be broken open to reveal a mold of the fish.

Petrifaction

Millions of years ago, trees, and bones of organisms that died were often covered with water that was rich in minerals. Minerals in the water diffused into the cells of the organism, forming rock that resembles the organic tissue that was replaced. In the United States, petrified forests can be found in Arizona and California.

Refrigeration

Refrigeration occurs when an organism is preserved by cold temperatures or by ice. In a part of Russia known as Siberia, mammoths (which are related to the modern-day elephant) have been found preserved in ice in remarkably good condition.

Amber

Amber is a form of fossilized resin from trees that lived millions of years ago. As the resin dripped down from the tree, it often trapped and surrounded an insect or part of a plant. As the resin hardened, the organism that was trapped inside was preserved. Most amber comes from the Dominican Republic on the island of Haiti and the Baltic region of Europe. Amber is prized by jewelers for its gem-like quality and by scientists for any organisms that might be trapped inside.

Tar Pits

A tar pit is an accumulation of oil or asphalt that seeped up to the earth's surface. Often animals became trapped in the sticky asphalt. The soft parts of the animal decayed, but the hard parts (such as bones and teeth) were preserved by the asphalt. In Los Angeles, the La Brea tar pits are famous for fossils of dire wolves, saber-toothed tigers, mammoths, horses, camels, and many other species. Other tar pits have been found in Iran, Peru, Poland, and Russia.

Example Problems

These problems review evolution and fossils.

1. Define evolution.

   Answer: Evolution is the change that has occurred in a species of organism with the passage of time.

2.   What is a fossil?

    Answer: A fossil is the remains of an organism that lived in the past.

3.   Why are the soft parts of organisms rarely fossilized?

    Answer: The soft parts of organisms are rarely fossilized because they decay rapidly after the death of the organism. The hard parts of the organism decay slowly and can form fossils.

4.   What is amber?

    Answer: Amber is a form of fossilized resin from trees that can contain trapped insects or plant material.

5.   How does refrigeration result in fossil formation?

    Answer: Refrigeration results in fossil formation because cold temperatures can dramatically slow down the process of decay.

Work Problems

Use these problems on evolution and fossils to give yourself additional practice.

1.   Why is the study of evolution important?

2.   Why is the fossil record strong evidence for evolution?

3.   How are imprints different from molds?

4.   How do tar pits help form fossils?

5.   Describe the process of petrifaction.

Worked Solutions

1.   Evolution shows how organisms have changed over time. Evolution accounts for the huge diversity of living things that exist on earth.

2.   Fossils are strong evidence in support of evolution because they are the tangible remains of organisms that have lived in the past. Fossils can be compared to current organisms so that any changes that have taken place can be determined.

3.   An imprint is an impression of a part of an organism. Examples of imprints are leaf prints and footprints. A mold is an outline of the organism that is left behind after the organism dies, becomes covered with sand or mud, and then decays.

4.   Tar pits contain asphalt that surrounds the bones of the organism preventing the bacteria of decay from breaking down the bones. These preserved bones eventually become fossils.

5.   Petrifaction occurs when minerals in water diffuse into the cells of a dead organism, forming rock.

Dating of Fossils

A dog brings home a bone that it found. Is this bone a fossil? Or, is this a bone leftover from someone's dinner? Two techniques are used to date fossils: relative dating and absolute dating.

Relative Dating

Relative dating is a method used to determine the age of a fossil by comparing its location *relative* to fossils in nearby rock layers. In relative dating an exact age is not assigned to a fossil. When a rock formation is examined, the oldest fossils are found in the bottom layer, and the youngest fossils are found in the top layer. Some species of organisms were once found all over the world, but they lived for a short time. Fossils of these organisms are known as *index fossils*. Their presence in a rock layer can be used to date the layer and the fossils of organisms that it contains.

Most fossils are found in sedimentary rock. *Sedimentary rock* is formed by the deposition of very small particles of rock, clay or silt. Sedimentary rock forms in layers, often in water, and takes millions of years to form. *Igneous rock* is formed from volcanic activity and never contains fossils. The high temperature associated with the formation of this rock incinerates any organism unfortunate enough to be caught up in the lava flow. *Metamorphic rock* is formed by tremendous heat and pressure applied to igneous and sedimentary rock. Fossils are not found in metamorphic rock.

Absolute Dating

Absolute dating is a method used to determine the approximate age of a fossil by relying on the radioactivity of certain elements and their half-lives. *Half-life* is the amount of time that it takes for an element to decay into half of its original amount. The older the fossil, the more the radioactive element has decayed. For example, carbon-14 decays into nitrogen-14 in 5,730 years. If at the time an organism died one of its bones had 1 milligram (mg) of carbon-14, and at the time the fossil bone was examined it had .25 milligrams of carbon-14, the bone is 11,460 (5,730 × 2) years old.

Carbon-14 Dating

$$C^{14} \xrightarrow{\quad 5730 \quad} N^{14} \; + \; C^{14} \xrightarrow{\quad 5730 \quad} N^{14} \; + \; C^{14}$$

1 mg   0.5 mg   0.25 mg

Fossils of organisms that lived up to 50,000–70,000 years ago can be dated using carbon-14. Beyond this, not enough carbon-14 remains to decay because the half-life of carbon-14 is short. Other elements used in absolute dating are uranium-238 (which decays to form lead-206 in 4.47 billion years) and potassium-40 (which decays into argon-40 in 1.25 billion years). Both of these elements can be used to date rocks that contain fossils.

Example Problems

These problems review the dating of fossils.

1. What kind of rock are fossils found in?

    Answer: Fossils are found in sedimentary rock.

2. Why are fossils dated?

    Answer: Fossils are dated to determine when the organism that formed the fossil was alive.

3. Why aren't fossils found in igneous rock?

    Answer: Igneous rock is formed by volcanic activity and lava flow. The high temperatures associated with this type of rock formation destroy any organisms, preventing the formation of fossils.

4. Why can't carbon-14 dating be used for fossils that are more than 50,000–70,000 years old?

    Answer: The half-life of carbon-14 is short (5,730 years), so after 50,000–70,000 years not enough carbon-14 is left in a fossil to be measured.

Work Problems

Use these problems on the dating of fossils to give yourself additional practice.

1. How does the relative dating of fossils work?

2. What is an index fossil?

3. What is absolute dating?

4. How does carbon-14 dating work?

5. Why can't carbon-14 be used to date a rock?

Worked Solutions

1. In relative dating, the age of a fossil is not determined. However, the age of one fossil relative to another can be determined. For example, fossils in layers near the top of a rock formation are younger than fossils near the bottom.

2. An index fossil represents a species of organism that was found all over the world but lived for a short period of time. If the age of an index fossil is known, it can be used to determine the age of other fossils in the same layer. In addition, fossils in layers above the index fossil are younger. Fossils in layers below the index fossil are older.

3. Absolute dating gives an approximate date of when the organism represented by a fossil died. Absolute dating relies on the radioactivity of certain elements and their half-life. The older the fossil, the more the radioactive element has decayed.

4. Carbon-14 dating is a form of absolute dating. Carbon-14 has a half-life of 5,730 years. The carbon-14 in a fossil sample decays into half its original amount every 5,730 years. A sample that has one-eighth the original amount of carbon-14 has gone through three half-lives and is $3 \times 5,730$ (or 17,190) years old.

5. Rocks generally don't contain carbon, so trying to use carbon-14 to date a rock is meaningless. Carbon-14 dating can be used to date fossils that contain carbon, such as bones and teeth (hard parts). To date rocks that might contain fossils, uranium 238 and the potassium-argon method can be used.

Additional Evidence of Evolution

The evolutionary relationship between many organisms can be traced back to a common ancestor. A *common ancestor* is an individual from which two or more related species could have evolved. With the passage of time, organisms change and diverge from their common ancestor to form new species. The following evidence for evolution demonstrates the concept of the common ancestor.

Comparative Anatomy

If the *anatomy* (body structure) of the human forelimb is compared with that of a bat, cat, and porpoise, a similarity of structure can immediately be noticed. This is to be expected because all these organisms are mammals, and all mammals at one time changed and diverged from a common ancestor. The wing of a bird is much different from the structures mentioned previously because mammals and birds are not in the same class and do not share a close common ancestor.

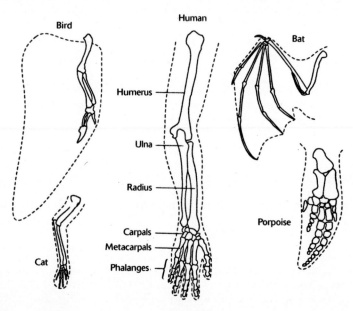

Evolutionary relationships based on comparative anatomy depend on homologous structures. *Homologous structures* are similar in construction and evolutionary development but dissimilar in function. Examples are the forelimb of a human and the wing of a bat. Analogous structures are not used in comparative anatomy. *Analogous structures* are similar in function but dissimilar in construction and evolutionary development. An example is the wing of a bird and the wing of an insect. Birds and insects use wings for locomotion, but the construction of their wings is dramatically different (as was their evolutionary development). A *vestigial structure* at one time had a function in the evolutionary history of an organism but now does not have a function. In humans, the appendix and the coccyx are examples of vestigial structures. The appendix is thought to have had a role in digestion, and the coccyx is what remains of a tail that human ancestors once possessed. The change from functionality to nonfunctionality of these structures is evidence of human evolution.

Comparative Embryology

An amazing similarity exists between vertebrate embryos. During their early stages of development, all vertebrate embryos have gill slits, a two-chambered heart and a tail. As the embryos develop, they begin to acquire the unique characteristics of their species. The similarity of embryological development supports the concept of the common ancestor.

Comparative Biochemistry

DNA, RNA, the genetic code and protein synthesis are similar in all organisms. The greater the genetic and molecular similarity between species, the closer their common ancestor. Humans and chimpanzees have 98% of their genes in common. The remaining 2% is what distinguishes these two species from each other. Humans did not descend from chimpanzees but from primitive humans. However, at some point in evolutionary history, primitive humans and primitive chimpanzees probably diverged from a common ancestor. Diabetics can use insulin from cows and pigs because insulin from these animals is almost identical to human insulin. In addition, hemoglobin in humans, which has almost 600 amino acids, is almost identical to hemoglobin in all other vertebrates. This similarity in chemical structure demonstrates that all vertebrates can be traced back to a common ancestor.

Example Problems

These problems review additional evidence of evolution.

1. What are homologous structures?

   Answer: Homologous structures are similar in construction and evolutionary development but dissimilar in function. Homologous structures are often used to show evolutionary relationships between organisms. Examples of homologous structures are the forelimb of a human and the wing of a bat.

2. What are analogous structures?

   Answer: Analogous structures are similar in function but dissimilar in construction and evolutionary development. Examples of analogous structures are the wing of a bird and the wing of an insect.

3. How do vestigial structures provide evidence of evolution?

   Answer: A vestigial structure at one time had a function but no longer has a function. The change from functionality to nonfunctionality demonstrates evolution. For example, at one time the human appendix had a digestive function but now does not. This change shows that humans have evolved.

Work Problems

Use these problems on additional evidence of evolution to give yourself more practice.

1. Explain the concept of common ancestry.

2. How does comparative anatomy provide evidence for evolution?

3. Scientists have found skeletons of camels dating back millions of years. Early skeletons of camels show that this animal was once smaller and different from modern-day camels. How can this information be used to provide evidence of evolution?

4. How does comparative embryology provide evidence for evolution?

5. How does comparative biochemistry provide evidence for evolution?

Worked Solutions

1. A common ancestor is an individual from which two or more related species could have evolved.

2. Comparative anatomy is a technique used to compare homologous body structures of organisms. The more the structures are similar to each other, the closer the evolutionary relationship is between the organisms.

3. If fossilized camel skeletons from different points in history are compared, the differences between them shows a pattern of change that proves evolution occurred.

4. The similarity between embryos during the early stages of development supports the concept of the common ancestor.

5. Comparative biochemistry provides strong evidence for evolution. The close similarity in complex molecules found between different species argues in favor of a common ancestor from which organisms could have evolved (changed). For example, human insulin differs from cow insulin in only 3 out of 51 amino acids. Cow insulin is close enough to human insulin that it can be given to diabetics.

Theories of Evolution

Charles Darwin was not the first person to propose a theory to explain evolution. In the early 1800s, naturalists tried to explain the changes they observed in the fossil record through the concepts of *acquired characteristics* and *use and disuse*. In 1859, Charles Darwin published his view of evolution in his book *On the Origin of Species* in which he proposed the theory of natural selection. Hugo De Vries (early 1900s) updated the theory of natural selection by suggesting that mutations are a source of variation in a population.

Inheritance of Acquired Characteristics

In Darwin's time many naturalists believed that the more an organism used a structure, the better and more highly developed the structure became. They also believed that the acquired improvement in the structure could be inherited by the offspring of the next generation. In addition, they said that the reverse was also true. If an organism doesn't use a structure, it is not inherited by the next generation. In other words, use it or lose it. A naturalist named Jean Baptiste Lamarck (who preceded Darwin) used the giraffe as a hypothetical example to explain this concept. Millions of years ago, the giraffe was a smaller animal with a short neck. The giraffe stretched its neck to reach leaves higher up in a tree. When the giraffe reproduced, the acquired characteristic of the stretched neck was inherited by its offspring. It is important to note that Lamarck was concerned with inheritance, not with evolution. However, the concepts of *acquired characteristics* and *use and disuse* were popular at the time because no alternatives existed.

Natural Selection

Darwin's voyage on the HMS *Beagle* provided him with the data that he needed to formulate his theory of natural selection. Natural selection explains how evolution takes place. The key concepts of this theory are as follows:

❑ **Overproduction:** Darwin observed that when members of a species reproduce, they produce more offspring than is necessary to maintain the species' numbers.

❑ **Struggle for survival:** Overproduction results in competition for scarce resources such as food, water, and territory, which are necessary for survival.

- **Variation:** Within every population, members of a species show variation (differences) in their traits.

- **Natural selection:** Nature selects those variations that are the most fit (survival of the fittest). Organisms with the best variation survive the struggle for existence and reproduce. As a result, these variations are passed on to the next generation.

- **Formation of new species:** With the passage of time and many generations, the most-fit variations become the norm within the population. The less-fit variations are eliminated. Eventually, a new species of organism evolves that is significantly changed from that of its distant ancestor.

Mutation

One problem with Darwin's theory of natural of selection is that it does not explain the source of variation within a *population* (all members of a species living together in an area). In 1901, Hugo De Vries (a Dutch botanist) noted sudden changes in the evening primrose. De Vries called these changes mutations. The modern approach to evolution proposes that mutations are a source of variation in a population that can lead to the formation of new species with the passage of time.

The Rate of Evolution

The rate of evolution and speciation (new species formation) can be explained by gradualism and punctuated equilibrium.

- **Gradualism:** Many biologists believe that evolution is a slow, continual and gradual process that proceeds in numerous small steps, taking many years and generations for new species to form.

- **Punctuated equilibrium:** Other biologists believe that evolutionary change occurs in sudden spurts during which many new species are formed, followed by long periods of stability with no speciation.

Example Problems

These problems review theories of evolution.

1. According to Darwin, what is the result of overproduction?

   Answer: The result of overproduction is competition for scarce resources such as food, water, and territory.

2. What is meant by variation?

   Answer: Variation refers to the different and diverse forms of a trait.

3. How are the most-fit or best variations in a population selected?

   Answer: According to Darwin, the best variations are selected by nature. The modern approach to evolution states that nature selects the organisms with alleles for the fittest variations; these organisms are the ones most likely to reproduce and pass on the alleles for these variations to the next generation.

4. Which theory of evolution is best described by gradualism?

Answer: Darwin's theory of natural selection is best described by gradualism, which states that evolution is a gradual, slow and continuous process that proceeds in numerous small steps over many generations.

Work Problems

Use these problems on the theories of evolution to give yourself additional practice.

1. Explain the concept of acquired characteristics.

2. What did Darwin mean by survival of the fittest?

3. How did Darwin account for the formation of new species?

4. How does the concept of mutations update Darwin's theory of natural selection?

5. What is the punctuated equilibrium explanation for the rate of evolution?

Worked Solutions

1. According to the concept of acquired characteristics, the more an organism uses a structure, the better and more highly developed the structure becomes. The acquired improvement can be inherited by the offspring.

2. To Darwin, survival of the fittest meant that those individuals that had the most-fit variations got the opportunity to reproduce and to pass these variations on to the next generation.

3. Darwin believed that with the passage of time, the most-fit variations became the norm within a population, eventually giving rise to a new species of organism.

4. The concept of mutations is used to explain how variations can occur in a species of organism.

5. The punctuated equilibrium explanation of evolution states that change occurs in sudden spurts during which many species are formed, followed by long periods of stability with no speciation.

The Hardy-Weinberg Law

In 1908, G. H. Hardy and W. Weinberg simultaneously and independently discovered a law that predicts how gene frequencies are maintained from one generation to the next. The Hardy-Weinberg law states that in a population the frequency of an allele remains constant from generation to generation, as long as five conditions are met. Any deviation from the allele frequencies predicted by the Hardy-Weinberg law is an indication that evolutionary forces are at work. The Hardy-Weinberg law can be stated in mathematical terms. If the two alleles for a trait in a population are p (dominant) and q (recessive), all the p alleles and all the q alleles equal 100% of the alleles for that trait in the population.

$p + q = 1$ (100% as a decimal is 1.)

A cross between two heterozygous individuals (pq × pq) produces all the possible genotypes in a population. Look at the following Punnett square.

The Hardy-Weinberg Law

pq × pq

	p	q
p	pp	pq
q	pq	qq

Results:

pp + 2pq + qq = 1 or 100% of the possible genotypes

or

$p^2 + 2pq + q^2 = 1$ or 100% of the possible genotypes

p^2 = frequency of the homozygous dominant allele

2pq = frequency of the heterozygous genotype

q^2 = frequency of the homozygous recessive allele

For the Hardy-Weinberg law to work, the following five conditions must be met:

❑ **The gene pool must be large** to provide statistical accuracy of the Hardy-Weinberg law.

❑ **No migrations** can be allowed into or out of the population because this changes the frequency of alleles.

❑ **Mating must be random** to ensure that the laws of probability work. The organisms decide who their mates are without restrictions or interference.

❑ **No mutations** can occur because mutations can change the frequency of an allele within a population.

❑ **No natural selection** can be present because it tends to favor individuals within the population that have more-fit alleles. This tends to skew the results of the Hardy-Weinberg law.

The reality is that all these conditions are rarely met in any population. Whenever actual gene frequencies differ from those predicted by the Hardy-Weinberg law, it is an indication that evolution is taking place. Two factors that cause changes in gene frequencies predicted by the Hardy-Weinberg law are geographic isolation and environmental factors.

Geographic and Reproductive Isolation

Geographic isolation is caused by geographical barriers such as mountains, deserts, oceans, and rivers. Geographic isolation results in the physical separation of individuals within a population, preventing random mating between individuals. Eventually the organisms become *reproductively isolated* and are no longer able to mate and produce fertile offspring. When this happens, two new species have developed. Darwin observed geographic and reproductive isolation in finches (a small bird) of the Galapagos Islands. Reproductive isolation also occurs when species reproduce during different times of the year, have different mating rituals, or have reproductive structures that prevent mating between the male and female.

Environmental Factors

Environmental factors can favor one variation of a trait over another, resulting in a change in the frequency of an allele within a population. The peppered moth (*Biston betularia*) provides an excellent example. Peppered moths come in two varieties—one has light wings, and the other has dark wings. Prior to the industrial revolution in Europe, the light-winged moths were more common despite the fact that the allele for dark wings was dominant. However, after the industrial revolution, the dark-winged moth became more common. How did this shift in gene frequency occur? The industrialization of Europe deposited a great deal of soot from coal-burning factories onto the trunks of trees. The lighter-colored moths stood out against the dark background and were more visible to the birds that preyed on them. Meanwhile, moths with dark wings blended in with the soot and were less likely to be preyed on by birds. The environment selected in favor of the moths with darker wings, and the frequency for the dark-winged allele in the population increased.

Example Problems

These problems review the Hardy-Weinberg law.

1. State the Hardy-Weinberg law.

    Answer: Within a population, the frequency of an allele remains constant from generation to generation, as long as certain conditions are met.

2. State the two mathematical equations used to express the Hardy-Weinberg law.

    Answer: The two mathematical equations used to express the Hardy-Weinberg law are $p + q = 1$ and $p^2 + 2pq + q^2 = 1$.

3. Why must the gene pool be large for the Hardy-Weinberg law to work?

    Answer: An increase in the size of the gene pool creates an increase in statistical accuracy.

4. How does migration into or out of a population affect the Hardy-Weinberg law?

    Answer: Migrations of individuals into or out of a population changes the frequency of alleles in a population.

5. In a population of fruit flies, the percentage of individuals with the dominant allele for red eyes is 80%. What is the percentage of the recessive allele for white eyes?

    Answer: Twenty percent of the fruit flies have the recessive allele for white eyes. The equation $p + q = 1$ can be used to solve the problem. Here, p is the percentage of individuals with the dominant allele (0.8 or 80%), and q is the percentage of individuals with the recessive allele.

    $p + q = 1$

    $0.8 + q = 1$

    $q = 0.2$ or 20%

Work Problems

Use these problems on the Hardy-Weinberg law to give yourself additional practice.

1. What is geographic isolation?

2. How does geographic isolation upset the Hardy-Weinberg law?

3. How can environmental factors upset the Hardy-Weinberg law?

4. How can the Hardy-Weinberg law be used to indicate that evolution is taking place?

Problems 5–7 are based on the following information. In a certain population of dogs, curly hair is dominant over straight hair. The percentage of the curly haired allele in this population is 60%.

5. What is the frequency or percentage of the straight-haired allele in this population?

6. What frequency or percentage of the population is heterozygous for curly hair?

7. What frequency or percentage of the population has straight hair?

Worked Solutions

1. Geographic isolation is the physical separation of individuals within a population and is caused by geographical barriers such as mountains, deserts, oceans, and rivers.

2. Geographic isolation upsets the Hardy-Weinberg law by reducing the size of the gene pool and interfering with random mating.

3. Environmental factors upset the Hardy-Weinberg law by selecting one variation within a population over another.

4. The Hardy-Weinberg law states that within a population the frequency of an allele remains constant. Any change in the frequency of an allele in a population might indicate that evolution is taking place.

5. Forty percent of the dogs have the recessive allele for straight hair. The equation $p + q = 1$ can be used to solve the problem. Here, p is the percentage of individuals with the dominant allele for curly hair (0.6 or 60%), and q is the percentage of individuals with the recessive allele for straight hair. Solve for q:

    $p + q = 1$

    $0.6 + q = 1$

    $q = 0.4$ or 40%

6. Forty-eight percent of the dogs are heterozygous for curly hair. To get the answer, substitute the values for p and q in 2pq, which is the frequency of the heterozygous genotype.

    $2pq$ = frequency of the heterozygous genotype

    $2(0.6 \times 0.4)$ = frequency of the heterozygous genotype

    0.48 or 48% = frequency of the heterozygous genotype

7. Sixteen percent of the dogs have straight hair. Straight hair is homozygous, so q^2 is the frequency of the homozygous recessive allele. Because q equals 0.4 (see the answer to question 5):

$q^2 = (0.4)^2$

$q^2 = 0.16$ or 16%

The Origin of Life on Earth

The *heterotroph hypothesis* was proposed by A. I. Oparin and J. B. S. Haldane to explain how the first cells might have originated on earth. This is a hypothesis or educated guess and not a fact; however, the key concepts and inferences are based on facts. According to the heterotroph hypothesis, the first cells on earth were *heterotrophs* (organisms that obtain food for energy production) and most likely originated in the ocean.

A modern explanation of this hypothesis states that the earth's early atmosphere was a reducing atmosphere (one with little or no oxygen). Among the gases that were probably present in this atmosphere were water vapor, nitrogen, carbon monoxide, carbon dioxide, and some hydrogen. Ultraviolet (UV) radiation, high temperature, and lightning were energy sources for chemical reactions between the gases. These reactions may have resulted in the production of *simple organic compounds* such as simple sugars, amino acids, fatty acids, glycerol, and nucleotides that were washed out of the atmosphere and into the oceans by rain. Some biologists feel that most of these reactions took place in the ocean and not in the atmosphere.

The formation of *complex organic compounds* took place in the oceans (perhaps at sea vents), producing complex carbohydrates, proteins, lipids, and nucleoproteins. Eventually these compounds grouped together, biochemical processes developed, and the first *primitive prokaryotic heterotrophic cells* appeared. These early heterotrophs used organic compounds in the ocean as a source of energy (ATP), producing alcohol and carbon dioxide as waste products. In this way, the early heterotrophs added to the levels of carbon dioxide in the atmosphere. *The first autotrophs* were primitive prokaryotic cells that used carbon dioxide and water to produce glucose and oxygen by the process of photosynthesis. The early autotrophs changed the earth's atmosphere by introducing oxygen. At some point, a nuclear membrane developed to enclose the genetic material of heterotrophic and autotrophic cells, giving rise to the first eukaryotic cells. *Modern heterotrophs* are eukaryotic and developed the ability to use oxygen for respiration. Aerobic respiration produces more energy than anaerobic respiration, allowing these heterotrophs to become the dominant organisms on earth.

The *endosymbiotic theory* states that heterotrophic and autotrophic eukaryotic cells developed from a symbiotic relationship between prokaryotic organisms. Prokaryotic organisms with the ability to produce energy from organic molecules were ingested (engulfed by endocytosis) by a larger prokaryote, were incorporated into the cell, and eventually evolved to become the mitochondria of modern heterotrophs. Other prokaryotic organisms with the ability of photosynthesis were ingested by a prokaryote and evolved into the chloroplasts of modern autotrophs. The endosymbiotic theory is supported by the fact that mitochondria and chloroplasts have their own DNA and reproduce independently of the cell by binary fission (as do primitive prokaryotic organisms such as bacteria). Eukaryotic cells reproduce by mitosis, and mitochondria and chloroplasts have ribosomes that are similar to prokaryotic organisms.

Example Problems

These problems review the origin of life on earth.

1. Identify the gases that were present in the earth's primitive atmosphere.

Answer: The gases that were probably present in this atmosphere were: water vapor, nitrogen, carbon monoxide, carbon dioxide, and some hydrogen.

2. Name the organic molecules that were the first to form on earth.

   Answer: Some of the first organic molecules to form on earth were simple sugars, amino acids, fatty acids, glycerol, and nucleotides.

3. What kind of cells were the first cells on earth?

   Answer: According to the heterotroph hypothesis, the first cells to originate on earth were primitive prokaryotic heterotrophs.

Work Problems

Use these problems on the origin of life on earth to give yourself additional practice.

1. How was carbon dioxide added to the earth's atmosphere?

2. How are modern-day heterotrophs different from the first heterotrophs on earth?

3. State the main concept of the endosymbiotic theory.

Worked Solutions

1. Carbon dioxide was added to the earth's atmosphere by anaerobic respiration of the first heterotrophs.

2. The first heterotrophs were prokaryotic, reproduced by binary fission, and produced energy by fermentation (anaerobic respiration). Modern-day heterotrophs are eukaryotic, reproduce by mitosis, have mitochondria, and produce energy by aerobic respiration.

3. The endosymbiotic theory states that eukaryotic cells incorporated primitive prokaryotic cells that eventually evolved to become mitochondria and chloroplasts.

Chapter Problems and Answers

Problems

The following is a brief paragraph based on evolution. For problems 1–10, fill in the missing terms.

One of the greatest discoveries of the 20th century was penicillin, the first antibiotic.

Antibiotics are important because they can destroy bacteria that cause disease and death

in humans. Bacteria are primitive cells that belong to the domain _____. Antibiotics kill
                                                              1

bacteria by destroying their cell walls or ribosomes. Unfortunately, many strains of bacteria

have become resistant to some of the best antibiotics. This can occur because in every

population of bacteria some individuals have _____ that make them resistant to the
                                                       2
antibiotic. The bacteria that survive the antibiotic treatment pass on their allele for

resistance to the next generation. In this example, the _____ is selecting in favor of the
                                                             3
resistant bacteria. Darwin would have called this _____. After many generations of
                                                        4
bacteria, a new _____ could develop that makes a particular antibiotic useless. The
                     5
theory that could explain the origin of the new bacteria is _____. The fact that a changed
                                                                  6
bacterial population can develop that is resistant to antibiotics demonstrates that the

process of _____ is occurring. Some biologists believe that sudden changes called _____
                7   8
can result in bacteria that are resistant to antibiotics. These sudden changes in a trait

were first discovered by _____, in the early 1900s. If the development of a new species
                               9
occurs in a slow, continual, and gradual process over many generations, this rate of

change is called _____.
                      10

For problems 11–15, fill in the missing answer.

11. The remains of an organism that lived in the distant past can be considered a
    _____.

12. Under the right climatic conditions, the _____ of an organism are able to
    resist the natural process of decay.

13. _____ is a form of fossilized resin from trees that lived millions of years ago.

14. The remains of an organism that lived in the distant past can be found in _____
    rock.

15. A(n) _____ is representative of a species of organism that was found all
    over the world but lived for a short time.

For problems 16–20, select the evidence for evolution from the following list that is associated
with each description. A choice can be used more than once or not at all.

        A. Comparative anatomy   B. Comparative biochemistry
        C. Comparative embryology   D. Fossils   E. Mutations

16. The study of evolutionary relationships based upon homologous structures. _____

17. Insulin, hemoglobin, DNA, and RNA have been used to establish evolutionary relationships
    between organisms. _____

18. Imprints, molds, and petrifaction formed remains of organisms that can be used to show
    change. _____

19. A comparison of vertebrates during their early stages of development shows that an
    amazing similarity exists between them. _____

20. Vestigial structures have often been used to demonstrate that organisms have changed with the passage of time. _____

For problems 21–25, select the item from the following list that is associated with each description. A choice can be used more than once or not at all.

A. Heterotroph hypothesis B. Inheritance of acquired characteristics
C. Mutations D. Natural selection E. The Hardy-Weinberg law

21. As an organism uses a structure and the structure becomes more developed, this improvement can be passed on to the next generation. _____

22. Gene frequencies remain the same from one generation to the next, provided evolution is not taking place. _____

23. Organisms with the best variations survive and pass those variations on to the next generation. _____

24. This concept is an attempt to explain how the first cells originated on earth. _____

25. Sudden changes in the expression of a trait can result in variation, which might lead to the formation of a new species. _____

Answers

1. **Bacteria.** In a six-kingdom system of classification, organisms in the domain Bacteria are classified in the kingdom Eubacteria.

2. **Variations.** Some bacteria in this population have a variation in their cell walls or ribosomes that a particular antibiotic cannot destroy.

3. **Antibiotic.** Usually nature selects the fittest variations. However, in this case the antibiotic selects for the resistant bacteria by killing off those bacteria that are not resistant.

4. **Survival of the fittest.** Individuals with the variation that can resist the antibiotic's action are the fittest. These bacteria survive, reproduce and pass the allele for this variation on to their offspring.

5. **Species.** Many species of bacteria can reproduce every 20 minutes. Therefore, in one 24-hour period, 72 (3 per hour multiplied by 24 hours) divisions can occur, resulting in over 26,000 generations being produced in one year. As a result, new species or strains of bacteria can originate in just a few years.

6. **The theory of natural selection.** Darwin's theory of natural selection explains how evolution takes place and new species form.

7. **Evolution.** Evolution is the change in a species of organism with the passage of time. Because bacteria reproduce rapidly, many generations can be produced in just a few years, and the time needed for new species to originate is reduced.

8. **Mutations.** Mutations are a source of variation in a population that can lead to the origin of new species.

9. **De Vries.** Most biologists today believe that mutations are an important source of variation in a population.

10. **Gradualism.** The alternative explanation for the rate of evolution is punctuated equilibrium, which states that evolutionary change occurs in sudden spurts during which many species are formed, followed by long periods of stability with no speciation.

11. **Fossil.** The existence of fossils is one indication that evolution has taken place.

12. **Hard parts.** The hard parts of an organism, such as bones and teeth, are resistant to decay. Soft parts, such as cartilage and organs, decay rapidly after the death of the organism. The bones are present in a skull, but the cartilage that makes up the nose and ears is gone.

13. **Amber.** Amber is important because as the resin drips down from a tree, it often traps and surrounds an insect or part of a plant. The enclosed organism, or part of the organism, is the actual fossil.

14. **Sedimentary.** Sedimentary rock forms in layers often taking millions of years to form. Organisms trapped during this rock formation process can sometimes become fossilized. Igneous rock and metamorphic rock do not contain fossils.

15. **Index fossil.** Index fossils are important for relative dating. Fossils in undisturbed rock layers above the one containing the index fossil are younger, and those in layers below the index fossil are older.

16. **A. Comparative anatomy.** Homologous structures are structures that are similar in construction and evolutionary development. By comparing these structures, the evolutionary relationships that exist between organisms can be determined.

17. **B. Comparative biochemistry.** The greater the similarity of biochemical compounds in cells between individuals of different species, the closer their common ancestry is.

18. **D. Fossils.** Imprints, molds and petrifaction are different methods of fossilization. Fossils can be used to show that evolution has taken place.

19. **C. Comparative embryology.** Vertebrates in their early stages of development are embryos. The later embryos become different from each other the closer the evolutionary relationship between them is.

20. **A. Comparative anatomy.** A vestigial structure is an anatomical feature that at one time had a function but now does not. Examples are the human appendix and coccyx.

21. **B. Inheritance of acquired characteristics.** This theory of evolution preceded Darwin's theory of natural selection.

22. **E. The Hardy-Weinberg law.** Hardy and Weinberg said that gene frequencies remain constant as long as the population is large, mating is random, mutations don't occur, migration doesn't occur, and natural selection does not take place.

23. **D. Natural selection.** Nature selects organisms with alleles for the fittest variations.

24. **A. Heterotroph hypothesis.** The heterotroph hypothesis is an educated guess stating that the first cells were heterotrophs.

25. **C. Mutations.** Mutations were first discovered by De Vries. Mutations are a source of variation in a population that can eventually lead to the origin of new species.

Supplemental Chapter Problems

Problems

For problems 1–25, select the *best* answer.

1. The great diversity of living things on earth today can best be explained by:

    A. Mendel's laws

    B. the concept that new species of organisms originated spontaneously at key times in the earth's history

    C. the inheritance of acquired characteristics

    D. biological evolution

2. Which of the following statements best describes evolution?

    A. Organisms change with the passage of time.

    B. With the passage of time, organisms change from simple to complex.

    C. With the passage of time, organisms change from complex to simple.

    D. Organisms develop any characteristics that they need to survive.

3. Fossils are often used as evidence for evolution because they:

    A. are found in metamorphic rock

    B. are old and date back to the formation of the earth

    C. can show how organisms changed with the passage of time

    D. decay at the same rate as radioactive elements

4. Fossils that have formed because of minerals in water diffusing into organisms and forming rock are called:

    A. hard parts   B. molds   C. imprints   D. petrified

5. Fossils of plants and animals are usually found in:

    A. metamorphic rock   B. sedimentary rock   C. igneous rock   D. lava rock

6. Which process can be used to determine the approximate age of a fossil bone?

    A. relative dating   B. absolute dating   C. couple dating   D. index dating

7. An index fossil is found in an undisturbed layer of rock. The fossil of an amphibian was found in a layer directly above the one containing the index fossil. Which statement about the two fossils is correct?

    A. The fossil of the amphibian is younger than the index fossil.

    B. The fossil of the amphibian is older than the index fossil.

    C. Both fossils are the same age.

    D. The organism represented by the index fossil evolved from the amphibian.

8. The half-life of carbon-14 is approximately 5,730 years. A fossilized tooth has one-quarter of the original amount of carbon-14. How many years ago did the animal from which the tooth came from die?

    A. 0   B. 5,730   C. 11,460   D. 17,190

9. The study of evolutionary relationships based on the similarity of structure and evolutionary development is known as:

    A. comparative selection   B. comparative embryology

    C. comparative biochemistry   D. comparative anatomy

10. Which evidence of evolution best illustrates that two species had a common ancestor?

    A. They live in the same environment.

    B. Their fossils are found in the same rock layer.

    C. Their body structures are analogous.

    D. They have the same hormones, enzymes and hemoglobin.

11. Several vertebrate embryos were studied during different stages of development. All the embryos had gill slits and a tail. This suggests that these organisms:

    A. might have evolved from a common ancestor

    B. are members of the same species

    C. are adapted to living in the same environment as adults

    D. all develop inside the uterus of the mother

12. A vestigial structure is one that:

    A. functions in vegetative propagation

    B. is present in an organism and no longer has a known function

    C. promotes mutations and variations

    D. is used to determine the relative age of a fossil

13. The inheritance of acquired characteristics is based on the concept of:

    A. natural selection   B. punctuated equilibrium

    C. use and disuse   D. mutations

14. Darwin's theory of evolution can best be described by:

    A. inheritance of acquired characteristics   B. the theory of mutations

    C. punctuated equilibrium   D. survival of the fittest and natural selection

15. Which of the following statements can be attributed to Darwin?

    A. Evolutionary change takes place in sudden spurts during which many new species are formed.

    B. Overproduction results in competition and a struggle for existence.

    C. Acquired characteristics are inherited by the offspring of the next generation.

    D. The strong beat up the weak.

16. Which of the following can be considered a source of variation within a population?

    A. half-life   B. asexual reproduction   C. mutation   D. use and disuse

17. According to Darwin, the rate of evolutionary change can best be explained by:

    A. gradualism   B. punctuated equilibrium
    C. the Hardy-Weinberg law   D. mutation

18. Evolutionary change that takes place in sudden spurts during which many new species are formed, followed by long periods where no new species are formed, is characteristic of:

    A. gradualism   B. punctuated equilibrium
    C. the Hardy-Weinberg law   D. mutation

19. Who is responsible for the following concept: Within a population, the frequency of an allele remains constant from generation to generation?

    A. Mendel   B. Darwin   C. Lamarck   D. Hardy and Weinberg

20. In a population of guinea pigs, the frequency of the recessive allele for white fur is 0.3. What is the frequency of the dominant allele for brown fur?

    A. 0.09   B. 0.42   C. 0.7   D. 1.0

21. Which of the following is *not* a condition of the Hardy-Weinberg law?

    A. no random mating   B. no migrations
    C. no mutations   D. no natural selection

22. Members of a population became separated from each other because of a newly formed river. This is an example of:

    A. genetic drift   B. natural selection
    C. geographic isolation   D. reproductive isolation

23. When organisms that were once members of the same species are no longer capable of mating with each other, they are said to be:

    A. genetically mutated   B. naturally selected
    C. geographically isolated   D. reproductively isolated

24. The increase in the frequency of the allele for dark wings in the peppered moth because of the industrial revolution demonstrates that:

    A. mutations cause variation

    B. environmental factors can tend to favor one variation over another

    C. acquired characteristics can be inherited

    D. the environment causes variation

25. The heterotroph hypothesis is an attempt to explain:

    A. evolution

    B. the theory of natural selection

    C. the inheritance of acquired characteristics

    D. how the first cells might have originated

Answers

1. **D.** "Evolution Defined," p. 275

2. **A.** "Evolution Defined," p. 275

3. **C.** "Fossil Evidence for Evolution," p. 275

4. **D.** "Fossil Evidence for Evolution," p. 275

5. **B.** "Fossil Evidence for Evolution," p. 275

6. **B.** "Dating of Fossils," p. 278

7. **A.** "Dating of Fossils," p. 278

8. **C.** "Dating of Fossils," p. 278

9. **D.** "Comparative Anatomy," p. 280

10. **D.** "Comparative Biochemistry," p. 281

11. **A.** "Comparative Embryology," p. 280

12. **B.** "Comparative Anatomy," p. 280

13. **C.** "Inheritance of Acquired Characteristics," p. 282

14. **D.** "Natural Selection," p. 282

15. **B.** "Natural Selection," p. 282

16. **C.** "Mutation," p. 283

17. **A.** "The Rate of Evolution," p. 283

18. **B.** "The Rate of Evolution," p. 283

19. **D.** "The Hardy-Weinberg Law," p. 284

20. **C.** "The Hardy-Weinberg Law," p. 284

21. **A.** "The Hardy-Weinberg Law," p. 284

22. **C.** "Geographic and Reproductive Isolation," p. 285

23. **D.** "Geographic and Reproductive Isolation," p. 285

24. **B.** "Environmental Factors," p. 286

25. **D.** "The Origin of Life on Earth," p. 288

Chapter 15

Ecology

Ecology is the branch of biology that studies living things, their interrelationships with each other and their interrelationship with the environment. The *environment* of an organism is its surroundings and is composed of abiotic (nonliving) and biotic (living) factors. The *abiotic* factors of the environment include climate, light, temperature, water, soil, and territory. The *biotic* factors of the environment include organisms, populations, communities, food chains, symbiotic relationships, competition between organisms, and succession. *Limiting factors* are any abiotic or biotic factors that determine the distribution and kinds of organisms that can survive in an area. In this chapter, we will review abiotic factors, biotic factors, major world biomes, and the role that humans have played in the environment.

Abiotic Factors

The nonliving factors in the environment determine the kinds of organisms (biotic factors) that can survive in an area. The plant life in an area determines which animals can survive in that area. All the abiotic factors are interrelated, and each one impacts the others.

Climate

Climate is the weather in an area over the period of a year. *Weather* is the day-to-day change in the atmosphere. Climate is concerned with light, temperature, and precipitation in a given location. Climate is determined by the latitude and geography of an area. *Latitude* is measured in degrees north and south of the equator. As a rule, the further a location is north or south of the equator, the colder the climate. At the poles, temperatures are so far below freezing that most plant and animal life cannot exist. The effects of latitude on living things will be discussed later in the chapter when we study the world's biomes. Geography can have a dramatic effect on climate. One side of a mountain range can have a lot of rainfall, while the other side is a desert. Higher elevations have colder climates. Locations close to water tend to be cooler in summer and warmer in winter. Climate is one abiotic factor that determines the kinds of plants, animals, and their distribution in an area.

Light

Light from the sun is the main source of energy for living things. Autotrophs (green plants) depend on light for photosynthesis to produce glucose and oxygen. Heterotrophs (animals) depend on green plants for food and oxygen. The amount of light (duration) and the intensity of light (strength) that an area receives determine the types of plants and their distribution in the area. Some plants such as those that make up the canopy in the tropical rain forest require light of long duration and high intensity. Plants at ground level require low-intensity light or shade.

Temperature

The temperature in an area limits the types of plants that can grow in the area, which in turn determines the kinds of animals that feed off the plants. All plants have a thermal minimum and maximum. The *thermal minimum* is the lowest temperature at which the plant can survive. The *thermal maximum* is the highest temperature at which the plant can survive. Palm trees can grow all year long in California and Florida. However, they cannot survive the winters of New York and Montana because the temperature is below the thermal minimum for this kind of plant. Similarly, plants that do well in Alaska cannot survive the summer heat of Arizona. Temperature also has an effect on rainfall, the rate of evaporation of water from soil and the loss of water by transpiration from leaves.

Water

The type of water in an area determines the distribution and types of plants and animals in an area. About 97% of the earth's water is found in the oceans, and the remaining 3% is fresh water. Plants and animals that live in the oceans cannot survive in fresh water. Plants and animals that live in fresh water cannot survive in the ocean. Elodea, a fresh water plant, dies if placed in the ocean. Tuna fish, which live in the ocean, die if put in fresh water.

Hydrophytes are plants that are adapted for growing in water; an example is the water lily. *Xerophytes* are plants that are adapted to growing in a dry environment; an example is the cactus. *Mesophytes* are land plants that require average amounts of rainfall during the course of a year. Mesophytes are plants whose water requirements fall between hydrophytes and xerophytes. Deciduous plants are mesophytes.

Soil

Land plants grow in soil. Many different types of soil exist, and they vary by texture, structure, organic content, and mineral content. One way to classify soil is by texture. Sandy soil contains large particles (0.02–2mm), clay soil has small particles (less than 0.002mm), and loam is a mixture of clay and sand. Sandy soils tend to be dry, while clay soil retains moisture. Soil also contains bacteria and fungi, which help decay plant and animal material forming *humus*. Humus returns nutrients and minerals to the soil. In addition, humus helps the soil retain water. The type of soil in an area determines the plants that can grow in that area.

Example Problems

These problems review abiotic factors.

1. Define ecology.

   Answer: Ecology is the branch of biology that studies living things, their interrelationships with each other and their interrelationship with the environment. Two interrelationships exist. The first is between living things. The second is between living things and their environment.

2. How are abiotic factors different from biotic factors?

   Answer: Abiotic factors are all the nonliving things in the environment such as climate, light, temperature, water, and soil. Biotic factors are all the living things that are found in the environment.

3. How is humus formed?

   Answer: Humus is the part of soil that is formed from decayed plant and animal material. Humus provides nutrients and minerals for plants and increases the capability of soil to retain water.

Work Problems

Use these problems on abiotic factors to give yourself additional practice.

1. How do latitude and geography affect climate?

2. How are light, temperature and water interrelated?

3. Why can't cactus plants survive outdoors in the northern United States and Canada?

Worked Solutions

1. Generally the further a location is north or south of the equator, the colder the climate. Geographical features such as mountains, proximity to water, and elevation affect climate. One side of a mountain range can have lots of rainfall, while the other side is a desert. Locations close to water tend to be cooler in summer and warmer in winter. Higher elevations tend to have colder climates.

2. The greater the duration and intensity of light in a given location, the higher the temperature. High temperatures in an area increase evaporation from bodies of water and increase transpiration, which is the loss of water from the leaves of a plant.

3. The abiotic conditions of the northern United States and Canada are very different from those found in desert regions where cactus plants grow. The cold winters, excess precipitation, different soil conditions, reduced duration of light, and reduced intensity of light are factors that prevent cactus plants from surviving in the northern United States and Canada.

Biotic Factors

The living factors in the environment interact with each other and with the abiotic factors. The environment has a hierarchy of biotic organization, beginning with species (the smallest category) and ending with biosphere (the largest category).

Species: Organisms that can mate with each other and produce fertile offspring.

Population: Organisms in the same species that live together in a specific location.

Community: Different populations that live together in a specific location.

Ecosystem: The interaction between the community and the abiotic factors in the environment.

Biosphere: The interaction between all the ecosystems on earth.

Some common biotic interrelationships between organisms are competition, predator-prey relationships, and symbiosis.

Competition

In the preceding chapter, we mentioned that Darwin observed that members of a species produce more offspring than is necessary to maintain the species. Overproduction results in competition for scarce resources such as food, water, and territory. Two types of competition exist: intraspecific competition and interspecific competition.

Intraspecific competition is competition between members of the same species for territory, food, water, and a mate. *Interspecific competition* is competition between members of different species for territory, food, and water. *Territory* is that piece of land that the organism is willing to fight for and defend against others. Territory is important because it can provide food and mating opportunities. Consider the following example: A male lion must have his own territory to hunt and to attract females for mating.

Habitat is where an organism lives in its environment. Think of habitat as the organism's address. *Niche* is the role or job that the organism has in its environment. The organism's niche is closely related to what it eats. As a rule, two organisms cannot occupy the same habitat if they have the same niche. This is true because it places the organisms in competition with each other. Birds in a tree in Africa and a lion in the shade of the tree have the same habitat, but they do not compete because their niches are different. However, the lion and a cheetah compete because they have the same niche: Both are predators hunting the same kinds of animals.

The Predator-Prey Relationship

A *predator* is an animal that hunts for food. Some examples of predators are lions, wolves and hawks. *Prey* are organisms that are hunted by predators. Some examples of prey are deer, rabbits, and mice. When food is plentiful for the prey population, its numbers increase. After a time this results in an increase in the predator population. As the predator population increases, the prey population decreases. The two populations are linked to each other, and both depend on the other. The following graph illustrates the predator-prey relationship.

A predator population keeps the prey population from increasing to the point where its numbers are so high that the prey organisms begin to die because of a lack of food. When this happens, the prey population has exceeded the *carrying capacity of the land*. Predators usually kill the old, diseased, and some young in a prey population. Consider the following example: In New York State, wolves (which are the predators of deer) have been eliminated. The deer population is now so high that they have exceeded the carrying capacity of the land. Many deer now forage for food in suburban yards and starve to death in the winter. To decrease the deer population, a hunting season was begun. However, unlike predators, hunters often kill the healthy adults in the prey population.

Example Problems

These problems review biotic factors.

1. How does a population differ from a community?

    Answer: A population contains members of the same species that live together in a specific location, while a community has several different populations that live together in a specific location.

2. What is an ecosystem?

    Answer: An ecosystem consists of a community and the abiotic factors in its environment.

3. How is habitat different from niche?

    Answer: Habitat is where an organism lives in its environment. Niche is the role or job that the organism has.

Work Problems

Use these problems on biotic factors to give yourself additional practice.

1. How does the carrying capacity of the land affect a prey population?

2. How does intraspecific competition differ from interspecific competition?

3. Explain the predator-prey relationship.

Worked Solutions

1. The carrying capacity of the land refers to the land's capability to supply enough food to feed a population. When the number of individuals in a population exceeds the carrying capacity of the land, some members of the population must migrate to a new location or some will die of starvation.

2. Intraspecific competition is competition between members of the same species. Interspecific competition is competition between members of different species. Organisms compete for territory, food and water. In the case of intraspecific competition, organisms also compete for a mate.

3. A predator is an animal that hunts and kills for food; the prey is the animal that is hunted. When the prey population increases, the predator population also increases. As the prey population decreases, the predator population decreases. Predators are important because they keep the prey population from increasing to the point where it exceeds the carrying capacity of the land.

Symbiosis

Symbiosis is a relationship between two organisms that live together where at least one of the organisms benefits from the association. The three kinds of symbiotic relationships are mutualism, commensalism and parasitism.

Mutualism

Mutualism is a relationship where both organisms benefit from living together. Lichen is a combination of two organisms that live together, a fungus and a green alga or blue-green bacteria. The alga or bacteria live inside the fungus and provide the fungus with food produced by photosynthesis. The fungus provides the alga or bacteria with a place to live and the moisture necessary for photosynthesis. Termites have a mutual relationship with animal-like protists that live in their digestive tract. Termites eat wood but lack the enzymes necessary to digest wood. The animal-like protists have the enzymes that digest wood for the termite. The termite benefits by getting food when the wood is digested, and the animal-like protists benefit by getting housing, protection, and food.

Commensalism

Commensalism is a relationship where one organism benefits from the association, and the other is neither helped nor harmed. An interesting relationship exists between the remora and the shark. The remora can attach itself to a shark and benefits by getting free transportation, food (it eats scraps left over by the shark), and protection (no organism picks a fight with a shark). The shark appears to be unaffected by the remora. Some orchids grow on trees and benefit from having a place to live. The trees are not affected by the orchids.

Parasitism

Parasitism is a relationship where one organism benefits from the association, and the other is harmed. The organism that benefits is the parasite, and the one that is harmed is the host. The parasite lives on or in the host. Ticks are arachnids that attach themselves to the skin of an animal and suck its blood. Many ticks carry disease-causing organisms, such as the deer tick, which is responsible for spreading the bacteria that causes Lyme disease. Tapeworms are parasites that live inside the digestive tract of an organism. The tapeworm benefits by getting food and a place to live but harms the host by causing pain, weight loss, and diarrhea.

Coevolution

For two species to develop a close symbiotic relationship, they must have coevolved. *Coevolution* is a process by which species adapt to changes in one another.

Example Problems

These problems review symbiosis.

1. Define symbiosis.

   Answer: Symbiosis is a relationship between two organisms that live together where at least one of the organisms benefits from the association.

2. Define commensalism.

   Answer: Commensalism is a relationship where one organism benefits from the association, and the other is neither helped nor harmed.

3. What is coevolution?

   Answer: Coevolution is a process by which species adapt to changes in one another.

Work Problems

Use these problems on symbiosis to give yourself additional practice.

1. Which type of symbiosis is beneficial to both organisms in a relationship?

2. Bees and other insects visit flowers for pollen, which they use as food. In the process of visiting a flower, insects help transfer pollen from the anther to the stigma. Identify the type of symbiosis described here.

3. Why is parasitism detrimental to the host?

Worked Solutions

1. Mutualism is a symbiotic relationship that is beneficial to both organisms.

2. Mutualism is the type of symbiosis described here. The insects benefit by gathering pollen for food. The plants benefit by having their flowers pollinated so that they can reproduce.

3. The parasite uses the host for a place to live and a source of food. The host is an unwilling participant in this relationship and is injured or killed by the parasite.

The Flow of Energy Through an Ecosystem

Autotrophs are green plants that can make their own food by the process of photosynthesis. During photosynthesis, green plants use light energy to convert carbon dioxide and water into glucose and oxygen. Green plants are often called producers. This is a misnomer because green plants don't produce energy; they convert light energy into the chemical energy of the glucose molecule. Plants and animals use glucose during respiration to produce energy in the form of ATP. *Heterotrophs* are organisms that eat to obtain food needed for energy production. Heterotrophs eat autotrophs or other heterotrophs. Heterotrophs can be classified by the kinds of organisms they eat.

The following chart lists the different kinds of heterotrophs and the organisms they eat.

Heterotroph Classification		
Heterotroph	*Type of Organism Eaten*	*Examples*
Herbivore	Plants	Cow, deer, rabbit, grasshopper
Carnivore	Meat	Lion, wolf, fox, owl
Omnivore	Plants and meat	Human, bear, raccoon
Predator	Meat that has been hunted	Lion, wolf, fox, owl
Scavenger	Meat that is left over by a predator	Vultures, hyenas
Saprobe	Dead plants and animals	Bacteria, fungi

The flow or transfer of energy through the ecosystem begins at the sun; autotrophs capture the sun's energy and produce glucose. Heterotrophs eat autotrophs and are eaten by other heterotrophs. The flow of energy through the ecosystem is unidirectional (one way). Energy is never recycled. Food chains, webs, and pyramids illustrate the transfer of energy from organism to organism.

Food Chain

A *food chain* shows how energy is transferred from one organism to another in an ecosystem. In the following food chain, the producers are autotrophs, the primary consumers are herbivores, and the secondary and tertiary consumers are carnivores. Sometimes a fourth consumer exists in a food chain. More than four consumers in a food chain is unusual because by the time the fourth consumer is reached, little energy remains to be transferred. When any of the organisms in a food chain die, they are broken down by organisms of decay called decomposers. Decomposers are saprobes, such as bacteria and fungi, and can be considered the final consumers.

Sun $\xrightarrow{\text{energy}}$ Producers $\xrightarrow{\text{energy}}$ Primary consumer $\xrightarrow{\text{energy}}$ Secondary consumer $\xrightarrow{\text{energy}}$ Tertiary consumer

(autotroph)   (herbivore)   (carnivore)   (carnivore)

Ecological Pyramid

An *ecological pyramid* is similar to a food chain because it shows the flow of energy from one group to the next. In addition, an ecological pyramid shows the relative numbers of organisms at each trophic (feeding) level. In the following pyramid, the most numerous organisms are producers, and the least numerous are tertiary consumers. The pyramid can also be used to illustrate the amount of energy available at each level. The group with the most energy is the producers, and the group with the least energy is the tertiary consumers. Each group has less energy to transfer to the next because each uses energy by performing life functions. Different ecological pyramids can be drawn to illustrate specific relationships (energy, biomass, numbers).

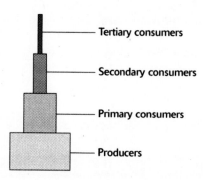

Tertiary consumers

Secondary consumers

Primary consumers

Producers

Food Web

A *food web* is made up of several interconnecting food chains. The disadvantage of a food chain is that it is only as strong as its weakest link. If a disease kills all the primary consumers, the secondary and tertiary consumers also die. A food web has the advantage of providing alternate sources of food.

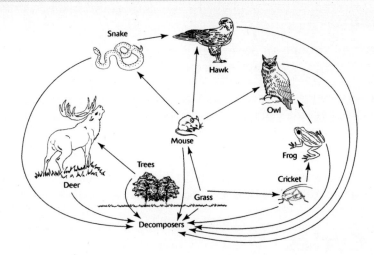

Example Problems

These problems review the flow of energy through an ecosystem.

1. What is the source of energy that flows through an ecosystem?

   Answer: The sun is the source of energy that flows through the ecosystem.

2. What does a food chain show?

   Answer: A food chain shows how energy is transferred from one organism to another in an ecosystem.

3. How is a predator different from a scavenger?

   Answer: A predator is a heterotroph that hunts and kills its prey to obtain food for energy. Predators are carnivores and are primary, secondary or tertiary consumers. A scavenger is a carnivore that eats food left over by predators. Scavengers do not kill the animals they eat.

Work Problems

Use these problems on the flow of energy through an ecosystem to give yourself additional practice.

1. Why is the flow of energy through an ecosystem considered unidirectional?

2. Why do food chains rarely exceed four consumers?

3. Why is it better for an organism to be part of a food web as opposed to a food chain?

Worked Solutions

1. The flow of energy through an ecosystem is considered unidirectional because energy is never recycled. Energy never returns to the sun.

2. As energy flows through the food chain from one group of organisms to the next, less energy is available because each group uses up energy for life functions such as locomotion, respiration, digestion, and reproduction.

3. Organisms in food webs have the opportunity to eat other foods if one food becomes unavailable. However, an organism in a food chain has only one choice of food. If that food is not available, the organism in the food chain dies.

Material Cycles

Unlike energy, some abiotic factors in the environment can be recycled. Water, carbon dioxide/oxygen and nitrogen are abiotic factors that can be recycled. *Recycling* means that these abiotic factors can be reused in one form or another.

The Water Cycle

The water cycle is made up of four components: evaporation (and transpiration), condensation, precipitation, and collection.

Evaporation: As the sun beats down on oceans, lakes, and rivers, water evaporates (changes from a liquid state to a gaseous state) to form water vapor. In addition, plants lose water through their stomates (transpiration) to form water vapor.

Condensation: As water vapor in the atmosphere cools, it changes back into a liquid, forming clouds.

Precipitation: Eventually, the clouds contain more water than they can hold, and they lose this water in the form of rain, sleet, or snow.

Collection: Water lost from clouds collects in oceans, lakes, rivers, and soil. The action of the sun starts the process all over again.

The Oxygen/Carbon Dioxide Cycle

Oxygen is released into the atmosphere by green plants during the process of photosynthesis. Animals use this oxygen during aerobic respiration to produce energy (ATP), water, and carbon dioxide. Plants use the water and carbon dioxide to produce glucose and oxygen. Theoretically, the amount of carbon dioxide released into the atmosphere should equal the amount of oxygen removed by respiration. However, this does not happen because additional carbon dioxide is produced by decomposers as they break down organic materials. In addition, carbon dioxide is added to the atmosphere by the burning of fossil fuels (oil, natural gas, and coal) and by the burning down of forests. Oxygen is the second most abundant gas in the earth's atmosphere at about 21%. Nitrogen makes up about 78% of the atmosphere, and carbon dioxide makes up about 0.036%.

The Nitrogen Cycle

Although nitrogen makes up about 78% of the earth's atmosphere, most organisms cannot use atmospheric nitrogen. Nitrogen is needed to make amino acids, nucleic acids, and proteins. Proteins are used for growth, repair, and reproduction of new cells, and they form many different molecules needed by organisms. For a plant to use nitrogen, it must be in the form of nitrates (NO_3^-). Plants get nitrogen from the soil by the action of nitrogen-fixing bacteria and nitrifying bacteria. In addition, lightning converts atmospheric nitrogen into nitrates. Animals get nitrogen by consuming plants or other animals. When plants and animals die, decomposers in the soil break them down, producing ammonia. Denitrifying bacteria take ammonia and release free nitrogen into the atmosphere. Nitrifying bacteria take ammonia and produce nitrites, which are converted into nitrates that are removed from the soil by the roots of plants.

Example Problems

These problems review material cycles.

1. Define recycling.

   Answer: In ecology, recycling refers to the reuse of important abiotic factors such as water, oxygen, carbon dioxide, and nitrogen.

2. Identify the abiotic factor from the environment that is never recycled.

   Answer: Light energy from the sun is never recycled. The flow of energy through an ecosystem is unidirectional.

3. How does water enter the atmosphere?

   Answer: Water can enter the atmosphere by evaporation from oceans, lakes, and rivers. Water also enters the atmosphere by transpiration from plants.

Work Problems

Use these problems on material cycles to give yourself additional practice.

1. How does the sun play an important role in the water cycle?

2. Identify the two biological processes that are involved with the oxygen/carbon dioxide cycle.

3. Explain why the amount of carbon dioxide that is introduced into the atmosphere exceeds the amount of oxygen used during aerobic respiration.

4. What are the different kinds of bacteria in the nitrogen cycle? Give the function of each.

Worked Solutions

1. The sun provides the energy needed for the evaporation of water from oceans, lakes, and rivers. This water condenses to form clouds, and precipitation returns the water to the oceans, lakes, and rivers for collection so that the process can begin again.

2. The two biological processes involved in the oxygen/carbon dioxide cycle are photosynthesis and respiration. Photosynthesis is a process that puts oxygen into the atmosphere and removes carbon dioxide. Respiration uses oxygen and puts carbon dioxide into the atmosphere. See Chapter 6 for a comparison of respiration and photosynthesis.

3. The amount of carbon dioxide that is introduced into the atmosphere exceeds the amount of oxygen used during aerobic respiration because additional carbon dioxide is added by decomposers, the burning of fossil fuels and the burning of forests.

4. Four different kinds of bacteria are involved in the nitrogen cycle:

   ❏ *Nitrogen-fixing bacteria* take free nitrogen from the atmosphere and produce nitrates.

   ❏ *Bacteria of decay* are responsible for breaking down dead plants and animals into ammonia.

   ❏ *Denitrifying bacteria* convert ammonia into free nitrogen.

   ❏ *Nitrifying bacteria* convert ammonia into nitrites.

Ecological Succession

Succession is a process by which one biotic community is replaced by another until a climax community develops. A *climax community* is the last community of plants and animals to occupy an area. A climax community is stable and self-perpetuating. Succession is often caused by changes in the environment brought about by the current community. These changes often make it impossible for the current community to survive and set the stage for its replacement by the succeeding (following) community. Succession can also occur because of changes in climate and natural catastrophes such as fire, hurricanes, floods, and volcanic eruptions. Humans cause succession by cutting down forests, polluting the environment and overdeveloping an area.

Succession from Rock to Forest

Primary succession begins in an area that never had life. Pioneer organisms such as microbes, lichens, and mosses are the first to inhabit a rock promoting soil formation. Eventually, enough soil is built up so that grasses can succeed the pioneer organisms. Grasses are replaced by shrubs, which are replaced by fast-growing trees. Fast-growing trees are replaced by slower-growing trees that are representative of the climax community. Each community changes the environment, making its replacement possible by the succeeding community. The types of grasses, fast-growing trees, and climax-community plants vary with the location and climate of an area. A climax community is not truly final; humans or nature can destroy it. For example, a forest fire can burn down most trees in an area. In this case, succession starts over with grasses. *Secondary succession* occurs in areas where life has previously existed.

Succession from Water to Land

Succession can also occur in lakes and ponds. Leaves, debris and silt that enter a lake lower the water level. Nitrates and phosphates that run off into a lake promote the growth of algae and deplete the water of oxygen, which kills fish and also causes a lowering of the water level. Algae are replaced by marsh plants, tall grasses that begin to fill in the lake. Eventually, the water is drained, and soil forms. At this point, succession proceeds with grasses as explained in the previous section. *Eutrophication* is a term used to describe the death of a lake by nitrates and phosphates.

Example Problems

These problems review succession.

1. What is succession?

    Answer: Succession is a process by which one biotic community is replaced by another until a climax community of plants and animals develops.

2. What are some of the causes of succession?

    Answer: The current community in an area often changes the physical conditions of its environment so that it is no longer favorable for it to continue living there. When this happens, it is replaced (succeeded) by a community that can survive under the altered conditions. A change in the climate of an area often makes conditions unfavorable for one community and more favorable for another.

3. What is a pioneer organism?

    Answer: A pioneer organism is the first organism to occupy an area. Lichens are an example of pioneer organisms.

Work Problems

Use these problems on succession to give yourself additional practice.

1. How does primary succession differ from secondary succession?

2. What is a climax community?

3. How can a climax community be destroyed?

Worked Solutions

1. Primary succession begins in an area that never had life. Secondary succession occurs in areas where life has previously existed. Secondary succession often follows a catastrophe that destroys an existing community.

2. A climax community is the last community to occupy an area. A climax community is stable and self-perpetuating.

3. A climax community can be destroyed by fires, volcanic eruptions, hurricanes, changes in climate, or human activities.

World Biomes

A *biome* is large area of the earth defined by a climax community of plants and specific abiotic factors such as light, precipitation, temperature, and soil type. Biomes do not have distinct boundaries; one merges into another. The biomes discussed in the following sections are presented in order of decreasing latitude. The most northern biomes are listed first.

Tundra

Location: The tundra is a biome that circles the Northern Hemisphere just above the Arctic Circle. Tundra can be found in northern Alaska, northern Canada, Greenland, Scandinavia, and Russia.

Abiotic factors: The tundra has a short growing season of approximately 60 days and is the coldest of all the biomes. Winter can last from 6 to 10 months. The ground in the tundra is permanently frozen, a condition known as *permafrost*. Very little precipitation falls in the tundra, which can be considered a frozen desert.

Biotic factors: The tundra is treeless and characterized by low-growing plants such as lichens, mosses, bearberry, small shrubs, and grasses. Animal life consists of insects, birds (snowy owl), lemmings, arctic hare, arctic fox, caribou, and polar bears. The tundra has the least biodiversity of all the world's biomes.

Taiga

Location: The taiga is located in the Northern Hemisphere below the Arctic Circle in a continuous belt around the earth. The taiga begins where the tundra ends and can be found in northern parts of the United States, southern Canada, Scandinavia, and vast areas of Russia. The taiga is the largest land biome. The taiga and the tundra are not found in the Southern Hemisphere because the landmass is not sufficient to create these biomes.

Abiotic factors: Winters are long and cold; summers are short and warm. The average temperature in the taiga is below freezing for more than half the year. Annual precipitation is low, and most of it occurs during the summer. Winter snows tend to remain on the ground due to cold temperatures. Many streams and rivers can be found in the taiga.

Biotic factors: Coniferous forests of spruce and fir trees are the main plants found in the taiga. These trees are evergreen and have needle-like leaves that are adapted to surviving cold winters. Many insects, birds, and mammals (moose, bears, deer, rabbits) make their home in the taiga.

Temperate Deciduous Forest

Location: The deciduous forest is found in the middle latitude regions such as the northeastern part of the United States, Europe, southwest Russia, eastern China, and Japan.

Abiotic factors: This biome has all four seasons and a moderate climate. Precipitation ranges from 75–150 cm per year. The soil is rich and contains a great deal of humus formed from decaying leaves.

Biotic factors: Deciduous trees (oak, maple, beech, hickory, hemlock) have broad leaves that they lose in the winter. Animal life includes insects, birds, deer, squirrels, and raccoons. Cold-blooded animals such as frogs and snakes are found in this biome.

Grassland

Location: In the United States, the grasslands are known as the prairie and are found in the Midwest (Oklahoma, Kansas, Nebraska, Iowa, Illinois, South Dakota, and North Dakota). In Canada, the provinces of Alberta, Saskatchewan, and Manitoba are called the Prairie Provinces. Grasslands are also found in Argentina, Africa, Russia, and New Zealand.

Abiotic factors: Prairies have hot summers, cold winters and moderate rainfall (50–90 cm). The soil in the grasslands is rich and fertile.

Biotic factors: Tall grasses, few trees, prairie dogs, birds, bison, coyotes, deer, and antelope live in grasslands. Today, much of the grassland has been replaced by farms that grow wheat and corn as well as ranches that raise cattle.

Desert

Location: About 20% of the earth's land surface area is desert. Deserts can be found on every continent except Europe. The United States has four deserts that are in the southwestern part of the country.

Abiotic factors: Deserts have very little precipitation (less than 25 cm per year) with low humidity (10%–20%). Some deserts are classified as hot deserts such as the Mojave Desert (southeastern California, parts of Nevada, Arizona, and Utah) or cold deserts such as the Great Basin desert (from the Sierra Nevada Mountains on the west to the Rocky Mountains on the east). The desert often has large changes in temperature between daytime and nighttime.

Biotic factors: Some desert plants are cactus, yucca, aloe, and ocotillo. Plants in the desert have evolved interesting methods of conserving and preventing water loss. Some cactus plants such as the Saguaro cactus have thick stems to store water and extensive root systems to capture water. The ocotillo plant has small leaves and few stomates to minimize transpiration. Others (like aloe) have a thick, waxy covering that helps conserve water. Animals in the desert include insects, snakes, lizards, birds, and rodents. Many animals are nocturnal, coming out only at night when the temperatures cool down. Most desert animals are small, but some are large like the fox and the coyote.

Tropical Rain Forest

Location: The rain forests are found near the equator. Tropical rain forests exist in Latin America, Brazil (Amazon), West Africa, Southeast Asia, and the Pacific islands.

Abiotic factors: The climate in a tropical rain forest is stable, with little variation in temperature (18°C–35°C). Rainfall can exceed 600 cm per year, with rain 200 days a year. The soil is acidic and not fertile.

Biotic factors: Tropical rain forests occupy 2% of the earth's land surface but contain more than 50% of the earth's plant and animal life. This land biome has the greatest degree of biodiversity. Trees are tall with broad evergreen leaves. Four different layers exist based on their need for light. The top layer is the canopy and requires the most light; below is the subcanopy and then the understory. The forest floor is the lowest layer and receives the least amount of light. Some animals found in various tropical rain forests are insects, amphibians, snakes, and birds. Mammals include gorillas, chimpanzees, and panthers.

Marine Biome

Location: The marine biome covers 71% of the earth's surface and is the largest and most stable biome on earth. The marine biome consists of five interconnected oceans, most of which are in the Southern Hemisphere below the equator.

Abiotic factors: The salt content of the ocean is about 3.5%. Temperatures vary from below freezing at the poles to about 30°C at the equator. The ocean is divided into different zones, each with its own unique characteristics. Climate has little effect on the oceans, but the oceans have a substantial effect on climate. Oceans tend to moderate climate by cooling land areas in summer and warming them in winter.

Biotic factors: The marine biome has lots of biodiversity. Kelp, algae, seaweed, and phytoplankton are the most abundant plant life. Some animals found in the ocean include jellyfish, octopus, bony fishes, and mammals (dolphins and whales).

Example Problems

These problems review the world biomes.

1. Define a biome.

   Answer: A biome is a large geographical area having one kind of climate and a climax community of unique plants and animals.

2. Why can't trees grow in the tundra?

   Answer: Trees cannot grow in the tundra because of permafrost, a condition where the ground is permanently frozen. The root system of trees cannot grow in frozen soil.

3. Which biome is the most stable and least changing?

   Answer: The most stable and least changing biome is the marine biome. Climate has very little effect on this biome. The temperature of the water at any given depth in the ocean varies within narrow limits. The salinity of water remains relatively constant at 3.5%.

Work Problems

Use these problems on the world biomes to give yourself additional practice.

1. How do abiotic factors such as climate determine the kinds and distribution of plants in a biome?

2. How are desert plants adapted for survival in a dry climate?

3. Why are the tropical rain forests ecologically important despite the fact that they occupy only 2% of the earth's land surface area?

Worked Solutions

1. The abiotic factors of the environment limit the kinds of plants and animals that are in a biome. As a result, each biome has its own unique plant and animal life. The trees of the taiga and the temperate deciduous forest cannot survive in the tundra because of permafrost. Tropical rain forest plants cannot survive in the desert because of insufficient rainfall. Rain forest plants can't grow in a marine biome.

2. Desert plants have adapted several ingenious methods for survival in a dry climate. Some plants have a thick, waxy cuticle that helps prevent the loss of water. Others have small or needle-like leaves with few stomates. Many desert plants have extensive underground root systems that help absorb any precipitation that might fall. Others have thick stems that are adapted for the storage of water.

3. The tropical rain forests are ecologically important because of their tremendous biodiversity. More than 50% of the earth's plant and animal life is found in this biome.

Humans and the Environment

Population growth is the main cause of environmental damage. The current population of the world is over 6.3 billion and is increasing at a rate of about 85 million new individuals each year. In 1950, the population of the United States was 151 million. Today, it is over 292 million. Increased population growth means more demands on the earth's limited resources of energy, food, water, and living space. More people means more pollution, waste, destruction of the environment, and decreased biodiversity. Of all the species on earth, humans have the greatest ability to alter the environment. This ability to change the environment can be used in a responsible manner with consideration for the environment and the organisms that share the biosphere, or it can be used irresponsibly and with indifference.

Human Population Growth

The change in population for a given location can be expressed as a simple mathematical equation: *Population change* = (births + immigration) – (deaths + emigration). Currently, the world population is increasing at a rate of about 1.4% annually. However, the rate of population growth is expected to decline to about 1% by 2015. The human population is increasing *exponentially;* this means it doubles every *X* number of years. In 1900, the world's population was 1.6 billion people, and in 2000 it was 6 billion people. By 2050, the world's population is expected to be about 10 billion people. As countries become more industrialized, death rates and birth rates decrease. Death rates decrease because of better medicine, food, nutrition, and sanitation. Birth rates decrease because of birth control, education, and fewer children being needed to help support families.

Scarcity of Resources

Most of the energy that is used by humans comes from fossil fuels such as oil, coal, and natural gas. These are *nonrenewable resources* because when they are used up, they cannot be replaced by the same kind of substance. Some researchers feel that the supply of oil will be completely depleted (used up) sometime before the end of this century. In the short term, increased fuel and energy efficiency are needed. In the long term, an increase is needed in the research and development of alternative sources of energy such as hydrogen fuel and wind and solar energy. In addition, increased funding is needed to develop new energy technologies. Along with industrialization comes increased urbanization. This results in the destruction of natural habitats for existing organisms and less land for food production. Increasing population density increases the need for food but also decreases available land and energy supplies. Efforts need to be concentrated on *renewable resources,* which are replaceable. Some examples of renewable resources are trees, wildlife, food crops, cattle, fish, and land. Conservation by recycling and reusing what is already here helps preserve and protect scarce resources.

Pollution

Pollution is the addition of chemical or biological substances into land, water, or air that changes the natural composition of these abiotic factors. The main sources of pollution are the burning of fossil fuels, industrial waste, pesticides, insecticides, fertilizer, and biological waste. Not all pollutants are man-made. For example, volcanic eruptions emit ash and other particles in the air that normally don't belong there. Pollutants are harmful to humans and other organisms in the environment. *Point source pollution* is pollution that can be traced back to its point of origin, such as a waste pipe from a factory or the aerial spraying of an insecticide. *Nonpoint source pollution* is defined as pollution coming from different sources that cannot be traced back to a specific point of origin. An example is the runoff of fertilizer from farms and suburban areas into water.

Air pollution: The burning of fossil fuels in cars and factories produces emissions that pollute the air. Some of these pollutants are carbon dioxide, carbon monoxide, nitrogen oxides, and sulfur oxides. These pollutants eventually return to the water and soil to contaminate these abiotic factors.

Water pollution: Sources of water pollution include wastes from factories, farms, suburban areas, sewage, and air pollutants.

Land (soil) pollution: Land is polluted by rain that washes out pollutants in the air. Landfill and toxic dumps also contaminate the land.

The degree of harm caused by pollutants depends on the chemical structure of the pollutant, its concentration in an area and its longevity. Air pollutants can cause respiratory problems such as bronchitis, asthma, and lung cancer. Pollutants in water and soil can cause cancer in humans. Pollutants that enter the food chain can cause illness in humans. Pollutants tend to become more concentrated in an animal's tissue the farther down the animal is in the food chain.

Global Warming

Global warming is caused by the greenhouse effect, which causes heat to be trapped by the earth's atmosphere. The major gases responsible for the greenhouse effect are water vapor and carbon dioxide. The concentration of carbon dioxide in the atmosphere has been rising since the start of the industrial revolution. This increase in carbon dioxide concentration is a direct result of the burning of fossil fuels. Possible effects of global warming include climate changes, rising sea levels, melting glaciers, and crop disasters. In addition, massive flooding because of a rise in sea level, droughts, wildfires, reduced water supplies, and reduced biodiversity can be expected. Reducing air pollution produced by burning fossil fuels eliminates most of the carbon dioxide emissions responsible for the greenhouse effect while at the same time improving the quality of air on earth for all organisms. This can be accomplished with alternative energy sources and by increased improvements in energy efficiency. At the same time, global warming can be reduced by decreasing deforestation and slowing population growth.

Example Problems

These problems review humans and the environment.

1. What is the effect of increased population growth on the environment?

   Answer: Increased population growth creates added demands for the earth's limited resources of energy, food, water, and living space. Furthermore, more people results in an increase in pollution and a decrease in biodiversity.

2. How are nonrenewable resources different from renewable resources?

   Answer: Nonrenewable resources are resources that cannot be replaced. Examples of nonrenewable resources are fossil fuels. Renewable resources are replaceable. Examples of renewable resources are trees, wildlife, food crops, cattle, fish, and land.

3. What are some of the major sources of man-made and natural pollution?

   Answer: The major sources of man-made pollution are the burning of fossil fuels, industrial waste, pesticides, insecticides, fertilizer, and biological waste. Volcanic eruptions that emit ash and other particles are an example of a source of natural pollution.

4. How is pollution harmful to a person's health?

      Answer: Air pollutants can cause respiratory problems such as bronchitis, asthma, and lung cancer. Pollutants in water and soil can cause cancer in humans.

Work Problems

Use these problems on humans and the environment to give yourself additional practice.

1. What is meant by exponential population growth?

2. How can scarce resources be preserved?

3. What threats are posed by global warming?

Worked Solutions

1. Exponential population growth refers to the doubling of the population every few years.

2. Scarce resources can be preserved by decreasing population growth, reducing the use of fossil fuels and through conservation of the environment. In addition, currently available resources can be reused and recycled.

3. Global warming has the potential of causing major disruptions in the world's ecosystems. An increase in temperature can result in melting glaciers and rising sea levels. Changes in climate might occur that produce droughts in some areas and flooding in others. Additional threats are wildfires, reduced water supplies, reduced biodiversity, and crop failures.

Environmental Protection

The best way to protect the environment is prevention. If a problem isn't created, it does not need to be fixed. If the air, water, and soil aren't polluted, they don't need to be cleaned up. However, for areas that are already polluted, expensive cleanup and control are the only solutions. In addition, if the tropical rain forests aren't burned down and wildlife habitats aren't reduced, expensive restorations don't need to be performed.

Conservation

Conservation is concerned with the protection and maintenance of ecosystems and their biological communities. Human population growth is the greatest danger that exists for any ecosystem. Controlling and reducing population is the best long-term method of protecting an ecosystem. In the meantime, use of nonrenewable resources such as fossil fuels needs to be reduced. Reducing the use of fossil fuels helps conserve these resources while at the same time reducing pollution of the air, water, and soil. Increased efforts to recycle paper, plastics, and metals help conserve these resources and reduce wastes that overburden landfills. Water conservation can be achieved by reducing pollution from agriculture. Runoff from fertilizers, pesticides, and animal wastes are the main causes of water pollution. Control of point source pollution such as pollution from sewage treatment plants and home septic tanks can help preserve the water supply. Soil conservation can be achieved by controlling erosion. *Erosion* is the loss of topsoil and other soil components from an area. Erosion is caused by the action of water and wind. Preserving forests, preserving vegetation, preventing overgrazing by livestock, and practicing good farming techniques conserve soil and prevent erosion.

Protecting the Tropical Rain Forests

The tropical rain forests are important for their biodiversity and because 50% of the earth's plant and animal life can be found there. In addition, the plants of the rain forest play a critical role in the carbon-oxygen cycle. More oxygen is produced in this biome than in any other land biome. The tropical rain forest plants are crucial in the removal of carbon dioxide from the atmosphere. Since 1950, more than half of the world's tropical rain forests have been destroyed. Many trees have been cut down for their lumber, and many acres have been burned down and used as farmland. The destruction of the rain forests is currently continuing.

Biological Pest Control

Biological pest control is the control of a pest population by nonchemical (not pesticides or insecticides) methods such as predators, parasites, animal-like protists, bacteria, and diseases, which destroy or control a pest population. Insects such as the ladybug feed on aphids. Egg parasites are small wasps that lay their eggs inside the eggs of moths and caterpillars. *Bacillus thuringiensis* (Bt) is a bacterium that can be sprayed on crops. These bacteria stop insects from eating, and in a few days they die. Nematodes are tiny roundworms that are parasitic on harmful insects such as Japanese grub beetles, flea beetles, fruit borers, potato tuber worms, and many other pests. The advantage of biological pest control is that it avoids the use of harmful insecticides and pesticides. Many of these chemicals kill helpful insects, birds, and fish and cause cancer and other diseases in humans. The use of biological pest controls is more expensive than the use of chemical controls and often requires repeated applications. Researchers are also producing genetically engineered plants that are resistant to certain insects and fungi.

Alternative Sources of Energy

One way to reduce dependency on fossil fuels is to utilize other sources of energy. Solar energy can be used to produce heat and electricity. New techniques of home building can make better use of the sun to heat a home. Solar panels can be used to generate electricity. Wind farms that utilize giant wind turbines can be used to generate electricity. Alcohol produced from plants can be combined with gasoline to produce gasohol, which can be burned in a car engine. Hydrogen, which can be obtained from a variety of sources, is being investigated as a future clean-burning fuel that will reduce dependence on fossil fuels. Currently the cost of producing energy from these alternatives is greater than the cost of energy from fossil fuels. As fossil fuels become scarcer and as technology improves, the production of alternative energy sources and their use will become more widespread. Nuclear energy does not produce air pollution. Despite this, nuclear energy is not an attractive alternative source of energy. The energy produced by a nuclear plant is expensive, removal and disposal of nuclear waste is dangerous, and no one wants to store the waste in his or her backyard. People don't want nuclear reactors in their neighborhood because they are afraid of an accident. Furthermore, nuclear wastes can potentially be used to produce weapons of mass destruction.

Example Problems

These problems review environmental protection.

1. What is conservation?

    Answer: Conservation is concerned with the protection and maintenance of an ecosystem and its biological communities.

2. What are the most important steps that can be taken to help conserve and protect the environment?

   Answer: The most important steps that can be taken to help conserve and protect the environment are reducing the rate of human population growth and striving toward negative population growth. In addition, reducing pollution and the use of fossil fuels, along with protecting open spaces and wildlife, helps to conserve resources.

3. What is biological pest control?

   Answer: Biological pest control is the control of a pest population by nonchemical means (predators, parasites, animal-like protists, bacteria, and diseases) to destroy or control a pest population.

Work Problems

Use these problems on environmental protection to give yourself additional practice.

1. How can the water supply be conserved and maintained?

2. Why is the destruction of the rain forest for use as farmland not a good idea?

3. What are the advantages and disadvantages of biological pest controls?

4. What are some alternative sources of energy?

Worked Solutions

1. Water conservation can be achieved by reducing pollution from agriculture. Runoff from fertilizers, pesticides, and animal wastes are the main causes of water pollution. Control of point source pollution from factories, sewage treatment plants, and home septic tanks can also help preserve the water supply.

2. The soil in rain forests is acidic, low in nutrients, and needs to be heavily fertilized to grow crops. However, fertilizer is expensive and not often used by farmers in land that was once rain forest. As a result, the farmers get a low yield on their crop production. In the meantime, this valuable resource (along with its diverse plant and animal life) is destroyed.

3. Biological pest controls do not pollute the environment and do not kill helpful insects and birds because they target a specific pest population. Biological controls are harmless to humans. The disadvantage of biological pest controls is that they are expensive and often require repeated applications. In addition, biological controls generally target only one type of pest, unlike chemical controls whose action is more widespread.

4. Some alternative sources of energy are solar energy, wind energy, hydrogen fuel, and nuclear energy.

Chapter Problems and Answers

Problems

The following is a brief paragraph based on the diagram shown. For problems 1–10, fill in the missing terms.

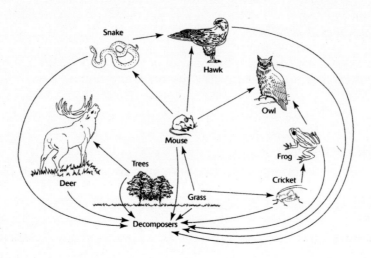

The preceding diagram shows a _____. The organisms represented by grass and trees
                            1

are known as the _____. Deer, mice, and crickets are the _____ consumers. These
                 2   3

organisms are also known as _____ because they eat plants. The snake and frog are the
                            4

_____ consumers. These organisms are also known as _____ because they eat animals.
   5   6

The hawk and the owl are _____ consumers. If a disease killed the frog, the owl would
                         7

survive because it can also eat the _____. The two organisms that have the least amount
                                    8

of energy available to them are _____. When any of these organisms die, their energy is
                                9

transferred to the _____.
                   10

For problems 11–15, fill in the missing answers.

11. Air, water, and soil are examples of _____ factors.

12. A distance north or south of the equator can be measured in degrees of _____.

13. _____ is the weather in an area over a period of a year.

14. Plants that live in water or that require large amounts of water such as those found in the
     tropical rain forest are called _____.

15. _____ is the part of the soil that is formed from decaying plant and animal material.

For problems 16–19, select the biotic relationship from the following list that is associated with each description. A choice can be used more than once or not at all.

A. Commensalism   B. Interspecific competition
C. Mutualism   D. Parasitism   E. Predator-prey relationship

16. Birds are often infected with lice that suck their blood. _____

17. Hawks and owls both feed on small rodents such as mice. _____

18. The red-billed oxpecker bird eats ticks and lice off the skin of a rhinoceros. _____

19. The clown fish swims among the stinging tentacles of the sea anemone where it is protected from being eaten by other fish. The sea anemone tolerates the presence of the clown fish. _____

For problems 20–22, provide the term that *best* completes each sentence.

20. _____ is a gas that is being released into the atmosphere, contributing to global warming.

21. _____ in the form of rain, sleet, or snow removes some of the pollutants from the air.

22. Oil and coal are examples of energy sources that are known as _____ resources.

For problems 23–25, select the biome from the following list that is associated with each description. A choice can be used more than once or not at all.

A. Deciduous forest   B. Taiga   C. Tropical rain forest   D. Tundra   E. Desert

23. Select the biome that is located in the most northerly latitude. _____

24. Which biome covers that largest geographical area? _____

25. Identify the biome with the most biodiversity. _____

Answers

1. **Food web.** This food web consists of several interconnecting food chains.

2. **Producers.** Green plants are producers. Green plants convert light energy from the sun into chemical energy stored in glucose molecules.

3. **Primary consumers.** In a food chain the first group of organisms to eat are called primary consumers.

4. **Herbivores.** Animals that eat plants are called herbivores.

5. **Secondary consumers.** The snake and frog are called secondary consumers because they are the second group of organisms to eat. The snake ate the mouse, which is a primary consumer. The frog ate the cricket, which is a primary consumer.

6. **Carnivores.** Animals that eat meat (other animals) are called carnivores.

7. **Tertiary consumers.** The hawk and owl are called tertiary consumers because they are the third group of organisms to eat.

8. **Mouse.** The advantage for organisms in a food web is that they have alternative sources of food in case one particular food is unavailable.

9. **The hawk and the owl.** In a food chain or a food web, the organisms that are at the end have the least amount of energy available to them because the organisms before them use energy performing their life functions.

10. **Decomposers.** The decomposers are the organisms of decay. Decomposers return minerals and nutrients to the soil. Sooner or later the decomposers are going to get you.

11. **Abiotic.** Abiotic factors are the nonliving factors in the environment. Climate, temperature, and light are examples of abiotic factors.

12. **Latitude.** The equator is the 0° line of latitude. Points on the earth north of the equator have positive values, while those south of the equator have negative values. The North Pole is at +90°, and the South Pole is at –90°. The prime meridian is the 0° line of longitude. Longitude is measured east and west of the prime meridian.

13. **Climate.** The climate of an area is its atmospheric conditions over a period of a year.

14. **Hydrophytes.** The prefix *hydro-* means water, and the suffix *-phyta* means plant.

15. **Humus.** Soil that is rich and fertile has lots of humus.

16. **D. Parasitism.** In parasitism, one organism benefits (the parasite), and the other is harmed (the host). In this example, the parasites are the lice, and the hosts are the birds.

17. **B. Interspecific competition.** Interspecific competition is competition between organisms of different species. The hawk and the owl are in different species, but they compete for many of the same kinds of food.

18. **C. Mutualism.** Mutualism is a symbiotic relationship where both organisms benefit. The bird in this example gets food, and the rhinoceros gets rid of parasites.

19. **A. Commensalism.** Commensalism is a symbiotic relationship where one organism benefits, and the other is not affected. The clown fish benefits from the protection of the sea anemone, which is not affected by the presence of the clown fish.

20. **Carbon dioxide.** Carbon dioxide is the main greenhouse gas that is responsible for global warming.

21. **Precipitation.** Unfortunately, pollutants that are removed from the air by precipitation pollute the soil and water.

22. **Nonrenewable.** Nonrenewable resources cannot be replaced. All fossil fuels were formed millions of years ago from organisms that died and decayed. Fossil fuels cannot be replaced by the same kind of fuel.

23. **D. Tundra.** The tundra is located just above the Arctic Circle.

24. **B. Taiga.** The taiga is located below the Arctic Circle in a continuous belt around the earth.

25. **C. Tropical rain forest.** This biome has 50% of the earth's plant and animal life.

Supplemental Chapter Problems

Problems

For problems 1–25, select the *best* answer.

1. Ecology can be defined as the study of:

    A. pollution

    B. climax communities

    C. the world biomes

    D. living things, their interrelationships with each other, and their interrelationship with the environment

2. Light, temperature, and climate are examples of:

    A. biotic factors   B. abiotic factors

    C. renewable resources   D. resources that can be recycled

3. Which of the following kinds of green plants are found in the desert?

    A. hydrophytes   B. mesophytes   C. xerophytes   D. endophytes

4. Soil that has large particles and is *not* able to retain water is called:

    A. clay   B. loam   C. sandy   D. humus

5. All the grass, trees, insects, birds, and squirrels living together in a park represent a:

    A. species   B. population   C. community   D. ecosystem

6. Which of the following best describes a population?

    A. all the mice living in a warehouse

    B. ants capable of reproducing with each other

    C. all the grass, grasshoppers, and frogs in a location

    D. all the trees, birds, cats, soil, and water in a location

7. In interspecific competition, organisms compete for all the following *except:*

    A. territory   B. food   C. water   D. a mate

8. The role that an organism has in its environment is called its:

    A. habitat   B. niche   C. territory   D. abiotic factor

9. The capability of an area to supply enough food to feed a given species of organism is known as:

    A. the carrying capacity of the land   B. primary production

    C. the flow of energy through the ecosystem   D. a food chain

10. Rabbits eat plants, and coyotes eat rabbits. What is the long-term effect on the rabbit population in an area if all coyotes are removed?

    A. The rabbit population decreases.

    B. The rabbit population increases.

    C. The rabbit population decreases and then increases.

    D. The rabbit population increases and then decreases.

11. Wolves, coyotes and hawks are organisms that hunt and kill other living things for food. The animals that are hunted and killed are called:

    A. decomposers   B. scavengers   C. prey   D. predators

12. Identify the following symbiotic relationship: Barnacles are small crustaceans that attach themselves to whales, and the whales are unaffected.

    A. commensalism   B. mutualism   C. parasitism   D. predation

13. Identify the following symbiotic relationship: Nitrogen-fixing bacteria live inside the roots of legumes such as the pea plant. The bacteria get a place to live, and the plant gets nitrogen, an important mineral.

    A. commensalism   B. mutualism   C. parasitism   D. predation

14. For two species to develop a close symbiotic relationship, they must:

    A. have coevolved   B. compete with each other

    C. have the same niche   D. have the same common ancestor

15. Identify the group of organisms that breaks down dead plant and animal tissue, returning minerals and nutrients to the soil.

    A. consumers   B. scavengers   C. omnivores   D. decomposers

16. To have a balanced and stable ecosystem:

    A. the number of predators must equal the number of prey

    B. the abiotic factors must equal the biotic factors

    C. materials such as water, nitrogen, carbon dioxide, and oxygen must cycle between organisms and their environment

    D. energy must be recycled between organisms and the environment

17. The energy that drives the water cycle comes from:

    A. producers   B. the sun   C. nuclear energy   D. hydrogen fuel

18. Which two processes are involved in the oxygen/carbon dioxide cycle?

    A. transpiration and evaporation   B. condensation and precipitation

    C. protein synthesis and nitrogen fixation   D. photosynthesis and respiration

19. Ecological succession can best be described as:

    A. the transfer of energy in a food chain from one group of organisms to the next

    B. a process by which one biotic community is replaced by another

    C. the transfer of materials from organisms to the environment

    D. the transfer of materials from the environment to organisms

20. The first organisms to inhabit an area that has not previously been inhabited are called:

   A. heterotrophs   B. climax community organisms
   C. pioneer organisms   D. decomposers

21. Which of the following biomes is the most stable?

   A. desert   B. tropical rain forest   C. tundra   D. ocean

22. Which biome in the United States has been replaced with farms and ranches?

   A. grasslands   B. taiga   C. tundra   D. desert

23. Select the factor that is most responsible for damaging the environment.

   A. nuclear energy   B. people   C. hurricanes   D. tornadoes

24. Select the best method to protect and preserve the environment.

   A. use of nuclear energy   B. conservation
   C. population increase   D. rain forest decrease

25. Which method of destroying the mosquitoes responsible for spreading the West Nile virus causes the least damage to the environment?

   A. spraying the adult mosquitoes with insecticides
   B. spraying oil on wet areas where mosquitoes reproduce
   C. introducing a predator that feeds only on these mosquitoes
   D. draining all wet areas where mosquitoes reproduce

Answers

1. **D.** "Chapter Introduction," p. 297

2. **B.** "Abiotic Factors," p. 297

3. **C.** "Abiotic Factors," p. 297

4. **C.** "Abiotic Factors," p. 297

5. **C.** "Biotic Factors," p. 299

6. **A.** "Biotic Factors," p. 299

7. **D.** "Competition," p. 300

8. **B.** "Competition," p. 300

9. **A.** "The Predator-Prey Relationship," p. 300

10. **D.** "The Predator-Prey Relationship," p. 300

11. **C.** "The Predator-Prey Relationship," p. 300

12. **A.** "Symbiosis," p. 301

13.   **B.** "Symbiosis," p. 301

14.   **A.** "Coevolution," p. 302

15.   **D.** "The Flow of Energy Through an Ecosystem," p. 303

16.   **C.** "Material Cycles," p. 306

17.   **B.** "Material Cycles," p. 306

18.   **D.** "Material Cycles," p. 306

19.   **B.** "Ecological Succession," p. 308

20.   **C.** "Ecological Succession," p. 308

21.   **D.** "World Biomes," p. 309

22.   **A.** "World Biomes," p. 309

23.   **B.** "Humans and the Environment," p. 313

24.   **B.** "Environmental Protection," p. 315

25.   **C.** "Environmental Protection," p. 315

Glossary

This glossary contains many of the key terms presented in this book.

Abiotic factors   Nonliving things in the environment such as climate, light, temperature, water, soil and territory.

Absolute dating   A method used to determine the approximate age of a fossil by relying on the radioactivity of certain elements and their half-lives.

Acetylcholine   A neurotransmitter produced by the terminal branches of a nerve cell.

Acid   A compound that forms hydrogen H^+ ions when dissolved in water and has a pH less than 7.

Addition   One type of addition is a gene mutation, where a nitrogen base is inserted into an mRNA molecule. Another type of addition is a chromosome mutation, where a segment of one chromosome breaks off and is added to its homologous chromosome.

Aerobic respiration   Cellular respiration that requires oxygen and takes place inside the mitochondria.

Allele   One of the two different kinds of genes for a trait. Alleles occupy the same location on a pair of homologous chromosomes.

Alveoli   Tiny air sacs, located at the end of bronchioles, where the exchange of oxygen for carbon dioxide in the blood takes place.

Amino acid   An organic compound that is the building block of proteins. Amino acids have an amino group (NH_2) on one end and a carboxyl group (COOH) on the other.

Anaerobic respiration   Cellular respiration that does not require oxygen and takes place in the cytoplasm outside the mitochondria.

Analogous structures   Structures that are similar in function but dissimilar in construction and evolutionary development.

Anther   Part of the stamen of a flower and the producer of pollen grains.

Anthophyta   A phylum in kingdom Plantae that includes flowering plants that usually reproduce by seeds hidden inside fruits.

Antibiotic   A substance produced by some bacteria or fungi that can destroy bacteria but not viruses.

Antibody   A protein in the blood that can help fight disease by destroying harmful substances such as antigens, bacteria and viruses.

Anticodon   The side of tRNA that is complementary to the codon on an mRNA molecule.

Antigen   A protein that stimulates the production of antibodies.

Archaea   A domain that includes prokaryotic unicellular organisms with cell walls lacking peptidoglycan. In a six-kingdom system of classification, organisms in the domain Archaea are classified in the kingdom Archaebacteria.

Artery   A blood vessel with thick muscular walls that has a pulse and carries blood away from the heart to the organs.

Asexual reproduction   The formation of a new individual from one parent.

ATP (adenosine triphosphate)   The energy molecule produced during cellular respiration.

Autosomes   The chromosomes in the nucleus of a cell that *do not* determine the sex of an individual.

Autotroph   An organism that makes its own food by the process of photosynthesis.

Axon   The part of a neuron that carries nerve impulses from the cyton to the terminal branches.

Bacteria   A domain that includes prokaryotic unicellular organisms with cell walls containing peptidoglycan. In a six-kingdom system of classification, organisms in the domain Bacteria are classified in the kingdom Eubacteria.

Bilateral symmetry   The right and left sides of an organism are similar.

Binary fission   A kind of asexual reproduction that results in the formation of two identical daughter cells, each with the same amount of cytoplasm and nuclear material (chromosomes).

Binomial nomenclature   A two-name naming system for organisms composed of their genus and specific epithet names.

Biological pest control   The control of a pest population by nonchemical methods (as opposed to pesticides and insecticides), such as the use of predators, parasites, animal-like protists, bacteria and diseases.

Biology   The science that studies living things, their structure and function.

Biome   A large area of the earth defined by a climax community of plants and specific abiotic factors such as light, precipitation, temperature and soil type.

Biosphere   The interaction between all the ecosystems on earth.

Biotic factors   Living things in the environment such as organisms, populations, communities, food chains, symbiotic relationships, competition between organisms and succession.

Blastocyst (blastula)   A stage of embryological development that takes place about five days after fertilization, characterized by a hollow ball of cells with fluid and an inner mass of cells at one end.

Bowman's capsule   The part of the nephron that receives wastes from blood in the glomerulus.

Bryophyta   A phylum in the kingdom Plantae characterized by the lack of vascular tissue. Bryophytes do not have true roots, stems or leaves.

Budding   A type of asexual reproduction that results in two unequal-sized cells. The larger cell is called the parent, and the smaller one is called the bud.

Cambium   The cells of a vascular bundle (vein) located between the phloem and the xylem. Cambium cells are *meristematic cells* and are capable of reproduction.

Capillary   Connects arteries to veins and is the smallest blood vessel (one cell in diameter).

Capsid   A protein coat that surrounds a virus.

Carcinogen   A chemical that causes mutations that result in cancer.

Carnivore   An animal that eats other animals.

Carpel The part of a flower that contains the stigma, style and ovary.

Carrier A female that has an allele for a sex-linked disease but does not have the disease.

Carrying capacity The ability of a land area to supply enough food to feed a population.

Cartilage A connective tissue that is flexible and softer than bone.

Cell The basic unit of structure and function in all living things.

Cell membrane The outer boundary of a cell that protects the cell, controls what goes in and out of the cell (*selective permeability*) and holds the cell together. This structure is composed of two layers of lipid with protein molecules suspended inside.

Cell wall Found only in plant cells surrounding the cell membrane. The cell wall is made of cellulose, is rigid and is nonliving. This organelle gives the plant cell support and shape.

Cellular respiration The process by which a cell converts food into energy (ATP).

Characteristic A version of a trait.

Chemistry The science that studies matter and changes in matter.

Chlorophyll A green pigment found inside the chloroplasts of cells located in the palisade and spongy mesophyll layers of a leaf. This pigment traps light energy that is used to split water during the process of photosynthesis.

Chloroplast A small, oval-shaped structure containing the green pigment chlorophyll that functions in photosynthesis.

Chordata A phylum in the animal kingdom characterized by a dorsal *notochord*, which is a flexible, rod-like structure found at some point during embryonic development.

Classification A system for organizing diverse living things into logical groupings that make it easier for biologists to study them.

Cleavage A special series of mitotic cell divisions where each time the cells divide they get smaller and smaller. This process results in embryological development.

Climate The weather in an area over a period of a year.

Climax community The last community to occupy an area. A climax community is a stable and self-perpetuating community of plant and animal life.

Clone DNA, genes, cells, tissues or organisms that are genetically identical to the ancestor they are derived from.

Codominance Both alleles for a trait are dominant, and organisms produced from these crosses have both characteristics of the trait.

Codon A set of three nitrogen bases on mRNA.

Coenzyme A molecule that transfers hydrogen and electrons from one reaction to another.

Coevolution The process by which species adapt to changes in one another.

Commensalism A symbiotic relationship where one organism benefits from the association, and the other is neither helped nor harmed.

Common ancestor An individual from which two or more related species could have evolved.

Community Different populations living together in a specific location.

Coniferophyta   A phylum in kingdom Plantae, which includes nonflowering plants that reproduce by seeds found in cones. These trees have needlelike leaves that in most species usually stay on the plant all year long.

Conservation   The protection and maintenance of an ecosystem and its biological communities.

Control   The comparison group in a scientific experiment.

Cotyledons   These are the seed leaves that contain endosperm, which is food for a germinating plant. Monocots have one cotyledon, and dicots have two.

Cuticle   A thin, waxy layer that covers the upper epidermis of a leaf and prevents the loss of water.

Cyton   The cell body of a neuron, consisting of the nucleus, cytoplasm and dendrites.

Cytoplasm   The liquid part of a cell found within the cell membrane.

Dark reactions   A series of reactions in photosynthesis that takes place in the absence of light.

Deamination   A process that takes place in the liver, where excess amino acids have their amino group (NH_2) removed, converted into ammonia and then into urea.

Dehydration synthesis   The formation of a large molecule from two smaller molecules by the removal of water.

Deletion   One type of deletion is a gene mutation, where a nitrogen base is lost from an mRNA molecule. Another type of deletion is a chromosome mutation, where a segment of a chromosome is lost, resulting in lost genes.

Deoxyribonucleic acid (DNA)   A nucleoprotein found in chromosomes that contains genes.

Deoxyribose   A five-carbon sugar that is part of the DNA molecule.

Dependent variable   In a scientific experiment, the variable that changes in response to changes in the independent variable.

Dicot (dicotyledon)   A plant that produces seeds that have two seed parts (cotyledons) and two seed leaves with netted veins.

Differentiation   A process that occurs at some point between the blastocyst stage and the gastrula stage and results in the cells of the developing embryo becoming different from each other.

Diffusion   The movement of *any* molecule (*except water*) from areas of high concentration to areas of low concentration.

Diploid number (2*n*)   The number of chromosomes found in a somatic cell.

Disjunction   The separation of homologous chromosomes during meiosis.

Domain   A taxon higher than a kingdom.

Dominant   In a hybrid, the characteristic of the trait that shows up.

Dormancy   In a seed, a period of inactivity where life functions have significantly slowed down or have been temporarily suspended.

Ecological pyramid   A pyramid that shows the flow of energy from one trophic level to the next and the relative numbers of organisms at each level.

Ecosystem   The interaction between a community and the abiotic factors in the environment.

Ectoderm   The outer cell layer of a hydra or the outer germ layer in the gastrula stage of embryological development.

Effector   A muscle or gland that can carry out a response to a stimulus.

Embryo   An organism during the early stages of development.

Emulsification   The breakdown of large fat molecules into small fat molecules by the action of bile.

Endocrine gland   A ductless gland that secretes hormones, which are *chemical messengers* that travel through the blood to regulate the activity of a target organ.

Endoderm   The inner cell layer of a hydra or the inner germ layer in the gastrula stage of embryological development.

Endoplasmic reticulum (ER)   A canal-like network within a cell that functions in the transport of materials.

Endoskeleton   A skeleton composed of bones and cartilage found inside the body.

Endosperm   The part of the cotyledon that provides food for a germinating seed.

Environment   The surroundings of an organism: its abiotic and biotic factors.

Enzyme   An organic catalyst that speeds up the rate of chemical reactions.

Epicotyl   The two leaves of a plant embryo that become the upper part of the stem and the leaves of the plant.

Erosion   The loss of topsoil and other soil components from one area and their movement to another.

Eubacteria   A kingdom that includes prokaryotic unicellular organisms with cell walls containing peptidoglycan.

Eukarya   A domain that includes eukaryotic organisms in the kingdoms Protista, Fungi, Plantae and Animalia.

Eukaryotic   A cell with a nuclear membrane surrounding its chromosomes.

Evaporation   The change of a liquid into a gas.

Evolution   The change that occurs in a species with the passage of time.

Excretion   The removal of cellular wastes such as water, carbon dioxide and urea.

Exocrine gland   A gland that has a duct.

Experiment   The procedures and materials used by a scientist to support or reject a hypothesis.

Exponential growth   The doubling of a population every few years.

Extracellular digestion   The breakdown of food outside the cells of an organism.

Fallopian tube   A narrow tube lined with cilia that connects the ovaries to the uterus.

Fermentation   A type of anaerobic respiration performed by yeast. Glucose is broken down to yield ethanol, carbon dioxide and ATP (energy).

Fibrinogen   A protein found in plasma that interacts with platelets and helps the blood to clot.

Flexor   A skeletal muscle that moves a bone toward the body.

Follicle   A cavity in an ovary where an egg cell matures.

Food chain   The transfer of energy from one organism to another in an ecosystem.

Food web   Several interconnecting food chains in an ecosystem.

Fossil   The remains of an organism that lived in the distant past.

Fossil fuel   Oil, natural gas or coal that formed millions of years ago from organisms that died and decayed.

Frameshift mutation   A type of mutation where one nitrogen base is deleted from or added to a codon on an mRNA molecule.

Free radical   An atom with unpaired electrons that can cause breaks in a DNA molecule.

Fungi   A kingdom of organisms with cells that do not have chloroplasts. Fungi are heterotrophic and obtain their food by absorption from decaying vegetation.

Gamete   A sperm or egg cell with the haploid number of chromosomes.

Gameteophyte   In organisms that reproduce by alternation of generations, this generation is characterized by the production of male and female haploid (n) gametes.

Gametogenesis   The production of gametes during meiosis.

Ganglion   A group of nerve cells and cytons combined.

Gastrula   A stage of embryological development characterized by the presence of three cell layers (ectoderm, mesoderm and endoderm).

Genetic engineering   A process by which genes from the chromosomes of one species are inserted into the chromosomes of another species.

Genomics   The study of genes in the chromosomes of an organism, their structure and function.

Genotype   The gene combination for a trait (homozygous or heterozygous).

Genus   The classification category that includes similar species.

Germination   The development of a seed into a new plant.

Global warming   An increase in the earth's temperature caused by the greenhouse effect.

Glucose   A simple sugar (monosaccharide) produced by green plants during the process of photosynthesis.

Glycogen   A large polysaccharide used to store glucose in animals.

Glycolysis   The breakdown of glucose during anaerobic respiration into two molecules of pyruvate with a gain of 2ATP.

Golgi apparatus   Flattened membranes in a cell that look like plates stacked one on top of the other. This organelle packages proteins.

Gonad   A sex gland; the ovary in a female or testis in a male.

Gradualism   The belief that evolution is a slow, continual and gradual process that proceeds in numerous small steps taking many years and generations to produce new species.

Guard cells   Cells containing chloroplasts found in the lower epidermis of a leaf that are responsible for forming the stomates.

Habitat   Where an organism lives in its environment.

Half-life   The time that it takes for an element to decay into half of its original amount.

Haploid number (n)   Half the diploid number, or the number of chromosomes in a gamete.

Hemoglobin   An iron-containing pigment that gives red blood cells their color and aids in the transport of oxygen to the cells of an organism.

Herbivore   An animal that eats plants, a primary consumer.

Heterotroph   An organism that eats to obtain food.

Heterozygous (hybrid)   Two different alleles for a trait.

Hilum   A scar on a seed that shows the point of attachment of the seed to the inside of the ovary.

Homologous chromosomes   Pairs of similar chromosomes found in diploid cells.

Homologous structures   Are similar in construction and evolutionary development but dissimilar in function.

Homozygous (pure)   Two identical alleles for a trait.

Hormone   A chemical messenger produced by an endocrine gland that travels through the blood and functions in regulation.

Humus   Soil that is rich, fertile and formed by the decay of dead plants and animals.

Hydrolysis   The breakdown of a large molecule into two or more smaller molecules by the addition of water.

Hydrophytes   Plants adapted for growth in water or in areas that have a great deal of precipitation.

Hypocotyl   The part of a seed embryo that develops into the roots and lower part of a stem.

Hypothesis   An educated guess that provides a possible answer to a scientific problem.

Igneous rock   A rock formed by volcanic activity.

Incomplete dominance   In genetics, the presence of two different alleles for a trait where neither one is dominant or recessive.

Independent assortment   In genetics, when two dihybrids are crossed, alleles for the different traits separate and are inherited independently.

Independent variable   In a science experiment, the experimenter controls the independent variable (the factor that is being tested). As the independent variable is changed, the dependent variable changes.

Index fossil   A fossil of a species of organism that is found all over the world but lived for a short period of time.

Interspecific competition   Competition between members of different species.

Intracellular digestion   The breakdown of food inside the vacuoles of a cell.

Intraspecific competition   Competition between members of the same species.

Invertebrate   An animal without a backbone.

Islets of Langerhans   Cells located in the pancreas that produce the hormones insulin and glucogon.

Karyotype   Chromosomes that are removed from the nucleus of a cell and arranged in decreasing size order.

Kingdom   A category in classification, including phylum, class, order, family, genus and species.

Latitude   The distance north or south of the equator measured in degrees.

Life functions   Various life processes that are performed by living things (for example, *nutrition, transport, respiration, excretion* and *reproduction*).

Ligament   A tissue that connects one bone to another.

Light reactions   A series of reactions that begins the process of photosynthesis. Light energy trapped by chlorophyll is used to split water molecules (*photolysis*).

Limiting factors   Any abiotic or biotic factors that determine the distribution and kinds of organisms that can survive in an area.

Lipids   Organic molecules composed of one molecule of glycerol combined with three molecules of fatty acid. Examples are fats, oils and waxes.

Loop of Henle   The *U*-shaped part of a nephron that reabsorbs food materials (glucose, amino acids) and water into the blood of the surrounding capillaries.

Lysosomes   Small, irregularly shaped structures in a cell that have their own membrane and contain digestive enzymes for the breakdown of food.

Malpighian tubules   A network of tubes in insects that functions in the excretion of nitrogen waste (uric acid).

Mechanical digestion   The physical breakdown of food into small particles.

Menstrual cycle   A hormonal cycle in women where an egg cell is matured and released, accompanied by the preparation of the uterine lining to receive a fertilized egg.

Menstruation   A stage in the menstrual cycle characterized by the shedding of the uterine lining.

Mesoderm   The middle germ layer found in the gastrula stage of embryological development.

Mesophytes   Land plants that require average amounts of rainfall during the course of a year.

Metamorphic rock   A rock formed from igneous or sedimentary rock by tremendous heat and pressure.

Metamorphosis   A change in body form from young to adult (for example, a caterpillar becoming a butterfly).

Micrometer   The unit used for cellular measurement. One micrometer (μm) equals one thousandth of a millimeter (mm).

Mitochondrion   This organelle is the energy factory of cells and is associated with cellular respiration. It has an elliptical shape with outer and inner membranes. Mitochondria contain their own DNA and reproduce independently of cells.

Mitosis   A type of cell division that results in the formation of daughter cells, which have the diploid number of chromosomes.

Molecular transport   The movement of molecules across a membrane by passive or active transport.

Monocot (monocotyledon)   A plant that produces seeds that have one seed part (cotyledon), which cannot be split, and one seed leaf with parallel veins.

Morula   The stage in embryological development characterized by a solid ball of cells.

Motile   Capable of movement.

Mutagenic agent   An environmental or man-made factor that increases the frequency of mutation.

Mutation   A change or error in a gene or chromosome.

Mutualism   A symbiotic relationship where both organisms benefit from living together.

Natural selection   Darwin's theory that nature selects those variations in a population that are the most fit. These variations are passed on to the next generation.

Negative feedback   A homeostatic control mechanism, where one action or change *inhibits* another. For example, increasing levels of thyroxin in the blood *inhibit* the pituitary gland from producing a thyroid-stimulating hormone (which in turn results in the thyroid gland producing less thyroxin).

Nephridia   A pair of tubes in an earthworm segment that is surrounded by capillaries and functions in excretion.

Nephron   A tiny tube found in the human kidney that filters waste products out of the blood.

Nerve   A bundle of axons held together by connective tissue.

Neuron   A nerve cell.

Neurotransmitter   A chemical secreted by the terminal branches of a neuron that carries a nerve impulse across a synapse to the dendrites of the next neuron.

Niche   The role or job of an organism in its environment.

Nondisjunction   The failure of homologous chromosomes to separate during meiosis.

Nonrenewable resources   Resources such as fossil fuels that cannot be replaced.

Nucleolus   A part of a cell found inside the nucleus that is associated with the production of ribosomes and RNA.

Nucleotide   A part of a DNA molecule that contains deoxyribose, a phosphate group and one of four possible nitrogen bases (adenine, guanine, cytosine or thymine).

Nucleus   The part of an atom that contains protons and neutrons. In a cell, this structure controls metabolism and reproduction. The nucleus contains *chromosomes*, which have genes that govern the heredity of the cell.

Omnivore   An animal that eats plants or other animals.

Oogenesis   The production of egg cells during meiosis.

Osmosis   The movement of water molecules from areas of high concentration to areas of low concentration.

Ovary   In an animal, the female gonad, which produces egg cells and the hormones estrogen and progesterone. In a flower, the part of the pistil that produces ovules and becomes the fruit.

Palisade layer　A leaf layer composed of long columnar cells that are packed tightly together. These cells contain chloroplasts for photosynthesis.

Pancreas　An exocrine gland that secretes pancreatic fluid, which aids in the chemical digestion of food, and an endocrine gland (the Islets of Langerhans), which produces the hormones insulin and glucogon.

Parasitism　A symbiotic relationship, where one organism benefits from the association, and the other is harmed.

Parthenogenesis　The development of an unfertilized egg into an embryo that eventually becomes a new individual.

Passive transport　The movement of molecules from areas of high concentration to areas of low concentration, without the use of energy. *Diffusion* and *osmosis* are two kinds of passive transport.

Pedigree chart　A chart used in genetics to track the inheritance of a trait in a family through several generations.

Peristalsis　A series of involuntary muscle contractions that forces food through the digestive system.

pH scale　A logarithmic scale (from 0 to 14) that is used to indicate the relative strength of acids and bases.

Phenotype　The characteristic (version) of a trait that is determined by observation.

Phloem　Cells in a vascular bundle (vein) that carry food downward in a stem.

Photolysis　The splitting of water molecules into hydrogen and oxygen during the light reactions of photosynthesis.

Photosynthesis　A process by which green plants take in carbon dioxide and water from their environment and produce glucose and oxygen.

Phylum　A classification category that includes similar classes.

Pioneer organism　In succession, the first organism to occupy an area.

Pistil　The female part of a flower.

Pituitary gland　An endocrine gland called the *master gland* because it produces hormones that regulate the other endocrine glands.

Placenta　In mammals, this structure is formed from a combination of tissues from the mother's uterine wall and the developing fetus.

Plantae　A kingdom of multicellular, autotrophic organisms that includes green plants.

Plasma　The straw-colored liquid part of blood that consists mostly of water and proteins.

Plasmid　The circular DNA found in bacteria.

Platelet　A cell fragment that interacts with fibrinogen and helps the blood to clot.

Point mutation　A type of mutation where one nitrogen base is replaced by another nitrogen base in an mRNA molecule.

Point-source pollution　Pollution that can be traced back to its point of origin.

Pollination   The transfer of pollen from an anther to a stigma.

Pollution   The addition of chemical or biological substances to land, water or air that changes the natural composition of these abiotic factors.

Population   Organisms in the same species that live together in a specific location.

Primary succession   Succession that begins from an area that never had life.

Prokaryotic   Cells that lack a nuclear membrane around their chromosomes.

Proteins   Organic molecules that contain nitrogen and are composed of 40 or more amino acids combined.

Protista   A kingdom of eukaryotic, unicellular organisms with animal-like, plant-like or fungus-like characteristics.

Pseudopod   A cellular extension in some animal-like protists that can change shape, capture food and help the organism move.

Punctuated equilibrium   The belief that evolutionary change occurs in sudden spurts during which many species are formed and followed by long periods of stability and no speciation.

Radial symmetry   A cut through the axis of an organism produces halves that are mirror images of each other.

Recessive   In a hybrid, the characteristic of the trait that we *don't* see.

Recombinant DNA   DNA produced when a gene from one chromosome is spliced into the chromosome of another species.

Red blood cell (RBC)   A blood cell that contains hemoglobin and transports oxygen to the cells and carbon dioxide from the cells.

Reflex   An automatic, involuntary, unlearned response to a stimulus.

Regeneration   The replacement or repair of a lost or damaged part of an organism.

Regulation   The ability of an organism to respond to a *stimulus* (a change in environment); the reaction of the organism is the *response*.

Relative dating   A method used to determine the age of a fossil by comparing its location *relative* to fossils in nearby rock layers.

Renewable resources   Resources that are replaceable.

Reproduction   The life function by which organisms produce new individuals of the same kind (species) or new cells necessary for growth and repair.

Respiration   The life function that provides an organism with energy (ATP) needed to carry out all the other life processes.

Response   The reaction of an organism to a stimulus.

Ribonucleic acid (RNA)   A nucleoprotein involved in the process of protein synthesis.

Ribose   A five-carbon sugar that is part of an RNA molecule.

Ribosome   A cell organelle that is the site for protein synthesis.

Root hair cells   Specialized root cells that have a large surface area adapted for the increased absorption of water.

Saprobe   An organism of decay that eats dead plants and animals.

Scavenger   A carnivore that eats food left over by a predator.

Scientific method   An organized series of steps used by biologists to solve problems. The steps of the scientific method include *problem, hypothesis, experiment, results* and *conclusion*.

Secondary succession   Succession that takes place in an area where life has already existed.

Sedimentary rock   A rock formed by the deposition of very small particles of rock, clay or silt.

Segregation   In genetics, when two hybrids are crossed, each gene segregates (separates) during gamete formation so that new gene combinations can be formed at fertilization.

Self-pollination   The transfer of pollen from anther to stigma within the same flower.

Semiconservative replication   The duplication of DNA where two molecules are formed, each composed of one original and one new strand.

Seminiferous tubules   Coiled tubes inside the testes that produce sperm cells.

Sensory neuron   A neuron that detects a stimulus and transmits a nerve impulse from a receptor to an interneuron in the spinal cord.

Sessile   Not capable of independent movement.

Setae   Small bristles on the segments of an earthworm that are used for locomotion.

Sex chromosomes   A pair of chromosomes that determines the sex of an individual.

Sex-linked trait   A trait that has an allele located on the X chromosome.

Sexual reproduction   The formation of a new individual from two parents (by the fertilization of gametes).

Somatic cell   A body cell with the diploid (2*n*) number of chromosomes.

Speciation   New species formation.

Species   Organisms that can mate with each other and produce fertile offspring.

Specific epithet   The second name in the scientific name of an organism that often describes a distinguishing characteristic of the organism.

Spermatogenesis   The production of sperm cells during meiosis.

Spinal cord   A nerve that begins at the base of the medulla and extends down through the vertebrae in the back.

Spiracles   Openings in the abdomen of an insect that function in respiration.

Spongy mesophyll layer   A leaf layer composed of cells that have chloroplasts.

Sporophyte   In organisms that reproduce by alternation of generations, this generation is characterized by the fertilization of gametes and the formation of a diploid (2n) organism.

Stamen   The male part of a flower.

Starch   A large polysaccharide used to store glucose in plants.

Stem cells   Undifferentiated cells inside the blastocyst stage of embryological development.

Stigma   Part of a flower's pistil (carpel) that is sticky and holds pollen grains.

Stimulus   A change in the environment of an organism.

Stomate   An opening in the lower epidermis of a leaf that allows carbon dioxide in and water and oxygen out.

Structural formula   Indicates the kinds of elements, number of atoms and arrangement of atoms in a compound.

Substrate   The substance that an enzyme acts on.

Succession   An ecological process by which one biotic community is replaced by another until a climax community develops.

Symbiosis   A relationship between two organisms that live together where at least one of the organisms benefits from the association.

Synapse   A space between two nerve cells.

Synapsis   A process in meiosis where two pairs of homologous chromosomes come together to form a tetrad.

Synthesis   A process by which two or more small molecules are combined to form a larger molecule.

Taxon   A classification group.

Taxonomy   The branch of biology that is concerned with the classification and naming of organisms.

Tendon   A connective tissue that attaches skeletal muscles to bones.

Terminal branches   The parts of a neuron that produce neurotransmitters.

Testes   The male gonads, which produce sperm cells and the hormone testosterone.

Tetrad   Two pairs of homologous chromosomes that have joined together by synapsis.

Tracheal tubes   Tiny tubes found in insects that carry oxygen to the cells and carbon dioxide from the cells.

Trait   A distinguishing feature or property of an organism.

Transcription   A process that produces ribonucleic acid (RNA) nucleoproteins.

Translation   A process by which ribosomes combine amino acids to produce proteins.

Translocation   The movement of food throughout a plant.

Transpiration   The loss of water through the stomates of a leaf.

Transport   The life function that distributes food, water and oxygen from the environment to all cells of an organism.

Umbilical cord   A structure that connects a fetus to the placenta.

Urea   A poisonous, nitrogenous waste produced from the breakdown of proteins.

Uterus (womb)   The muscular organ in the female reproductive system where an embryo develops into a fetus.

Vacuoles   Organelles of a cell that are storage sites for food or water.

Vas deferens (sperm duct)   A tube that carries sperm cells from the epididymis to the ejaculatory duct and into the urethra.

Vegetative propagation   Asexual reproduction in plants that results in the production of a new plant from the root, stem or leaf of an existing plant.

Vein   In animals, a blood vessel with a thin, muscular wall that has valves and carries blood back to the heart. In plants, vascular tissue specialized for the transport of materials.

Vertebrates   Animals with backbones.

Vestigial structure   A structure that at one time had a function in the evolutionary history of an organism but no longer has a function.

Weather   The day-to-day change in the atmosphere.

White blood cell (WBC)   A cell that helps the body fight disease. Some white blood cells engulf and kill bacteria, others produce *antibodies*, which are proteins that can destroy *antigens* (foreign substances in the body).

Xerophytes   Plants adapted to growing in a dry environment, such as a desert.

Xylem   Cells in a vascular bundle (vein) that carry food upward in a stem.

Zygote   A fertilized egg cell.

Customized Full-Length Exam

This exam will help you determine the areas you still need to focus on.

For problems 1–150, answer the question or fill in the blank.

1. Which life function provides an organism with the energy needed to carry out all its life processes? _____

   Answer: Respiration

   If you answered *correctly*, go to problem 3.
   If you answered *incorrectly*, go to problem 2.

2. A term that includes all the life functions and chemical reactions that take place within an organism is _____.

   Answer: Metabolism

   If you answered *correctly*, go to problem 3.
   If you answered *incorrectly*, review "Life Functions" on pages 15–17.

3. The diameter of a microscope field is 1,200 μm, and 20 equal-sized cells fit across the field. What is the size of one cell?

   Answer: 60 μm

   If you answered *correctly*, go to problem 5.
   If you answered *incorrectly*, go to problem 4.

4. A cell viewed with the high-power lens of a microscope (1,000X) measures 200 μm. How long is this cell when viewed with the low-power lens (100X)?

   Answer: 200 μm

   If you answered *correctly*, go to problem 5.
   If you answered *incorrectly*, review "Cell Measurement" on pages 21–22.

5. The _____ is a tool used to spin materials at very high speeds. The heaviest and densest materials settle to the bottom, and lighter ones are at the top.

   Answer: Centrifuge

   If you answered *correctly*, go to problem 7.
   If you answered *incorrectly*, go to problem 6.

6. _____ are small tools used to remove or add parts of cells.

   Answer: Microdissection tools

   If you answered *correctly*, go to problem 7.
   If you answered *incorrectly*, review "Other Tools" on page 23.

7. An educated guess that provides a possible answer to a scientific problem is called a
    _____.

    Answer: Hypothesis

    If you answered *correctly,* go to problem 9.
    If you answered *incorrectly,* go to problem 8.

8. An organized series of steps used by biologists to solve problems is known as the
    _____.

    Answer: Scientific method

    If you answered *correctly,* go to problem 9.
    If you answered *incorrectly,* review "The Scientific Method" on page 23.

9. The _____ is the basic unit of structure and function in living things.

    Answer: Cell

    If you answered *correctly,* go to problem 11.
    If you answered *incorrectly,* go to problem 10.

10. New _____ arise only from previously existing cells.

    Answer: Cells

    If you answered *correctly,* go to problem 11.
    If you answered *incorrectly,* review "Cell Theory" on page 33.

11. Name the organelles that are the storage sites of a cell. _____.

    Answer: Vacuoles

    If you answered *correctly,* go to problem 13.
    If you answered *incorrectly,* go to problem 12.

12. What part of the cell is selectively permeable? _____.

    Answer: Cell membrane

    If you answered *correctly,* go to problem 13.
    If you answered *incorrectly,* review "Cell Structure and Function" on page 35.

13. The _____ is an organelle that can engulf particles of food that are too
    large to pass through the membrane.

    Answer: Endocytotic vesicle or pinocytotic vesicle

    If you answered *correctly,* go to problem 15.
    If you answered *incorrectly,* go to problem 14.

14. Name the cell structure that contains digestive enzymes. _____.

    Answer: Lysosome

    If you answered *correctly,* go to problem 15.
    If you answered *incorrectly,* review "The Animal Cell As Seen with an Electron Microscope"
    on page 38.

15. The movement of water molecules from areas of high concentration to areas of low concentration is called _____.

Answer: Osmosis

If you answered *correctly*, go to problem 17.
If you answered *incorrectly*, go to problem 16.

16. The movement of molecules from areas of low concentration to areas of high concentration is called _____.

Answer: Active transport

If you answered *correctly*, go to problem 17.
If you answered *incorrectly*, review "Molecular Transport" on page 39.

17. The branch of biology that is concerned with the classification and naming of organisms is called _____.

Answer: Taxonomy

If you answered *correctly*, go to problem 19.
If you answered *incorrectly*, go to problem 18.

18. Organisms that can mate with each other and produce fertile offspring belong to a category of classification called a _____.

Answer: Species

If you answered *correctly*, go to problem 19.
If you answered *incorrectly*, review "Classification Subdivisions" on pages 47–48.

19. *Escherichia coli* is a bacterium that can cause illness in humans. What is the genus name for this organism?

Answer: *Escherichia*

If you answered *correctly*, go to problem 21.
If you answered *incorrectly*, go to problem 20.

20. The system of naming organisms that was developed by Linnaeus is known as _____.

Answer: Binomial nomenclature

If you answered *correctly*, go to problem 21.
If you answered *incorrectly*, review "Binomial Nomenclature" on page 48.

21. Unicellular prokaryotic organisms whose cell walls contain peptidoglycan are considered primitive living things and belong to the domain _____.

Answer: Bacteria

If you answered *correctly*, go to problem 23.
If you answered *incorrectly*, go to problem 22.

22. Organisms that lack a nuclear membrane are called _____.

Answer: Prokaryotic

If you answered *correctly*, go to problem 23.
If you answered *incorrectly*, review "Domain Bacteria" on page 51.

23. Eukaryotic, heterotrophic organisms that are decomposers belong the kingdom
_____.

Answer: Fungi

If you answered *correctly*, go to problem 25.
If you answered *incorrectly*, go to problem 24.

24. Ringworm and athlete's foot are examples of parasitic diseases caused by
_____.

Answer: Fungi

If you answered *correctly*, go to problem 25.
If you answered *incorrectly*, review "Kingdom Fungi" on pages 54–55.

25. Members of the kingdom Plantae probably evolved from _____.

Answer: Algae

If you answered *correctly*, go to problem 27.
If you answered *incorrectly*, go to problem 26.

26. Plants that lack vascular tissue, true roots, stems and leaves belong to the kingdom
Plantae and the phylum _____.

Answer: Bryophyta

If you answered *correctly*, go to problem 27.
If you answered *incorrectly*, review "Kingdom Plantae" on pages 56–57.

27. Animals without backbones are called _____.

Answer: Invertebrates

If you answered *correctly*, go to problem 29.
If you answered *incorrectly*, go to problem 28.

28. Which class in the animal kingdom has the largest number of species and is in the phylum
Arthropoda?

Answer: Insecta

If you answered *correctly*, go to problem 29.
If you answered *incorrectly*, review "Kingdom Animalia" on pages 58–61.

29. List the four elements that are found in all living things.

Answer: Hydrogen, oxygen, nitrogen and carbon (HONC)

If you answered *correctly*, go to problem 31.
If you answered *incorrectly*, go to problem 30.

30. A _____ formula indicates the number of molecules, elements, number of atoms and their arrangement for each element in a compound.

Answer: Structural

If you answered *correctly*, go to problem 31.
If you answered *incorrectly*, review "Drawing Organic Compounds" on pages 72–73.

31. In a carbohydrate, the ratio of hydrogen to oxygen is _____.

Answer: 2:1

If you answered *correctly*, go to problem 33.
If you answered *incorrectly*, go to problem 32.

32. Two glucose molecules can be combined by the process of _____.

Answer: Dehydration synthesis

If you answered *correctly*, go to problem 33.
If you answered *incorrectly*, review "Carbohydrates" on pages 75–76.

33. When one molecule of glycerol is combined with three molecules of fatty acid, one molecule of _____ and three molecules of water are produced.

Answer: Lipid

If you answered *correctly*, go to problem 35.
If you answered *incorrectly*, go to problem 34.

34. A lipid molecule can be broken down by the process of _____.

Answer: Hydrolysis

If you answered *correctly*, go to problem 35.
If you answered *incorrectly*, review "Lipids" on page 78.

35. A peptide bond is a bond that combines two _____ molecules together.

Answer: Amino acid

If you answered *correctly*, go to problem 37.
If you answered *incorrectly*, go to problem 36.

36. A molecule that has an (NH_2) group on one end and a (COOH) group on the other end is called a(n) _____.

Answer: Amino acid

If you answered *correctly*, go to problem 37.
If you answered *incorrectly*, review "Proteins" on pages 79–80.

37. A solution that has a large number of H^+ ions is _____.

Answer: Acidic

If you answered *correctly*, go to problem 39.
If you answered *incorrectly*, go to problem 38.

38. What is the pH of a molecule with the formula HOH?

Answer: Neutral or pH 7

If you answered *correctly,* go to problem 39.
If you answered *incorrectly,* review "Acids and Bases" on pages 82–83.

39. A(n) _____ is an organic catalyst that can speed up the rate of a chemical reaction.

Answer: Enzyme

If you answered *correctly,* go to problem 41.
If you answered *incorrectly,* go to problem 40.

40. The lock-and-key model is used to explain _____ action.

Answer: Enzyme

If you answered *correctly,* go to problem 41.
If you answered *incorrectly,* review "Enzyme Action" on pages 84–85.

41. Long columnar cells in a leaf that are specialized for photosynthesis can be found in the _____ layer.

Answer: Palisade

If you answered *correctly,* go to problem 43.
If you answered *incorrectly,* go to problem 42.

42. The _____ is a thin, waxy layer that covers the upper epidermis of a leaf and prevents the loss of water.

Answer: Cuticle

If you answered *correctly,* go to problem 43.
If you answered *incorrectly,* review "Leaf Structure and Function" on pages 95–96.

43. The splitting of water into hydrogen and oxygen during the light reactions of photosynthesis is called _____.

Answer: Photolysis

If you answered *correctly,* go to problem 45.
If you answered *incorrectly,* go to problem 44.

44. Phosphoglyceraldehyde (PGAL) is formed during the _____ of photosynthesis.

Answer: Dark reactions or Calvin cycle

If you answered *correctly,* go to problem 45.
If you answered *incorrectly,* review "The Photosynthetic Process" on pages 98–99.

45. The _____ cells of a vein or vascular bundle are specialized for the downward transport of materials.

Answer: Phloem

If you answered *correctly,* go to problem 47.
If you answered *incorrectly,* go to problem 46.

46. _____ cells are reproductive cells that separate xylem and phloem cells in a vascular bundle.

   Answer: Cambium

   If you answered *correctly*, go to problem 47.
   If you answered *incorrectly*, review "Stem Structure and Function" on pages 101–102.

47. The root cap or root tip contains _____ cells that reproduce to form new cells, causing the root to grow deeper into the ground.

   Answer: Meristematic

   If you answered *correctly*, go to problem 49.
   If you answered *incorrectly*, go to problem 48.

48. In the center of the root, we find the _____, which contains xylem and phloem cells that function in the transport of materials.

   Answer: Vascular cylinder

   If you answered *correctly*, go to problem 49.
   If you answered *incorrectly*, review "Root Structure and Function" on pages 103–104.

49. In humans, the large intestine reabsorbs _____ into the blood.

   Answer: Water

   If you answered *correctly*, go to problem 51.
   If you answered *incorrectly*, go to problem 50.

50. In humans, the chemical digestion of food is completed in the _____.

   Answer: Small intestine

   If you answered *correctly*, go to problem 51.
   If you answered *incorrectly*, review "Nutrition and Digestion" on pages 113–115.

51. The organs of the body contain tiny blood vessels called _____ that are one cell in diameter and connect arteries to veins.

   Answer: Capillaries

   If you answered *correctly*, go to problem 53.
   If you answered *incorrectly*, go to problem 52.

52. Human red blood cells contain _____, which aids in the transport of oxygen to the cells and carbon dioxide from the cells.

   Answer: Hemoglobin

   If you answered *correctly*, go to problem 53.
   If you answered *incorrectly*, review "Circulation" on pages 116–117.

53. The _____ is a muscle located under the lungs that aids in respiration.

   Answer: Diaphragm

   If you answered *correctly*, go to problem 55.
   If you answered *incorrectly*, go to problem 54.

54. Tiny air sacs in the lungs called _____ are surrounded by capillaries and function in the exchange of oxygen for carbon dioxide.

Answer: Alveoli

If you answered *correctly*, go to problem 55.
If you answered *incorrectly*, review "Organism Respiration" on pages 119–120.

55. During glycolysis, one molecule of glucose is broken down to yield two molecules of _____ and four molecules of ATP.

Answer: Pyruvate or pyruvic acid

If you answered *correctly*, go to problem 57.
If you answered *incorrectly*, go to problem 56.

56. To break a glucose molecule, the energy from _____ molecules of ATP are needed.

Answer: Two

If you answered *correctly*, go to problem 57.
If you answered *incorrectly*, review "Cellular Respiration" on pages 121–122.

57. In aerobic respiration, _____ is the final hydrogen acceptor.

Answer: Oxygen

If you answered *correctly*, go to problem 59.
If you answered *incorrectly*, go to problem 58.

58. Aerobic respiration takes place in the _____ of the cell.

Answer: Mitochondria

If you answered *correctly*, go to problem 59.
If you answered *incorrectly*, review "Cellular Respiration" on pages 121–122.

59. Wastes filtered out of the blood by the kidneys are stored in the _____.

Answer: Urinary bladder

If you answered *correctly*, go to problem 61.
If you answered *incorrectly*, go to problem 60.

60. Wastes from the glomerulus enter a part of the nephron known as _____.

Answer: Bowman's capsule

If you answered *correctly*, go to problem 61.
If you answered *incorrectly*, review "Structure and Function of the Kidneys" on pages 125–126.

61. Any change in the environment of an organism is called a _____.

Answer: Stimulus

If you answered *correctly*, go to problem 63.
If you answered *incorrectly*, go to problem 62.

62.   The muscles and glands of the body are examples of _____, which help carry out a response.

Answer: Effectors

If you answered *correctly*, go to problem 63.
If you answered *incorrectly*, review "Regulation" on pages 132–133.

63.   The parts of a neuron that can detect a stimulus are called the _____.

Answer: Dendrites

If you answered *correctly*, go to problem 65.
If you answered *incorrectly*, go to problem 64.

64.   Acetylcholine and cholinesterase are produced by the _____ of a neuron.

Answer: Terminal branches

If you answered *correctly*, go to problem 65.
If you answered *incorrectly*, review "The Structure of a Neuron" on pages 134–135.

65.   The _____ neuron detects a stimulus and transmits a nerve impulse from a receptor to an interneuron.

Answer: Sensory

If you answered *correctly*, go to problem 67.
If you answered *incorrectly*, go to problem 66.

66.   In a reflex arc, the _____ neuron serves as a relay neuron.

Answer: Interneuron

If you answered *correctly*, go to problem 67.
If you answered *incorrectly*, review "The Reflex Arc" on page 137.

67.   Endocrine responses are slow because hormones must travel through the _____ to reach their target organ(s).

Answer: Blood or circulatory system

If you answered *correctly*, go to problem 69.
If you answered *incorrectly*, go to problem 68.

68.   The pituitary gland controls other endocrine hormones through the production of _____ hormones.

Answer: Stimulating

If you answered *correctly*, go to problem 69.
If you answered *incorrectly*, review "Structure and Function of the Endocrine System" on pages 138–139.

69.   The nitrogen waste product produced by animal-like protists is _____.

Answer: Urea

If you answered *correctly*, go to problem 71.
If you answered *incorrectly*, go to problem 70.

70. Pseudopods, cilia and flagella are structures used by animal-like protists for the life function of _____.

Answer: Locomotion

If you answered *correctly,* go to problem 71.
If you answered *incorrectly,* review "Animal-Like Protists" on pages 153–154.

71. The hydra excretes nitrogenous waste in the form of _____.

Answer: Ammonia

If you answered *correctly,* go to problem 73.
If you answered *incorrectly,* go to problem 72.

72. The hydra has a _____ between its ectoderm and endoderm cells but lacks a brain or nerve cord to coordinate its responses.

Answer: Nerve net

If you answered *correctly,* go to problem 73.
If you answered *incorrectly,* review "Cnidaria" on pages 155–156.

73. Locomotion in an earthworm is accomplished by the interaction of muscles and paired _____.

Answer: Setae

If you answered *correctly,* go to problem 75.
If you answered *incorrectly,* go to problem 74.

74. Oxygen in the air diffuses through the moist _____ of an earthworm into capillaries, which carry the oxygen to the dorsal and ventral blood vessels.

Answer: Skin

If you answered *correctly,* go to problem 75.
If you answered *incorrectly,* review "Annelida" on pages 157–158.

75. An organism containing Malpighian tubules most likely possesses openings in its abdomen called _____, which function in respiration.

Answer: Spiracles

If you answered *correctly,* go to problem 77.
If you answered *incorrectly,* go to problem 76.

76. _____ are organisms that have an open circulatory system.

Answer: Insects or grasshoppers

If you answered *correctly,* go to problem 77.
If you answered *incorrectly,* review "Arthropoda" on pages 159–160.

77. In mitosis, the two halves of the replicated chromosome, called _____, are joined to each other by a structure called a centromere.

Answer: Chromatids

If you answered *correctly,* go to problem 79.
If you answered *incorrectly,* go to problem 78.

78. During the _____ stage of mitosis, the nuclear membrane and nucleolus disappear, and the centrioles begin to move to opposite sides of the cell. Also, for the first time, astral rays, spindle fibers and chromatids are visible.

Answer: Prophase

If you answered *correctly,* go to problem 79.
If you answered *incorrectly,* review "Mitosis" on pages 171–172.

79. Fruit flies have a diploid chromosome number of 8. What is the chromosome number in a sperm cell of the fruit fly?

Answer: 4

If you answered *correctly,* go to problem 81.
If you answered *incorrectly,* go to problem 80.

80. The union of a sperm cell and an egg cell to form a zygote is known as
_____.

Answer: Fertilization

If you answered *correctly,* go to problem 81.
If you answered *incorrectly,* review "Sexual Reproduction" on pages 173–174.

81. Synapsis, tetrad formation and disjunction are processes that take place during
_____.

Answer: Meiosis

If you answered *correctly,* go to problem 83.
If you answered *incorrectly,* go to problem 82.

82. Meiosis is often referred to as reduction division because the cytoplasm divides
_____ times.

Answer: Two

If you answered *correctly,* go to problem 83.
If you answered *incorrectly,* review "Meiosis" on pages 175–176.

83. In the human male reproductive system, sperm cells leave the epididymis by moving through a tube called the _____.

Answer: Vas deferens

If you answered *correctly,* go to problem 85.
If you answered *incorrectly,* go to problem 84.

84. In the human male reproductive system, the testes are located in a sac outside the body cavity known as the _____.

Answer: Scrotum

If you answered *correctly*, go to problem 85.
If you answered *incorrectly*, review "The Male Reproductive System" on pages 183–184.

85. The human female gamete is produced in the _____.

Answer: Ovary (follicle)

If you answered *correctly*, go to problem 87.
If you answered *incorrectly*, go to problem 86.

86. Immediately after ovulation, an egg cell enters the _____.

Answer: Fallopian tube

If you answered *correctly*, go to problem 87.
If you answered *incorrectly*, review "The Female Reproductive System" on pages 185–186.

87. In an embryo, the skeleton and muscles of the body develop from the _____ layer of the gastrula stage.

Answer: Mesoderm

If you answered *correctly*, go to problem 89.
If you answered *incorrectly*, go to problem 88.

88. Embryological development involves a special form of mitosis called _____.

Answer: Cleavage

If you answered *correctly*, go to problem 89.
If you answered *incorrectly*, review "Embryological Development" on pages 189–190.

89. A root, stem or leaf is detached from an existing plant and put into water, moist sand or soil. After several weeks, a new plant develops. This type of vegetative propagation is called _____.

Answer: Cutting

If you answered *correctly*, go to problem 91.
If you answered *incorrectly*, go to problem 90.

90. _____ is a type of artificial propagation where a stem is cut off a plant (scion) and attached to the stem of a plant that is rooted in the ground (stock).

Answer: Grafting

If you answered *correctly*, go to problem 91.
If you answered *incorrectly*, review "Artificial Propagation" on pages 202–203.

91. The anther and filament make up part of a flower called the _____.

Answer: Stamen or male part of the flower

If you answered *correctly*, go to problem 93.
If you answered *incorrectly*, go to problem 92.

92. The _____ is the part of a flower that produces female gametes.

   Answer: Ovary or pistil

   If you answered *correctly*, go to problem 93.
   If you answered *incorrectly*, review "Flower Structure and Function" on pages 205–206.

93. After a pollen grain lands on a stigma, its protective wall breaks down and a
   _____ begins to form through the style.

   Answer: Pollen tube

   If you answered *correctly*, go to problem 95.
   If you answered *incorrectly*, go to problem 94.

94. During plant reproduction, two sperm nuclei enter the ovule through an opening called
   the _____.

   Answer: Micropyle

   If you answered *correctly*, go to problem 95.
   If you answered *incorrectly*, review "Pollination and Fertilization" on pages 207–208.

95. The _____ is a thin, outer covering that protects the embryo and
   endosperm of the seed.

   Answer: Seed coat

   If you answered *correctly*, go to problem 97.
   If you answered *incorrectly*, go to problem 96.

96. The _____ is the part of a seed that forms from the cell of the ovule with
   3*n* chromosomes. Its function is to provide food for the germinating seed.

   Answer: Endosperm

   If you answered *correctly*, go to problem 97.
   If you answered *incorrectly*, review "Seed Structure and Function" on pages 209–210.

97. The type of gene combination for a trait (homozygous or heterozygous) that an organism
   has is called its _____.

   Answer: Genotype

   If you answered *correctly*, go to problem 99.
   If you answered *incorrectly*, go to problem 98.

98. One of the two different kinds of genes for a trait that occupy the same location on a pair
   of homologous chromosomes is called a(n) _____.

   Answer: Allele

   If you answered *correctly*, go to problem 99.
   If you answered *incorrectly*, review "Genetic Vocabulary" on pages 221–222.

99.   What is the probability of getting a homozygous tall (TT) pea plant from a cross between a heterozygous tall plant (Tt) and a homozygous short plant (tt)?

Answer: Zero

If you answered *correctly*, go to problem 101.
If you answered *incorrectly*, go to problem 100.

100.   Mendel's Law of _____ can be demonstrated by a cross between a homozygous tall pea plant and a homozygous short pea plant.

Answer: Dominance

If you answered *correctly*, go to problem 101.
If you answered *incorrectly*, review "Mendel's Law of Dominance" on pages 223–224.

101.   In mice, the allele for brown fur (B) is dominant over the allele for white fur (b). What is the probability of getting a mouse with white fur if two hybrid brown mice are crossed?

Answer: 25%

If you answered *correctly*, go to problem 103.
If you answered *incorrectly*, go to problem 102.

102.   In mice, the gene for brown fur (B) is dominant over the allele for white fur (b). What is the probability of getting a hybrid brown mouse if two hybrid brown mice are crossed?

Answer: 50%

If you answered *correctly*, go to problem 103.
If you answered *incorrectly*, review "Mendel's Law of Segregation" on pages 224–225.

103.   Two pea plants, heterozygous for both height and seed color, are crossed with each other. This cross illustrates Mendel's Law of _____.

Answer: Independent Assortment

If you answered *correctly*, go to problem 105.
If you answered *incorrectly*, go to problem 104.

104.   The phenotype ratio that results from the dihybrid cross is _____.

Answer: 9:3:3:1

If you answered *correctly*, go to problem 105.
If you answered *incorrectly*, review "Mendel's Law of Independent Assortment" on page 227.

105.   In a _____, the chromosomes are removed from the nucleus of a cell, arranged in decreasing size order and photographed.

Answer: Karyotype

If you answered *correctly*, go to problem 107.
If you answered *incorrectly*, go to problem 106.

106. What combination of sex chromosomes produces a female?

Answer: XX

If you answered *correctly*, go to problem 107.
If you answered *incorrectly*, review "Sex Determination" on page 237.

107. People with type O blood have antibodies _____ and
_____.

Answer: A and B.

If you answered *correctly*, go to problem 109.
If you answered *incorrectly*, go to problem 108.

108. What is the probability that two parents whose genotypes are I^AI^B and I^Bi can produce a child with blood type O?

Answer: 0%

If you answered *correctly*, go to problem 109.
If you answered *incorrectly*, review "Multiple Alleles" on pages 238–240.

109. A woman with a gene for hemophilia is married to a man who is a hemophiliac. What percentage of their children can be expected to have hemophilia?

Answer: 50%

If you answered *correctly*, go to problem 111.
If you answered *incorrectly*, go to problem 110.

110. Traits that are controlled by genes located on the X chromosome are considered
_____.

Answer: Sex linked

If you answered *correctly*, go to problem 111.
If you answered *incorrectly*, review "Sex Linkage" on page 242.

111. The Himalayan rabbit has white fur *except* for its extremities where the fur color tends to be black. This difference can be explained by the effect of _____ on the expression of the allele that controls fur color.

Answer: Temperature or environment

If you answered *correctly*, go to problem 113.
If you answered *incorrectly*, go to problem 112.

112. The phenotype of an organism is determined by the influence of the
_____ on gene expression.

Answer: Environment

If you answered *correctly*, go to problem 113.
If you answered *incorrectly*, review "The Influence of Environment on Heredity" on page 246.

113. Both DNA and RNA molecules are composed of repeating units known as _____.

Answer: Nucleotides

If you answered *correctly*, go to problem 115.
If you answered *incorrectly*, go to problem 114.

114. One strand of a DNA molecule has the following nitrogen base sequence: A-T-C. What is the nitrogen base sequence on the complementary strand?

Answer: T-A-G

If you answered *correctly*, go to problem 115.
If you answered *incorrectly*, review "DNA Structure" on pages 253–254.

115. Amino acid molecules are combined at the _____ to make proteins.

Answer: Ribosome

If you answered *correctly*, go to problem 117.
If you answered *incorrectly*, go to problem 116.

116. Which molecule brings amino acids from the cytoplasm to the mRNA molecule?

Answer: Transfer RNA (tRNA)

If you answered *correctly*, go to problem 117.
If you answered *incorrectly*, review "Translation" on pages 256–257.

117. Sickle-cell anemia is a genetic disease caused by a gene mutation in the _____ molecule of red blood cells.

Answer: Hemoglobin

If you answered *correctly*, go to problem 119.
If you answered *incorrectly*, go to problem 118.

118. Most human genetic diseases are caused by recessive alleles. Therefore, to have the disease an individual must be _____ and have two defective alleles.

Answer: Homozygous

If you answered *correctly*, go to problem 119.
If you answered *incorrectly*, review "Genetic Diseases" on pages 262–263.

119. _____ DNA is produced when a gene from one chromosome is spliced into the chromosome of another species.

Answer: Recombinant

If you answered *correctly*, go to problem 121.
If you answered *incorrectly*, go to problem 120.

120. The circular DNA found in a bacterium is called a _____.

Answer: Plasmid

If you answered *correctly*, go to problem 121.
If you answered *incorrectly*, review "Genetic Engineering" on page 265.

121. Fossils are used as evidence for evolution because they often show a pattern of consecutive _____.

   Answer: Changes

   If you answered *correctly*, go to problem 123.
   If you answered *incorrectly*, go to problem 122.

122. The _____ of an organism (bones, teeth or shells) often become fossilized because they can resist the natural process of decay.

   Answer: Hard parts

   If you answered *correctly*, go to problem 123.
   If you answered *incorrectly*, review "Fossil Evidence for Evolution" on pages 275–276.

123. Fossils are usually found in _____ rock.

   Answer: Sedimentary

   If you answered *correctly*, go to problem 125.
   If you answered *incorrectly*, go to problem 124.

124. The use of carbon-14 to date a fossil is an example of _____ dating.

   Answer: Absolute

   If you answered *correctly*, go to problem 125.
   If you answered *incorrectly*, review "Dating of Fossils" on page 278.

125. Two organisms that have many homologous structures in common probably have a common _____.

   Answer: Ancestor

   If you answered *correctly*, go to problem 127.
   If you answered *incorrectly*, go to problem 126.

126. The similarities of hormones and nucleoproteins between organisms provide evidence for evolution. This is an example of comparative _____.

   Answer: Biochemistry

   If you answered *correctly*, go to problem 127.
   If you answered *incorrectly*, review "Additional Evidence for Evolution" on pages 280–281.

127. An early theory of evolution proposed that organisms could inherit acquired _____ based on need.

   Answer: Characteristics

   If you answered *correctly*, go to problem 129.
   If you answered *incorrectly*, go to problem 128.

128. Modern concepts of evolution are based on Darwin's theory of _____.

   Answer: Natural selection

   If you answered *correctly*, go to problem 129.
   If you answered *incorrectly*, review "Theories of Evolution" on pages 282–283.

129. Modern concepts of evolution propose that _____ are a cause of variations within a population.

Answer: Mutations

If you answered *correctly*, go to problem 131.
If you answered *incorrectly*, go to problem 130.

130. A sudden and unexpected change in a trait of an organism can be called a

_____.

Answer: Mutation

If you answered *correctly*, go to problem 131.
If you answered *incorrectly*, review "Mutations" on page 283.

131. The Hardy-Weinberg law states that within a population, the frequency of an allele _____ from generation to generation, as long as certain conditions are met.

Answer: Remains constant or stays the same

If you answered *correctly*, go to problem 133.
If you answered *incorrectly*, go to problem 132.

132. A condition necessary for the Hardy-Weinberg law to apply is that a population must be _____ in size.

Answer: Large

If you answered *correctly*, go to problem 133.
If you answered *incorrectly*, review "The Hardy-Weinberg Law" on pages 284–285.

133. Light, temperature and water are all examples of _____ factors in the environment.

Answer: Abiotic

If you answered *correctly*, go to problem 135.
If you answered *incorrectly*, go to problem 134.

134. _____ factors are any abiotic or biotic factors that determine the distribution and kinds of organisms that can survive in an area.

Answer: Limiting

If you answered *correctly*, go to problem 135.
If you answered *incorrectly*, review "Chapter Introduction" on page 297.

135. The members of the species *Acer rubrum* living in a certain location make up a

_____.

Answer: Population

If you answered *correctly*, go to problem 137.
If you answered *incorrectly*, go to problem 136.

136. All the different populations that live together in a specific location make up a
_____.

Answer: Community

If you answered *correctly,* go to problem 137.
If you answered *incorrectly,* review "Biotic Factors" on pages 299–300.

137. Ants feed on secretions from the acacia tree, and at the same they protect the tree from harmful insects. This type of symbiotic relationship is known as _____.

Answer: Mutualism

If you answered *correctly,* go to problem 139.
If you answered *incorrectly,* go to problem 138.

138. A tapeworm can live in the digestive system of a dog eating foods that the dog eats. However, the tapeworm harms the dog in this symbiotic relationship, which is called _____.

Answer: Parasitism

If you answered *correctly,* go to problem 139.
If you answered *incorrectly,* review "Symbiosis" on page 301.

139. Organisms of decay such as bacteria and fungi are called _____.

Answer: Decomposers or saprobes

If you answered *correctly,* go to problem 141.
If you answered *incorrectly,* go to problem 140.

140. In a food chain, the primary consumers are always _____.

Answer: Herbivores or plant eaters

If you answered *correctly,* go to problem 141.
If you answered *incorrectly,* review "Food Chain" on page 304.

141. Evaporation and precipitation are two components of the _____ cycle.

Answer: Water

If you answered *correctly,* go to problem 143.
If you answered *incorrectly,* go to problem 142.

142. _____ bacteria take free nitrogen from the atmosphere and produce nitrates.

Answer: Nitrogen-fixing

If you answered *correctly,* go to problem 143.
If you answered *incorrectly,* review "Material Cycles" on pages 306–307.

143. _____ succession begins in an area that never had life.

Answer: Primary

If you answered *correctly,* go to problem 145.
If you answered *incorrectly,* go to problem 144.

144. A stable, self-perpetuating group of plant and animal species in an area is known as a
_____.

Answer: Climax community

If you answered *correctly*, go to problem 145.
If you answered *incorrectly*, review "Succession from Rock to Forest" on page 308.

145. The _____ is a biome characterized by trees that shed their leaves in the fall. This biome has four seasons: Winter is cold, summer is warm and precipitation is evenly distributed throughout the year.

Answer: Temperate, deciduous forest

If you answered *correctly*, go to problem 147.
If you answered *incorrectly*, go to problem 146.

146. The _____ is the land biome that has the greatest amount of biodiversity.

Answer: Tropical rain forest

If you answered *correctly*, go to problem 147.
If you answered *incorrectly*, review "World Biomes" on pages 309–311.

147. The greatest danger to the environment is _____.

Answer: Increased population growth or people

If you answered *correctly*, go to problem 149.
If you answered *incorrectly*, go to problem 148.

148. Which species has the greatest ability to alter the environment?

Answer: *Homo sapiens* or humans

If you answered *correctly*, go to problem 149.
If you answered *incorrectly*, review "Humans and the Environment" on pages 313–314.

149. The _____ movement is concerned with the protection and maintenance of an ecosystem and its biological communities.

Answer: Conservation

If you answered *correctly*, you have successfully completed the exam. Congratulations!
If you answered *incorrectly*, go to problem 150.

150. _____ is the control of a pest population by nonchemical means (such as the use of predators, parasites, bacteria and diseases) that destroy the pest population.

Answer: Biological pest control

If you answered *correctly*, you have successfully completed the exam. Congratulations!
If you answered *incorrectly*, review "Environmental Protection" on pages 315–316.

Index

D

E